N INDOLOGY

2003

NARSIDASS

SHERS

# VEDIC MATHEMATICS

*The Author*
Jagadguru Śaṅkarācārya
Śrī Bhāratī Kṛṣṇa Tīrthajī Mahārāja
(1884—1960)

# Vedic Mathematics

*by Jagadguru Swāmī Śrī*
BHĀRATĪ KṚṢṆA TĪRTHAJĪ MAHĀRĀJA
Śaṅkarācārya of Govardhana Maṭha, Puri

*General Editor*
DR. V.S. AGRAWALA

MOTILAL BANARSIDASS PUBLISHERS
PRIVATE LIMITED • DELHI

*Revised Edition: Delhi, 1992*
*Reprint: Delhi, 1994, 1995, 1997, 1998, 2000, 2001, 2002, 2003*



ISBN: 81-208-0163-6 (Cloth)
ISBN: 81-208-0164-4 (Paper)

*Also available at:*
MOTILAL BANARSIDASS
41 U.A. Bungalow Road, Jawahar Nagar, Delhi 110 007
8 Mahalaxmi Chamber, 22 Bhulabhai Desai Road, Mumbai 400 026
236, 9th Main III Block, Jayanagar, Bangalore 560 011
120 Royapettah High Road, Mylapore, Chennai 600 004
Sanas Plaza, 1302 Baji Rao Road, Pune 411 002
8 Camac Street, Kolkata 700 017
Ashok Rajpath, Patna 800 004
Chowk, Varanasi 221 001

Printed in India
BY JAINENDRA PRAKASH JAIN AT SHRI JAINENDRA PRESS,
A-45 NARAINA, PHASE-I, NEW DELHI 110 028
AND PUBLISHED BY NARENDRA PRAKASH JAIN FOR
MOTILAL BANARSIDASS PUBLISHERS PRIVATE LIMITED,
BUNGALOW ROAD, DELHI 110 007

# General Editor's Note

The work, VEDIC MATHEMATICS or 'Sixteen Simple Mathematical Formulae from the Vedas', was written by His Holiness Jagadguru Śaṅkarācārya Śrī Bhāratī Kṛṣṇa Tīrthajī Mahārāja of Govardhana Maṭha, Puri (1884-1960). It forms a class by itself not pragmatically conceived and worked out as in the case of other scientific works, but is the result of the intuitional visualisation of fundamental mathematical truths and principles during the course of eight years of highly concentrated mental endeavour on the part of the author and therefore appropriately has been given the title of "mental" mathematics appearing more as a miracle than the usual approach of hard-baked science, as the author has himself stated in the Preface.

Swāmī Śaṅkarācārya was a gifted scholar on many fronts of learning including science and humanities but his whole milieu was something of a much higher texture in that he was a Ṛṣi fulfilling the ideals and attainments of those seers of ancient India who discovered the cosmic laws embodied in the Vedas. Swāmī Bhāratī Kṛṣṇa Tīrtha had the same reverential approach towards the Vedas. The question naturally arises as to whether the *Sūtras* which form the basis of this treatise exist anywhere in the Vedic literature as known to us. But this criticism loses all its force if we inform ourselves of the definition of Veda given by Śrī Śaṅkarācārya himself as quoted below:

> "The very word 'Veda' has this derivational meaning, i.e. the fountain-head and illimitable store-house of all knowledge. This derivation, in effect, means, connotes and implies that the Vedas *should contain* (italics mine) within themselves all the knowledge needed by mankind relating not only to the so-called 'spiritual' (or other-worldly) matters but also to those usually described as purely 'secular', 'temporal', or 'worldly' and also to the means required by humanity as such for the achievement of all-round, complete and perfect success in all conceivable directions and that there can be no adjectival or restrictive epithet calculated (or tending) to limit that knowledge down in any sphere, any direction or any respect whatsoever.

"In other words, it connotes and implies that our ancient Indian Vedic lore *should be* (italics mine) all-round , complete and perfect and able to throw the fullest necessary light on all matters which any aspiring seeker after knowledge can possibly seek to be enlightened on."

It is the whole essence of his assessment of Vedic tradition that it is not to be approached from a factual standpoint but from the ideal standpoint viz., as the Vedas, as traditionally accepted in India as the repository of all knowledge, *should be* and not what they are in human possession. That approach entirely turns the tables on all critics, for the authorship of Vedic mathematics then need not be laboriously searched in the texts as preserved from antiquity. The Vedas are well-known as four in number Ṛg, Yajur, Sāma and Atharva, but they have also the four Upavedas and the six Vedāṅgas all of which form an indivisible corpus of divine knowledge as it once was and as it may be revealed. The four Upavedas are as follows:

| *Veda* | *Upaveda* |
|---|---|
| Ṛgveda | Āyurveda |
| Sāmaveda | Gāndharvaveda |
| Yajurveda | Dhanurveda |
| Atharvaveda | Sthāpatyaveda |

In this list the Upaveda of Sthāpatya or engineering comprises all kinds of architectural and structural human endeavour and all visual arts. Swāmījī naturally regarded mathematics or the science of calculations and computations to fall under this category.

In the light of the above definition and approach must be understood the author's statement that the sixteen *Sūtras* on which the present volume is based form part of a *Pariśiṣṭa* of the Atharvaveda. We are aware that each Veda has its subsidiary apocryphal texts some of which remain in manuscripts and others have been printed but that formulation has not closed. For example, some *Pariśiṣṭas* of the Atharvaveda were edited by G.M. Bolling and J. Von Negelein, Leipzig, 1909-10. But this work of Śrī Śaṅkarācāryajī deserves to be regarded as a new *Pariśiṣṭa* by itself and it is not surprising that the *Sūtras* mentioned herein do not appear in the hitherto known *Pariśiṣṭas*.

A list of these main 16 *Sūtras* and of their sub-*sūtras* or corollaries is prefixed in the beginning of the text and the style of language also points to their discovery by Śrī Śwāmījī himself. At any rate, it is needless to dwell longer on this point of origin since the vast merit of these rules should be a matter of discovery for each intelligent reader. Whatever is written here by the author stands on its own merits and is presented as such to the mathematical world.

Swāmījī was a marvellous person with surpassing qualities and was a prolific writer and eloquent speaker. I had the good fortune of listening to his discourses for weeks together on several occasions when he used to visit Lucknow and attracted large audiences. He could speak for several hours at a stretch in Sanskrit and English with the same felicity and the intonation of his musical voice left a lasting impression on the minds of his hearers. He was an ardent admirer of Bhartṛhari, the great scientific thinker of the Golden Age of Indian history in a different field viz., that of philosophy of grammar.

Swāmījī had planned to write 16 volumes on all aspects and branches of mathematical processes and problems and there is no doubt that his mental powers were certainly of that calibre, but what has been left to us is this introductory volume which in itself is of the highest merit for reason of presenting a new technique which the author styles as "mental" mathematics different from the orthodox methods of mathematicians all over the world. Arithmetical problems usually solved by 18, 28 or 42 steps in case of such vulgar fractions as 1/19, 1/29, 1/49 are here solved in one simple line and even young boys can do it. The truth of these methods was demonstrated by this saintly teacher before many University audiences in India and in the U.S.A. including learned Professors and everyone present was struck with their originality and simplicity.

We are told in his Preface by Swāmī Śaṅkarācārya that he contemplated to cover all the different branches of mathematics such as arithmetic, algebra, geometry (plane and solid), trigonometry (plane and spherical), conics—geometrical and analytical, astronomy, calculus—differential and integral etc., with these basic *Sūtras*. That comprehensive application of the *Sūtras* could not be left by him in writing but if someone has the patience and the genius to pursue the method and implications of these formulae he may probably be able to bring these various branches within the orbit of this original style.

A full-fledged course of his lecture-demonstrations was organised by the Nagpur University in 1952 and some lectures were delivered by Swāmījī at the B.H.U. in 1949. It is, therefore, in the fitness of things and a happy event for the B.H.U. to be given the opportunity of publishing this book by the courtesy of Srimati Manjula Devi Trivedi, a disciple of Śrī Swāmījī who agreed to make over this manuscript to us through the efforts of Dr. Pandit Omkarnath Thakur. The work has been seen through the press mainly by Dr. Prem Lata Sharma, Dean, Faculty of Music & Fine Arts in the University. To all of these our grateful thanks are due. Dr. Brij Mohan, Head of the Department of Mathematics, B.H.U., took the trouble, at my request, to go through the manuscript and verify the calculations for which I offer him my best thanks. I also express gratitude to Sri Lakshmidas, Manager, B.H.U. Press, for taking great pains in printing this difficult text.

We wish to express our deepest gratitude to Śrī Swāmī Pratyagātmānanda Saraswatī for the valuable foreword that he has written for this work. Today he stands pre-eminent in the world of Tantric scholars and is a profound mathematician and scientific thinker himself. His inspiring words are like fragrant flowers offered at the feet of the ancient Vedic Ṛṣis whose spiritual lineage was revealed in the late Śaṅkarācārya Śrī Bhāratī Kṛṣṇa Tīrtha. Swāmī Pratyagātmānandajī has not only paid a tribute to Śrī Śaṅkarācāryajī but his ambrosial words have showered blessings on all those who are lovers of intuitional experiences in the domain of metaphysics and physics. Swāmījī, by a fortunate chance, travelled from Calcutta to Varanasi to preside over the Tantric Sammelan of the Varanaseya Sanskrit University (8th to 11th March 1965) and although he is now 85 years of age, his innate generosity made him accept our request to give his foreword.

I am particularly happy that I am able to publish this work under the Nepal Endowment Hindu Vishvavidyalaya Publication Series for I entertained an ardent desire to do so since our late President Dr. Rajendra Prasadji spoke to me about its existence when I once met him in New Delhi in the lifetime to Śrī Swāmījī.

V. S. AGRAWALA, M.A., PH.D., D. LITT.
General Editor,
Hindu Vishvavidyalaya
Nepal Rajya Sanskrit Granthamala Series

*Banaras Hindu University*
*Varanasi*
March 17, 1965

# Foreword

*Vedic Mathematics* by the late Śaṅkarācārya (Bhāratī Kṛṣṇa Tīrtha) of Govardhana Pīṭha is a monumental work. In his deep-layer explorations of cryptic Vedic mysteries relating specially to their calculus of shorthand formulae and their neat and ready application to practical problems, the late Śaṅkarācārya shows the rare combination of the probing insight and revealing intuition of a Yogī with the analytic acumen and synthetic talent of a mathematician. With the late Śaṅkarācārya we belong to a race, now fast becoming extinct, of die-hard believers who think that the Vedas represent an inexhaustible mine of profound wisdom in matters both spiritual and temporal; and that this store of wisdom was not, as regards its assets of fundamental validity and value at least, gathered by the laborious inductive and deductive methods of ordinary systematic enquiry, but was a direct gift of revelation to seers and sages who in their higher reaches of Yogic realization were competent to receive it from a source, perfect and immaculate. But we admit, and the late Śaṅkarācārya has also practically admitted, that one cannot expect to convert or revert criticism, much less carry conviction, by merely asserting one's staunchest beliefs. To meet these ends, one must be prepared to go the whole length of testing and verification by accepted, accredited methods. The late Śaṅkarācārya has, by his comparative and critical study of Vedic mathematics, made this essential requirement in Vedic studies abundantly clear. So let us agree to gauge Vedic mysteries not as we gauge the far-off nebulae with the poet's eyes or with that of the seer, but with the alert, expert, scrutinizing eye of the physical astronomer, if we may put it as that.

That there is a consolidated metaphysical background in the Vedas of the objective sciences including mathematics as regards their basic conceptions is a point that may be granted by a thinker who has looked broadly and deeply into both the realms.

In our paper recently published, 'The Metaphysics of Physics', we attempted to look into the mysteries of creative emergence as contained in the well-known cosmogonic Hymn (Ṛg. X. 190) with a view to unveiling the metaphysical background where both ancient wisdom and modern physics may meet on a common basis of logical understanding, and compare notes, discovering, where possible, points of significant or suggestive parallelism between the two sets of concepts, ancient and modern. That metaphysical background includes mathematics also; because physics as ever pursued is the application of mathematics to given or specified space-time-event situations. There we examined *tapas* as a fundamental creative formula whereby the Absolute emerges into the realms of measures, variations, limits, frameworks and relations. And this descent follows a logical order which seems to lend itself, within a framework of conditions and specifications, to mathematical analysis. *Rātri* in the Hymn represents the Principle of Limits, for example, *ṛtañca satyañca* stand for Becoming (*calana-kalana*) and Being (*vartana-kalana*) at a stage where limits or conditions or conventions do not yet arise or apply. The former gives the unconditioned, unrestricted *how* or *thus* of cosmic process, the latter, *what or that* of existence. *Tapas*, which corresponds to *Ardhamātrā* in Tantric symbolism, negotiates, in its role specially of critical variation, between what is, *ab initio*, unconditioned and unrestricted, and what appears otherwise, as for instance, in our own universe of logico-mathematical appreciation.

This is, necessarily, abstruse metaphysics, but it is, nevertheless, the starting background of both physics and mathematics. But for all practical purposes we must come down from mystic nebulae to the *terra firma* of our actual apprehension and appreciation. That is to say, we must descend to our own pragmatic levels of time-space-event situations. Here we face actual problems, and one must meet and deal with these squarely without evasion or mystification. The late Śaṅkarācārya has done this masterly feat with an adroitness that compels admiration.

It follows from the fundamental premises that the universe we live in must have a basic mathematical structure, and consequently, to know a fact or obtain a result herein, to any required degree of precision, one must obey the rule of mathematical measures and relations. This, however, one may do consciously or semi-consciously, systematically or haphazardly. Even some species of lower animals are by instinct gifted mathematicians; for example, the migratory bird which flies thousands of miles off from its nest and after a period, unerringly returns. This implies a subconscious mathematical talent that works wonders. We may cite the case of a horse who was a mathematical prodigy and could 'tell' the result of a cube root (requiring 32 operations, according to M. Materlink in his 'Unknown Quest') in a twinkle of the eye. This sounds like magic, but it is undeniable that the feat of mathematics *does* sometimes assume a magical look. Man, undoubtedly, has been given his share of this magical gift. And he can improve upon it by practice and discipline, by *Yoga* and allied methods. This is also undeniable. Lately, he has devised the 'automatic brain' for complicated calculations of science, that look like magic.

But apart from this 'magic', there is and has been, the 'logic' of mathematics also. Man works from instinct, talent, or even genius. But ordinarily he works as a logical entity requiring specified data or premises to start from, and more or less elaborate steps of reasoning to arrive at a conclusion. This is his normal process of induction and deduction. Here formulae (*Sūtras*) and relations (e.g. equations) must obtain as in mathematics. The magic and logic of mathematics in some cases get mixed up; but it is sane to keep them apart. You can get a result by magic, but when you are called upon to *prove* it, you must have recourse to logic.

Even in this latter case, your logic, your formulae and applications may be either simple and elegant or complicated and cumbersome. The former is the ideal to aim at. We have classical instances of master mathematicians whose methods of analysis and solution have been regarded as marvels of cogency, compactness and elegance. Some have been 'beautiful' as a poem, e.g. Lagrange's 'Analytical Mechanics'.

The late Śaṅkarācārya has claimed, and rightly we may think, that the Vedic *Sūtras* and their applications possess these virtues to a degree of eminence that cannot be challenged. The outstanding merit of his work lies in his actual proving of this contention.

Whether or not the Vedas be believed as repositories of perfect wisdom, it is unquestionable that the Vedic race lived *not* as merely pastoral folk possessing a half-or-quarter-developed culture and civilization. The Vedic seers were, again, not mere 'navel-gazers' or 'nose-tip-gazers'. They proved themselves adepts in all levels and branches of knowledge, theoretical and practical. For example, they had their varied objective science, both pure and applied.

Let us take a concrete illustration. Suppose in a time of drought we require rains by artificial means. The modern scientist has his own theory and art (or technique) for producing the result. The old seer scientist had his both also, but different from these now availing. He had his science and technique, called *Yajña*, in which *Mantra*, *Yantra* and other factors must co-operate with mathematical determinateness and precision. For this purpose, he had developed the six auxiliaries of the Vedas in each of which mathematical skill and adroitness, occult or otherwise, play the decisive role. The *Sūtras* lay down the shortest and surest lines. The correct intonation of the *Mantra*, the correct configuration of the *Yantra* (in the making of the *Vedī* etc., e.g. the quadrature of a circle), the correct time or astral conjugation factor, the correct rhythms etc., all had to be perfected so as to produce the desired result effectively and adequately. Each of these required the calculus of mathematics. The modern technician has his logarithmic tables and mechanics manuals; the old Yājñika had his *Sūtras*. How were the *Sūtras* obtained—by magic or logic or both—is a vital matter we do not discuss here. The late Śaṅkarācārya has claimed for them cogency, compactness and simplicity. This is an even more vital point, and we think, he has reasonably made it good.

*Varanasi*
March 22, 1965

Swāmī Pratyagātmānanda Saraswatī

# Contents

# Conventional to Unconventionally Original

*Vedic Mathematics* deals mainly with various Vedic mathematical formulae and their applications for carrying out tedious and cumbersome arithmetical operations, and to a very large extent, executing them mentally. In this field of mental arithmetical operations the works of the famous mathematicians Trachtenberg and Lester Meyers (High Speed Maths) are elementary compared to that of Jagadguruji.

Some people may find it difficult, at first reading, to understand the arithmetical operations although they have been explained very lucidly by Jagadguruji. It is not because the explanations are lacking in any manner but because the methods are totally unconventional. Some people are so deeply rooted in the conventional methods that they probably, subconsciously reject to see the logic in unconventional methods.

An attempt has been made in this note to explain the unconventional aspects of the methods. Once the reader gets used to the unconventional in the beginning itself, he would find no difficulty in the later chapters. Therefore, the explanatory notes are given for the first few chapters only.

### Chapter One

Chapter One deals with a topic that has been dealt with comprehensively in Chapter Twenty-six, viz. 'Recurring Decimal'. Gurudeva has discussed the recurring decimals of 1/19, 1/29, etc. in Chapter One to arouse curiosity and create interest. In conversion of vulgar fractions into their decimal equivalents Gurudeva has used very unconventional methods of multiplication and division.

In calculation of decimal equivalent of 1/19, first method of the *Ekādhika Sūtra* requires multiplication of 1 by 2 by a special and unconventional process. In conventional method product of 1, the multiplicand, by 2 the multiplier, is 2 and that is the end of multiplication process. It is not so in the unconventional *Ekādhika* method. In this method, in the above example, 1 is the first multiplicand and its product with multiplier '2' is 2 which in this special process becomes the second multiplicand. This when multiplied by the multiplier (which remains the same) 2 gives the product as 4 which becomes the third multiplicand. And the process of multiplication thus goes on till the digits start recurring.

Similarly in the second method of the *Ekādhika Sūtra* for calculating the decimal equivalent of 1/19, it is required to divide 1 by 2 by an unconventional and special process. In the conventional method when 1, the dividend, is to be divided by the divisor '2', the quotient is 0.5 and the process of division ends. In the special method of *Ekādhika Sūtra* for calculating decimal equivalents, the process starts by putting zero as the first digit of the quotient, 1 as the first remainder. A decimal point is put after the first quotient digit which is zero. Now, the first remainder digit '1' is prefixed to the first quotient digit '0' to form '10' as the second dividend. Division of 10 by the divisor 2 (which does not change) gives 5 as the second quotient digit which is put after the decimal point. The second remainder digit '0' is prefixed to the second quotient digit 5 to form 5 as the third dividend digit. Division of 5 by 2 gives 2 as the third quotient digit and 1 as the third remainder digit which when prefixed to the third quotient digit '2' gives 12 as the fourth dividend and so the process goes on till the digits start recurring.

### Chapter Three

*Vinculum* is an ingenious device to reduce single digits larger than 5, thereby facilitating multiplication specially for the mental one-line method. *Vinculum* method is based on the fact that 18 is same as (20−2) and 76 as (100−24) or 576 as (600−24). Gurudeva has made this arithmetical fact a powerful device by writing 18 as $2\bar{2}$, 76 as $1\ \bar{2}\ 4$ and 576 as $6\ \bar{2}\,\bar{4}$. This device is specially useful in Vedic division method.

A small note on 'aliquot' may facilitate the study for some. Aliquot part is the part contained by the whole an integral number of times, e.g. 12 is contained by the whole number 110, 9 times, or in simple words it is the quotient of that fraction.

### Chapter Four

In the division by the *Nikhilaṃ* method the dividend is divided into two portions by a vertical line. This vertical line should have as many digits to its right as there can be in the highest possible remainder. In general the number of such digits are the same as in the figure which is one less than the divisor. Needless to state that the vertical and horizontal lines must be drawn neatly when using this method.

Wing. Com. Vishva Mohan Tiwari

॥ॐ श्रीः॥

# Vedic Mathematics

## Or Sixteen Simple Mathematical Formulae from the Vedas

### Sixteen Sūtras and Their Corollaries

| *Sūtras* | *Sub-Sūtras or Corollaries* |
|---|---|
| 1. एकाधिकेन पूर्वेण<br>*Ekādhikena Pūrveṇa* (also a corollary) | 1. आनुरूप्येण<br>*Ānurūpyeṇa* |
| 2. निखिलं नवतश्चरमं दशतः<br>*Nikhilaṃ Navataścaramaṃ Daśataḥ* | 2. शिष्यते शेषसंज्ञः<br>*Śiṣyate Śeṣasaṃjñaḥ* |
| 3. ऊर्ध्वतिर्यग्भ्याम्<br>*Ūrdhva-tiryagbhyām* | 3. आद्यमाद्ये नान्त्यमन्त्येन<br>*Ādyamādyenāntyamantyena* |
| 4. परावर्त्य योजयेत्<br>*Parāvartya Yojayet* | 4. केवलैः सप्तकं गुण्यात्<br>*Kevalaiḥ Saptakaṃ Guṇyāt* |
| 5. शून्यं साम्यसमुच्चये<br>*Śūnyaṃ Sāmyasamuccaye* | 5. वेष्टनम्<br>*Veṣṭanaṃ* |
| 6. (आनुरूप्ये) शून्यमन्यत्<br>*(Ānurūpye) Śūnyamanyat* | 6. यावदूनं तावदूनम्<br>*Yāvadūnaṃ Tāvadūnaṃ* |
| 7. संकलनव्यवकलनाभ्याम्<br>*Saṅkalana-vyavakalanābhyām* (also a corollary) | 7. यावदूनं तावदूनीकृत्य वर्गं च योजयेत्<br>*Yāvadunaṃ Tāvadūnīkṛtya Vargañca Yojayet* |
| 8. पूरणापूरणाभ्याम्<br>*Pūraṇāpūraṇābhyām* | 8. अन्त्ययोर्दशकेऽपि<br>*Antyayordaśake'pi* |
| 9. चलनकलनाभ्याम्<br>*Calana-kalanābhyām* | 9. अन्त्ययोरेव<br>*Antyayoreva* |
| 10. यावदूनम्<br>*Yāvadūnam* | 10. समुच्चयगुणितः<br>*Samuccayaguṇitaḥ* |
| 11. व्यष्टिसमष्टिः<br>*Vyaṣṭisamaṣṭiḥ* | 11. लोपनस्थापनाभ्याम्<br>*Lopanasthāpanābhyām* |

12. शेषाण्यङ्केन चरमेण
*Śeṣāṇyaṅkena Carameṇa*
13. सोपान्त्यद्वयमन्त्यम्
*Sopāntyadvayamantyam*
14. एकन्यूनेन पूर्वेण
*Ekanyūnena Pūrveṇa*
15. गुणितसमुच्चयः
*Guṇitasamuccayaḥ*
16. गुणकसमुच्चयः
*Guṇakasamuccayaḥ*

12. विलोकनम्
*Vilokanam*
13. गुणितसमुच्चयः समुच्चयगुणितः
*Gunitasamuccayaḥ*
*Samuccayaguṇitaḥ*

[*Note*: This list has been compiled from stray references in the text.—EDITOR]

॥ ॐ श्रीः ॥

# Prolegomena

In our 'Descriptive, Prefatory Note on the Astounding Wonders of Ancient Indian Vedic Mathematics', we have again and again, so often and at such great length and with such wealth of detail, dwelt on the almost incredible simplicity of the Vedic Mathematical *Sūtras* (aphorisms or formulae) and the indescribable ease with which they can be understood, remembered and applied (even by little children) for the solution of the wrongly-believed-to-be 'difficult' problems in the various branches of Mathematics that we need not, at this point, traverse the same ground and cover the same field once again here.

Suffice it, for our present immediate purpose, to draw the earnest attention of every scientifically-inclined mind and re-searchward-attuned intellect, to the remarkably extraordinary and characteristic—nay, unique fact that the Vedic system does not academically countenance (or actually follow) any automatical or mechanical rule even in respect of the correct sequence or order to be observed with regard to the various subjects dealt with in the various branches of Mathematics (pure and applied) but leaves it entirely to the convenience and the inclination, the option, the temperamental predilection and even the individual idiosyncrasy of the teachers and even the students themselves as to what particular order or sequence they should actually adopt and follow!

This manifestly out-of-the-common procedure must doubtless have been due to some special kind of historical background, background which made such a consequence not only natural but also inevitable under the circumstances in question.

Immemorial tradition has it and historical research confirms the orthodox belief that the sages, seers and saints of ancient India who are accredited with having observed, studied and meditated in the *Araṇya* (i.e. in forest-solicitude)—on physical nature around them and deduced their grand *Vedantic Philosophy* therefrom as the result not only of their theoretical reasonings but also of what may be more fittingly described as True Realisation by means of Actual Visualisation seem to have similarly observed, studied and meditated on the mysterious workings of numbers, figures etc. of the mathematical world (to wit, Nature) around them and deduced their *Mathematical Philosophy* therefrom by a similar process of what one may, equally correctly, describe as processes of True Realisation by means of Actual Visualisation.

And, consequently, it naturally follows that, inasmuch as, unlike human beings who have their own personal prejudices, partialities, hatreds and other such *subjective* factors distorting their visions, warping their judgements and thereby contributing to their inconsistent or self-contradictory decisions and discriminatory attitudes, conducts etc. *numbers* in Mathematics labour under no such handicaps and disadvantages based on personal prejudices, partialities, hatreds etc. They are, on the contrary, strictly and purely *impersonal and objective* in their behaviour etc., follow the same rules uniformly, consistenly and invariably with no question of outlook, approach, personal psychology etc. involved therein and are therefore absolutely reliable and dependable.

This seems to have been the real historical reason why, barring a few unavoidable exceptions in the shape of elementary, basic and fundamental first principles (of a preliminary or prerequisite character), almost all the subjects dealt with in the various branches of Vedic Mathematics are explicable and expoundable on the basis of those very 'basic principles' or 'first principles', with the natural consequence that no particular subject or subjects (or chapter or chapters) need necessarily precede or follow some other particular subject or subjects (or chapter or chapters).

Nevertheless, it is also undeniable that, although *any* particular sequence is quite possible, permissible and feasible, yet, *some* particular sequence will actually have to be adopted by a teacher (and, much more, therefore, by an author). And so, we find that subjects like analytical conics and even calculus differential and integral (which is usually the bugbear and terror of even the advanced students of mathematics under the present system all the world over) are found to figure and fit in at a very early stage in our Vedic Mathematics because of their being expounded and worked out on basic first principles. And they help thereby to facilitate mathematical study especially for the children.

And, with our more-than-half-a-century's actual personal experience of the very young mathematics-students and their difficulties, we have found the Vedic sequence of subjects and chapters the most suitable for our purpose, namely, the eliminating from the children's minds of all fear and hatred of mathematics and the implanting therein of a positive feeling of exuberant love and enjoyment thereof! And we fervently hope and trust that other teachers too will have a similar experience and will find us justified in our ambitious description of this volume as "Mathematics without tears".

From the hereinabove described historical background to our Vedic Mathematics, it is also obvious that, being based on basic and fundamental principles, this system of mathematical study cannot possibly come into conflict with any other branch, department or instrument of science and scientific education. In fact, this is the exact reason why all other sciences have different *Theories* to propound but Mathematics has only theorems to expound!

And, above all, we have our Scriptures categorically laying down the wholesome dictum:

युक्तियुक्तं वचो ग्राह्यं बालादपि शुकादपि।
युक्तिहीनं वचस्त्याज्यं वृद्धादपि शुकादपि।।

i.e. whatever is consistent with right reasoning should be accepted, even though it comes from a boy or even from a parrot; and whatever is inconsistent therewith ought to be rejected, although emanating from an old man or even from the great sage Śrī Śuka himself.

In other words, we are called upon to enter on such a scientific quest as this, by divesting our minds of all pre-conceived notions, keeping our minds ever open and, in all humility (as humility alone behoves and befits the real seeker after truth), welcoming the light of knowledge from whatever direction it may be forthcoming. Nay, our scriptures go so far as to inculcate that even their expositions should be looked upon by us not as "teachings" or even as advice, guidance etc. but as acts of "thinking aloud" by a fellow student.

It is in this spirit and from this view-point that we now address ourselves to the task before us, in this series of volumes[1] (i.e. a sincere exposition of the mathematical *Sūtras* under discussion, with what we may call our "running comments" just as in a blackboard demonstration or a magic lantern lecture or a cricket match etc.

In conclusion, we appeal to our readers (as we always, appeal to

[1]Unfortunately, only one volume has been left over by His Holiness.
—Editor

our hearers) to respond hereto from the same standpoint and in the same spirit as we have just hereinabove described.

We may also add that, inasmuch as we have since long promised to make these volumes[2] "self-contained", we shall make our explanations and expositions as full and clear as possible. Brevity may be the soul of wit; but certainly not at the expense of clarity (and especially in mathematical treatises like these).

॥ ॐ तत् सत् ॥

---

[2]Unfortunately, only one volume has been left over by His Holiness. —Editor

# My Beloved Gurudeva

ŚRĪ BHĀRATĪ KṚṢṆA TĪRTHA

[In the lines that follow the writer gives a short biographical sketch of the illustrious author of Vedic Mathematics and a short account of the genesis of his work based on intimate personal knowledge. —EDITOR]

Very few persons can there be amongst the cultured people of India who have not heard about HIS HOLINESS JAGADGURU ŚAṄKARĀCĀRYA ŚRĪ BHĀRATĪ KṚṢṆA TĪRTHAJĪ MAHĀRĀJA the magnificent and divine personality that gracefully adorned the famous Govardhan Math, Puri, his vast and versatile learning, his spiritual and educational attainments, his wonderful research achievements in the field of Vedic Mathematics and his consecration of all these qualifications to the service of humanity as such.

His Holiness, better known among his disciples by the beloved name 'Jagadguruji' or 'Gurudeva', was born of highly learned and pious parents in March, 1884. His father, late Sri P. Narasimha Shastri, was then in service as a Tahsildar at Tinnivelly (Madras Presidency) who later retired as a Deputy Collector. His uncle, late Sri Chandrashekhar Shastri, was the Principal of the Maharaja's College, Vizianagaram and his great-grandfather was late Justice C. Ranganath Shastri of the Madras High Court.

Jagadguruji, named as Venkatraman in his early days, was an exceptionally brilliant student and invariably won the first place in all the subjects in all the classes throughout his educational career. During his school days, he was a student of National College, Trichanapalli; Church Missionary Society College, Tinnivelli and Hindu College, Tinnivelli. He passed his matriculation examination from the Madras University in January, 1899, topping the list as usual.

He was extraordinarily proficient in Sanskrit and oratory and on account of this he was awarded the title of 'Saraswati' by the Madras Sanskrit Association in July, 1899 when he was still in his 16th year. One cannot fail to mention at this stage the profound impression left on him by his Sanskrit Guru Sri Vedam Venkatrai Shastri whom Jagadguruji always remembered with deepest love, reverence and gratitude, with tears in his eyes.

After winning the highest place in the B.A. Examination, Sri Venkatraman Saraswati appeared at the M.A. Examination of the American College of Sciences, Rochester, New York, from Bombay Centre in 1903; and in 1904 at the age of just twenty he passed M.A. Examination in seven subjects simultaneously securing the highest honours in all, which is perhaps the all-time world-record of academic brilliance. His subjects included Sanskrit, Philosophy, English, Mathematics, History and Science.

As a student Venkatraman was marked for his splendid brilliance, superb retentive memory and ever-insatiable curiosity. He would deluge his teachers with myriads of piercing questions which made them uneasy and forced them frequently to make a frank confession of ignorance on their part. In this respect, he was considered to be a terribly mischievous student.

Even from his University days Sri Venkatraman Saraswati had started contributing learned articles on religion, philosophy, sociology, history, politics, literature etc., to late W.T. Stead's "Review of Reviews" and he was specially interested in all the branches of modern science. In fact study of the latest researches and discoveries in modern science continued to be Sri Jagadguruji's hobby till his very last days.

Sri Venkatraman started his public life under the guidance of late Hon'ble Sri Gopal Krishna Gokhale, C.I.E. in 1905 in connection with the National Education Movement and the South African Indian issue. Although, however, on the one hand, Prof. Venkatraman Saraswati had acquired an endless fund of learning and his desire to learn ever more was still unquenchable and on the other hand the urge for selfless service of humanity swayed his heart mightly, yet the undoubtedly deepest attraction that Venkatraman Saraswati felt was that towards the study and practice of the science of sciences—the holy ancient Indian spiritual science or *Adhyātma-Vidyā*. In 1908, therefore, he proceeded to the Sringeri Math in Mysore to lay himself at the feet of the renowned late Jagadguru Shankaracharya Maharaj Sri Satcidānanda Sivābhinava Nrisimha Bhārati Swami.

But he had not stayed there long, before he had to assume the post of the first Principal of the newly started National College at Rajmahendri under a pressing and clamant call of duty from the nationalist leaders. Prof. Venkatraman Saraswati continued there for three years but in 1911 he could not resist his burning desire for spiritual knowledge, practice and attainment any more and, therefore, tearing himself off suddenly from the said college he went back to Sri Satcidānanda Sivābhinava Nrisimha Bhārati Swami at Sringeri.

The next eight years he spent in the profoundest study of the most advanced Vedanta Philosophy and practice of the Brahma-sādhana. During these days Prof. Venkatraman used to study Vedānta at the feet of Sri Nrisimha Bhārati Swami, taught Sanskrit and Philosophy in schools there, and practise the highest and most vigorous Yoga-sādhana in the nearby forests. Frequently, he was also invited by several institutions to deliver lectures on philosophy; for example, he delivered a series of sixteen lectures on Shankaracharya's Philosophy at Shankar Institute of Philosophy, Amalner (Khandesh) and similar lectures at several other places like Poona, Bombay etc.

After several years of the most advanced studies, the deepest meditation, and the highest spiritual attainment Prof. Venkatraman Saraswati was initiated into the holy order of Saṃnyāsa at Banaras (Varanasi) by his Holiness Jagadguru Shankaracharya Sri Trivikram Tirthaji Maharaj of Shāradāpeeth on the 4th July 1919 and on this occasion he was given the new name, Swami Bhāratī Kṛṣṇa Tīrtha.

This was the starting point of an effulgent manifestation of Swāmiji's real greatness. Within two years of his stay in the holy order, he proved his unique suitability for being installed on the pontifical throne of Shārada Peeṭha Śaṅkarācārya and accordingly in 1921, he was so installed with all the formal ceremonies despite his reluctance and active resistance. Immediately on assuming the pontificate Sri Jagadguruji started touring India from corner to corner and delivering lectures on Sanātana Dharma and by his scintillating intellectual brilliance, powerful oratory, magnetic personality, sincerity of purpose, indomitable will, purity of thought, and loftiness of character he took the entire intellectual and religious class of the nation by storm.

Jagadguru Śaṅkarācārya Sri Madhusudan Tirtha of Govardhan Maṭh, Puri was at this stage greatly impressed by Jagadguruji and when the former was in failing health he requested Jagadguruji to succeed him on Govardhan Math Gadi. Sri Jagadguruji continued to resist his importunate requests for a long time but at last when Jagadguru Sri Madhusudan Tirtha's health took a serious turn in 1925 he virtually forced Jagadguru Śrī Bhāratī Kṛṣṇa Tīrthajī to accept the Govardhan Math's Gadi and accordingly Jagadguruji installed Sri Swarupanandji on the Shāradapeeth Gadi and himself assumed the duties of the ecclesiastical and pontifical head of Sri Govardhan Math, Puri.

In this capacity of Jagadguru Śaṅkarācārya of Govardhan Maṭh, Puri, he continued to disseminate the holy spiritual teachings of Sanātana Dharma in their pristine purity all over the world the rest of his life for

35 years. Month after month and year after year he spent in teaching and preaching, talking and lecturing, discussing and convincing millions of people all over the country. He took upon himself the colossal task of the renaissance of Indian culture, spreading of Sanātana Dharma, revival of the highest human and moral values and enkindling of the loftiest spiritual enlightenment throughout the world and he dedicated his whole life to this lofty and noble mission.

From his very early days Jagadguruji was aware of the need for the right interpretation of "*Dharma*" which he defined as "the sum total of all the means necessary for speedily making and permanently keeping all the people, individually as well as collectively superlatively comfortable, prosperous, happy, and joyous in all respects (including the physical, mental, intellectual, educational, economic, social, political, psychic, spiritual etc. *ad infinitum*)". He was painfully aware of the "escapism" of some from their duties under the garb of spirituality and of the superficial modern educational varnish of the others, divorced from spiritual and moral standards. He, therefore, always laid great emphasis on the necessity of harmonising the 'spiritual' and the 'material' spheres of daily life. He also wanted to remove the false ideas, on the one hand, of those persons who think that *Dharma* can be practised by exclusively individual spiritual Sādhanā coupled with more honest bread-earning, ignoring one's responsibility for rendering selfless service to the society and on the other hand of those who think that the Sādhanā can be complete by mere service of society even without learning or practising any spirituality oneself. He wanted a happy blending of both. He stood for the omnilateral and all-round progress simultaneously of both the individual and society towards the speedy realisation of India's spiritual and cultural ideal, the lofty Vedāntic ideal of 'Pūrṇatva' (perfection and harmony all-round).

With these ideas a gitating his mind for several decades he continued to carry on a laborious, elaborate, patient and day-and-night research to evolve finally a splendid and perfect scheme for all-round reconstruction first of India and through it of the world. Consequently Sri Jagadguruji founded in 1953 at Nagpur an institution named Sri Vishwa Punarnirmana Sangha (World Reconstruction Association). The Administrative Board of the Sangha consisted of Jagadguruji's disciples, devotees and admirers of his idealistic and spiritual ideals for humanitarian service and included a number of high court judges, ministers, educationists, statesmen and other personage of the highest calibre in Indian public life. It was, however, after a long and incessant search that Guruji had found his General Secretary Sri Chimanlal Trivedi whom he called his *Scipio Africanus* and who truly thought, worked, planned and dreemt unceasingly for the Sangha's welfare and progress. Although this Sangha could not function very effectively in the beginning on account of Jagadguruji's failing health, various pre-occupations and other unforeseen hurdles, it is actively engaged now in disseminating Jagadguruji's message and teachings with Justice B.P. Sinha, the Chief Justice of India as its President and Dr. C. D. Deshmukh, (I.C.S.) the ex-Finance Minister of India and ex-Chairman, University Grants Commission as its Vice-President.

With a view to promote the cause of world peace and to spread the lofty Vedāntic spiritual ideals even outside India Sri Jagadguru went on a tour to America in February, 1958, the first tour outside India by a Śaṅkarācārya in the history of the said Order. The tour was sponsored by Self Realisation Fellowship of Los Angeles, the Vedāntic Society founded by Paramhansa Yoganandji in America. Jagadguruji stayed there for about three months and during this period addressed rapt audiences in hundreds of colleges of universities, churches and other public institutions. He was also invited to give talks and mathematical demonstrations on the television. In fact, he released an exceptionally powerful current of moral and spiritual enlightenment, peace and harmony throughout America during his tour which proved a phenomenal success comparable perhaps with that of Swami Vivekananda only. A request was also received by him from Dr. Hornday, the Minister for Church of Religious Science to open a branch of Sri Vishwa Punarnirmana Sangha in America with a view to establish one religion all over the world. The suggestion, however, could not materialise at that time for certain reasons. On his way back Jagadguruji gave some lectures in U.K. also and returned to India in May, 1958.

Guruji had been undergoing a terrific strain for more than five decades in devoting his body, mind, heart and soul to the cause of service of humanity, spreading of spiritual enlightenment and revival of Vedāntic ideals. This had already undermined his health but still Guruji never devoted any attention to his personal comforts. The excessive strain of the vast hurricane tour abroad came as a severe blow to his health but still he refused to take rest and incessantly continued to pursue his studies, talks, lectures and writings with unabated and youthlike vigour and enthusiasm. In fact it required a great vigilance and heroic effort to prevent him from giving '*darshan*', advice and talks to his devotees and disciples even when he could hardly speak on account of strain. As a result he fell seriously ill in November, 1959 and despite the best available treatment shed off his mortal frame and took Mahāsamādhi at Bombay on 2nd February, 1960.

From the very day of his assuming the throne of Jagadguru Śaṅkarācārya, Śrī Bhāratī Kṛṣṇa Tīrthajī had become the cynosure of all eyes. His winning personality, his charming innocence, his eager thirst for knowledge, his religious zeal, his earnest belief in the "*śāstras*", his universal kindness, his retentive memory, all these attracted towards him every living soul that came in contact with him. People flocked to him in crowds and waited at his doors for hours together just to get a glimpse of that divine countenance. It was nothing but the divine lustre that shone in his face. It was nothing but the marvellous superhuman milk of kindness that flowed from his heart.

He was always perfectly impartial. Every one was equal in his eyes. He cared not for riches. He cared not for position. Nothing but *Bhakti* could attract people to him, rich or poor, high or low, everybody had to go through the portals of *Bhakti* to approach his august presence. Exhibiting his divinity, he loved as himself everyone came to him. Everyone who had even two minute's conversation with him went out with the full conviction that he was the object of some special love of His Holiness.

Of such a divine personality it is impossible to draw a sketch. His activities were many-sided. To hear him was a pleasure. To see him was a privilege. To speak to him was a real blessing and to be granted a special interview—Ah! that was the acme of happiness which people coveted most in all earnestness. The magnetic force of his wonderful personality was such that one word, one smile, or even one look was quite enough to convert even the most sceptic into his most ardent and obedient disciple. He belonged to all irrespective of caste or creed and he was a real Guru to the whole world.

People of all nationalities, religions and climes, Brahmins and non-Brahmins, Hindus and Mahomedans, Parsis and Christians, Europeans and Americans received equal treatment at the hands of His Holiness. That was the secret of the immense popularity of this great Mahātmā.

He was grand in his simplicity. People would give anything and everything to get his blessings and he would talk words of wisdom as freely without fear or favour. He was most easily accessible to all. Thousands of people visited him and prayed for the relief of their miseries. He had a kind word to say to each, after attentively listening to his or her tale of woe and then give them some "*prasad*" which would cure their malady whether physical or mental. He would actually shed tears when he found people suffering and would pray to God to relieve their suffering.

He was mighty in his learning and voracious in his reading. A sharp intellect, a retentive memory and a keen zest went to mark him as the most distinguished scholar of his day. His leisure moments he would never spend in vain. He was always reading something or repeating something. There was no branch of knowledge which he did not know and that also '*śāstrically*'. He was equally learned in Chandaḥśāstra, Āyurveda and Jyotish Śāstra. He was a poet of uncommon merit and wrote a number of poems in Sanskrit in the praise of his guru, gods and godesses with a charming flow of *Bhakti* so conspicuous in all his writings.

I have a collection of over three thousand *ślokas* forming part of the various eulogistic poems composed by Gurudeva in adoration of various Devas and Devis. These Ślokas have been edited and are being translated into Hindi. They are proposed to be published in three volumes along with Hindi translation.

The book on "Sanātana Dharma" by H.H Swāmī Bhāratī Kṛṣṇa Tīrtha Mahārāja has been published by Bharatiya Vidya Bhavan, Bombay.

Above all, his *Bhakti* towards his Vidyāguru was something beyond description. He would talk for days together about the greatness of his Vidyāguru. He would be never tired of worshipping the Guru. His Guru also was equally attached to him and called our Swāmījī as the own son of the Goddess of Learning, Śrī Śāradā. Everyday he would first worship his guru's sandals. His "Gurupādukā Stotra" clearly indicates the qualities he attributed to the sandals of his guru.

Śrī Bhāratī Kṛṣṇa Tīrtha was a great *Yogin* and a "*Siddha*" of a very high order. Nothing was impossible for him. Above all he was a true Samnyāsin. He held the world but as a stage where every one had to play a part. In short, he was undoubtedly a very great Mahātmā but without any display of mysteries or occultisms.

I have not been able to express here even one millionth part of what I feel. His spotless holiness, his deep piety, his endless wisdom, his childlike peacefulness, sportiveness and innocence and his universal affection are beyond all description. His Holiness has left us a noble example of simplest living and highest thinking. May all the world benefit by the example of a life so nobly and so simply, so spiritually and so lovingly lived.

### Introductory Remarks on the Present Volume

I now proceed to give a short account of the genesis of the work published here Revered Guruji used to say that he had reconstructed the sixteen mathematical formulae (given in this text) from the Atharvaveda after assiduous research and 'Tapas' for about eight years in the forests surrounding Sringeri. Obviously these formulae are not to be found in the present recensions of Atharvaveda; they were actually reconstucted, on the basis of intuitive revelation, from materials scattered here and there in the Atharvaveda. Revered Gurudeva used to say that he had written sixteen volumes on these *Sūtras*, one for each *Sūtra* and that the manuscripts of the said volumes were deposited at the house of one of his disciples. Unfortunately, the said manuscripts were lost irretrievably from the place of their deposit and this colossal loss was finally confirmed in 1956. Revered Gurudeva was not much perturbed over this irretrievable loss and used to say that everything was there in his memory and that he could re-write the 16 volumes!

My late husband Sri C.M. Trivedi, Hon. Gen. Secretary V.P. Sangh noticed that while Śrī Jagadguru Mahārāja was busy demonstrating before learned people and societies Vedic Mathematics as discovered and propounded by him, some persons who had grasped a smattering of the new *Sūtras* had already started to dazzle audiences as prodigies claiming occult powers without knowledging indebtedness of the *Sūtras* of Jagadguruji. My husband, therefore, pleaded earnestly with Gurudeva and persuaded him to arrange for the publication of the *Sūtras* in his own name.

In 1957, when he had decided finally to undertake a tour of the U.S.A. he re-wrote from memory the present volume, giving an introductory account of the sixteen formulae reconstructed by him. This volume was written in his old age within one month and a half with his failing health and weak eyesight. He had planned to write subsequent volumes, but his failing health (and cataract developed in both eyes) did not allow the fulfilment of his plans. Now the present volume is the only work on Mathematics that has been left over by Revered Guruji; all his other writings on Vedic Mathematics have, alas, been lost for ever.

The typescript of the present volume was left over by Revered Gurudeva in U.S.A. in 1958 for publication. He had been given to understand that he would have to go to the U.S.A. for correction of proofs and personal supervision of printing. But his health deteriorated after his return to India and finally the typescript was brought back from the U.S.A. after his attainment of Mahāsamādhi, in 1960.

### Acknowledgements

I owe a deep debt of gratitude to Justice N.H. Bhagwati, the enlightened Vice-Chancellor of the Banaras Hindu University and other authorities of the B.H.U. who have readily undertaken the publication of this work which was introduced to them by Dr. Pt. Omkarnath Thakur. I am indebted to Dr. Thakur for this introduction. My hearty and reverent thanks are due to Dr. V.S. Agrawala (Professor, Art & Architecture,

B.H.U.) the veteran scholar, who took the initiative and throughout kept up a very keen interest in this publication. It is my pleasant duty to offer my heartfelt gratitude to Dr. Prem Lata Sharma, Dean, Faculty of Music and Fine Arts, B.H.U. who voluntarily took over the work of press-dressing of the typescript and proof-reading of this volume after a deadlock had come to prevail in the process of printing just at the outset. But for her hard labour which she has undertaken out of a sheer sense of reverence for the noble and glorious work of Revered Gurudeva this volume would not have seen the light of the day for a long time. I trust that Revered Gurudeva's Holy Spirit will shower His choicest blessings on her. My sincere thanks are also due to Sri S. Nijabodha of the Research Section under the charge of Dr. Sharma, who has ably assisted her in this onerous task.

The Humblest of His Disciples
MANJULA TRIVEDI
*Honorary General Secretary*
Sri Vishwa Punarnirmana Sangha,
Nagpur

*Nagpur*
16 March, 1965

# Author's Preface

## A. A Descriptive Prefatory Note on the Astounding Wonders of Ancient Indian Vedic Mathematics

1. In the course of our discourses on manifold and multifarious subjects (spiritual, metaphysical, philosophical, psychic, psychological, ethical, educational, scientific, mathematical, historical, political, economic, social etc., from time to time and from place to place during the last five decades and more, we have been repeatedly pointing out that the Vedas (the most ancient Indian scriptures, nay, the oldest "Religious" scriptures of the whole world) claim to deal with all branches of learning (spiritual and temporal) and to give the earnest seeker after knowledge all the requisite instructions and guidance in full detail and on scientifically—nay, mathematically—accurate lines in them all and so on.

2. The very word "Veda" has this derivational meaning, i.e. the fountain-head and illimitable store-house of all knowledge. This derivation, in effect, means, connotes and implies that the Vedas should contain within themselves all the knowledge needed by mankind relating not only to the so-called 'spiritual' (or other-worldly) matters but also to those usually described as purely "secular", "temporal", or "wordly"; and also to the means required by humanity as such for the achievement of all-round, complete and perfect success in all conceivable directions and that there can be no adjectival or restrictive epithet calculated (or tending) to limit that knowledge down in any sphere, any direction or any respect whatsoever.

3. In other words, it connotes and implies that our ancient Indian Vedic lore should be all-round, complete and perfect and able to throw the fullest necessary light on all matters which any aspiring seeker after knowledge can possibly seek to be enlightened on.

4. It is thus in the fitness of things that the Vedas include (i) *Āyurveda* (anatomy, physiology, hygiene, sanitary science, medical science, surgery etc.) not for the purpose of achieving perfect health and strength in the after-death future but in order to attain them *here and now* in our

present physical bodies; (ii) *Dhanurveda* (archery and other military sciences) not for fighting with one another after our transportation to heaven but in order to quell and subdue all invaders from abroad and all insurgents from within; (iii) *Gandharva Veda* (the science and art of music); and (iv) *Sthāpatya Veda* (engineering, architecture etc., and all branches of mathematics in general). All these subjects, be it noted, are inherent parts of the Vedas, i.e. are reckoned as "spiritual" studies and catered for as such therein.

5. Similar is the case with regard to the *Vedāṅgas* (i.e. grammar, prosody, astronomy, lexicography etc.,) which, according to the Indian cultural conceptions, are also inherent parts and subjects of *Vedic (i.e. Religious) study*.

6. As a direct and unshirkable consequence of this analytical and grammatical study of the real connotation and full implications of the word "Veda" and owing to various other historical causes of a personal character (into details of which we need not now enter), we have been from our very early childhood, most earnestly and actively striving to study the Vedas critically from this stand-point and to realise and prove to ourselves (and to others) the correctness (or otherwise) of the derivative meaning in question.

7. There were, too, certain personal historical reasons why in our quest for the discovering of all learning in all its departments, branches, sub-branches etc., in the Vedas, our gaze was riveted mainly on ethics, psychology and metaphysics on the one hand and on the "positive" sciences and especially mathematics on the other.

8. And the contemptuous or, at best patronising attitude adopted by some so-called Orientalists, Indologists, antiquarians, research-scholars etc., who condemned, or light heartedly, nay irresponsibly, frivolously and flippantly dismissed, several abstruse-looking and recondite parts of the Vedas as "sheer-non-sense"—or as "infant-humanity's prattle", and so on, merely added fuel to the fire (so to speak) and further confirmed and strengthened our resolute determination to unravel the too-long hidden mysteries of philosophy and science contained in ancient India's Vedic lore, with the consequence that, after eight years of concentrated contemplation in forest-solitude, we were at long last able to recover the long lost keys which alone could unlock the portals thereof.

9. And we were agreeably astonished and intensely gratified to find that exceedingly tough mathematical problems (which the mathematically most advanced present day Western scientific world had spent huge lots of time, energy and money on and which even now it solves with the utmost difficulty and after vast labour involving large numbers of difficult, tedious and cumbersome "steps" of working) can be easily and readily

solved with the help of these ultra-easy Vedic *Sūtras* (or mathematical aphorisms) contained in the *Pariśiṣṭa* (the Appendix-portion) of the *Atharvaveda* in a few simple steps and by methods which can be conscientiously described as mere "mental arithmetic".

10. Ever since (i.e. since several decades ago), we have been carrying on an incessant and strenuous campaign for the India-wide diffusion of all this scientific knowledge, by means of lectures, blackboard-demonstrations, regular classes and so on in schools, colleges, universities etc., all over the country and have been astounding our audiences everywhere with the wonders and marvels not to say, miracles of Indian Vedic mathematics.

11. We were thus able to succeed in attracting the more-than-passing attention of the authorities of several Indian universities to this subject. And, in 1952, the Nagpur University not merely had a few lectures and blackboard-demonstrations given but also arranged for our holding regualr classes in Vedic mathematics (in the University's Convocation Hall) for the benefit of all in general and especially of the University and college professors of mathematics, physics etc.

12. And, consequently, the educationists and the cream of the English educated section of the people including the highest officials (e.g. high-court judges, ministers etc.,) and the general public as such were all highly impressed nay, thrilled, wonder-struck and flabbergasted! And not only the newspapers but even the University's official reports described the tremendous sensation caused thereby in superlatively eulogistic terms; and the papers began to refer to us as "the Octogenarian Jagadguru Śaṅkarācārya who had taken Nagpur by storm with his Vedic mathematics", and so on !

13. It is manifestly impossible, in the course of a short note (in the nature of a "trailer"), to give a full, detailed, thorough-going, comprehensive and exhaustive description of the unique features and startling characteristics of all the mathematical lore in question. This can and will be done in the subsequent volumes of this series (dealing seriatim and in extenso with all the various portions of all the various branches of mathematics).

14. We may, however, at this point, draw the earnest attention of everyone concerned to the following salient items thereof:

(*i*) The *Sūtras* (aphorisms) apply to and cover each and every part of each and every chapter of each and every branch of mathematics (including arithmetic, algebra, geometry—plane and solid, trigonometry —plane and spherical, conics—geometrical and analytical, astronomy, calculus—differential and integral etc.). In fact, there is no part of mathematics, pure or applied, which is beyond their jurisdiction;

(*ii*) The *Sūtras* are easy to understand, easy to apply and easy to remember; and the whole work can be truthfully summarised in one word "mental" !

(*iii*) Even as regards complex problems involving a good number of mathematical operations (consecutively or even simultaneously to be performed), the time taken by the Vedic method will be a third, a fourth, a tenth or even a much smaller fraction of the time required according to modern Western methods;

(*iv*) And, in some very important and striking cases, sums requiring 30, 50, 100 or even more numerous and cumbersome "steps" of working (according to the current Western methods) can be answered in a single and simple step of work by the Vedic method! And children of even 10 or 12 years of age merely look at the sums written on the blackboard (on the platform) and immediately shout out and dictate the answers from the body of the convocation hall (or other venue of the demonstration). And this is because, as a matter of fact, each digit automatically yields its predecessor and its successor! and the children have merely to go on tossing off (or reeling off) the digits one after another (forwards or backwards)by mere mental arithmetic (without needing pen or pencil, paper or slate etc.)!

(*v*) On seeing this kind of work actually being performed by children, the doctors, professors and other "big-guns" of mathematics are wonder-struck and exclaim : "Is this mathematics or magic"? And we invariably answer and say: "It is both. It is magic until you understand it; and it is mathematics thereafter"; and then we proceed to substantiate and prove the correctness of this reply of ours !

(*vi*) As regards the time required by the students for mastering the whole course of Vedic mathematics as applied to all its branches, we need merely state from our actual experience that 8 months (or 12 months) at an average rate of 2 or 3 hours per day should suffice for completing the whole course of mathematical studies on these Vedic lines instead of 15 or 20 years required according to the existing systems of the Indian and also of foreign universities

15. In this connection, it is a gratifying fact that unlike some so-called Indologists (of the type herein above referred to) there have been some great modern mathematicians and historians of mathematics (like Prof. G.P. Halstead, Professor Ginsburg, Prof. De Moregan, Prof. Hutton etc.,) who have, as truth-seekers and truth-lovers, evinced a truly scientific attitude and frankly expressed their intense and whole-hearted appreciation of ancient India's grand and glorious contributions to the progress of mathematical knowledge (in the Western hemisphere and elsewhere).

16. The following few excerpts from the published writings of some universally acknowledged authorities in the domain of the history of mathematics, will speak eloquently for themselves:

(*i*) On page 20 of his book "On the Foundation and Technique of Arithmetic", we find Prof. G.P. Halstead saying "The importance of the creation of the zero mark can never be exaggerated. This giving of airy nothing not merely a local habitation and a name, a picture but helpful power is the characteristic of the Hindu race whence it sprang. It is like coining the Nirvāṇa into dynamos. No single mathematical creation has been more potent for the general on-go of intelligence and power".

(*ii*) In this connection, in his splendid treatise on "The present mode of expressing numbers" (the *Indian Historical Quarterly,* Vol. 3, pages 530-540) B.B. Dutta says : "The Hindus adopted the decimal scale vary early. The numerical language of no other nation is so scientific and has attained as high a state of perfection as that of the ancient Hindus. In symbolism they succeeded with ten signs to express any number most elegantly and simply. It is this beauty of the Hindu numerical notation which attracted the attention of all the civilised peoples of the world and charmed them to adopt it".

(*iii*)In this very context, Prof. Ginsburg says:

"The Hindu notation was carried to Arabia about 770 A.D. by a Hindu scholar named Kaṅka who was invited from Ujjain to the famous Court of Baghdad by the Abbaside Khalif Al-Mansur. Kaṅka taught Hindu astronomy and mathematics to the Arabian scholars; and, with his help, they translated into Arabic the *Brahma-Sphuṭa-Siddhānta* of Brahma Gupta. The recent discovery by the French savant M.F. Nau proves that the Hindu numerals were well-known and much appreciated in Syria about the middle of the seventh centuary A.D." (Ginsburg's "New Light on our numerals", *Bulletin of the American Mathematical Society,* Second series, Vol. 25, pages 366-369).

(*iv*) On this point, we find B.B. Dutta further saying:

"From Arabia, the numerals slowly marched towards the West through Egypt and Northern Arabia; and they finally entered Europe in the eleventh century. The Europeans called them the Arabic notations, because they received them from the Arabs. But the Arabs themselves, the Eastern as well as the Western, have unanimously called them the Hindu figures (*Al-Arqan-Al-Hindu*)".

17. The above-cited passages are, however, in connection with and in appreciation of India's invention of the "Zero" mark and her contributions of the 7th century A.D. and later to world mathematical knowledge.

In the light, however, of the hereinabove given detailed description of the unique merits and characteristic excellences of the still earlier Vedic *Sūtras* dealt with in the 16 volumes of this series,* the conscientious (truth-loving and truth-telling) historians of Mathematics (of the lofty eminence of Prof. De Morgan etc.) have not been guilty of even the least exaggeration in their candid admission that "even the highest and farthest reaches of modern Western mathematics have not yet brought the Western world even to the threshold of Ancient Indian Vedic Mathematics".

18. It is our earnest aim and aspiration, in these 16 volumes,* to explain and expound the contents of the Vedic mathematical *Sūtras* and bring them within the easy intellectual reach of every seeker after mathematical knowledge.

### B. Explanatory Exposition of Some Salient, Instructive and Interesting Illustrative Specimens by Way of Comparison and Contrast

*Preliminary Note*:

1. With regard to every subject dealt with in the Vedic Mathematical *Sūtras*, the rule generally holds good that the *Sūtras* have always provided for what may be termed the 'General Case' by means of simple processes which can be easily and readily—nay, instantaneously applied to any and every question which can possibly arise under any particular heading.

2. But, at the same time, we often come across special cases which, although classifiable under the general heading in question, yet present certain additional and typical characteristics which render them still easier to solve. And, therefore, special provision is found to have been made for such special cases by means of special *Sūtras*, sub-*Sūtras*, corollaries etc., relating and applicable to those particular types alone.

3. And all that the student of these *Sūtras* has to do is to look for the special characteristics in question, recognise the particular type before him and determine and apply the special formula prescribed therefor.

4. And, generally speaking it is only *in case* no special case is involved, that the general formula has to be resorted to. And this process is naturally a little longer. But it need hardly be pointed out that, even

* Only one volume has been bequeathed by His Holiness to posterity cf. p. xxxviii—General Editor

then, the longest of the methods according to the Vedic system comes nowhere in respect of length, cumbrousness and tediousness etc., near the corresponding process according to the system now current everywhere.

5. For instance, the conversion of a vulgar fraction (say $\frac{1}{19}$ or $\frac{1}{29}$ or $\frac{1}{49}$ etc.,) to its equivalent recurring decimal shape involves 18 or 28 or 42 or more steps of cumbrous working (according to the current system) but requires only one single and simple step of mental working (according to the Vedic *Sūtras*))!

6. This is not all. There are still other methods and processes (in the latter system) whereby even that very small (mental) working can be rendered shorter still ! This and herein is the beatific beauty of the whole scheme.

7. To start with, we should naturally have liked to begin this explanatory and illustrative exposition with a few processes in arithmetical computations relating to multiplications and divisions of huge numbers by big multipliers and big divisors respectively and then go on to other branches of mathematical calculation.

8. But, as we have just hereinabove referred to a particular but wonderful type of mathematical work wherein 18, 28, 42 or even more steps of working can be condensed into a single-step answer which can be written down immediately by means of what we have been describing as straight, single-line-mental arithmetic; and, as this statement must naturally have aroused intense eagerness and curiosity in the minds of the students, and the teachers too and especially as the process is based on elementary and basic fundamental principles and requires no previous knowledge of anything in the nature of an indispensable and inescapable prerequisite chapter, subject and so on, we are beginning this exposition here with an easy explanation and a simple elucidation of that particular illustrative specimen.

9. And then we shall take up the other various parts, one by one, of the various branches of mathematical computation and hope to throw sufficient light thereon to enable the students to make their own comparison and contrast and arrive at correct conclusions on all the various points dealt with.

## C. Illustrative Samples: Comparison and Contrast Specimens of Arithmetical Computations

I. *Multiplication:*

*The Sanskrit Sūtra (Formula) is:*

।। ऊर्ध्वतिर्यग्भ्याम्।।

(i) Multiply 87265 by 32117

*By current method:*

```
       87265
       32117
    --------
      610855
      87265
     87265
    174530
   261795
  ----------
  2802690005
  ----------
```

*By Vedic mental one-line method:*

```
    87265
    32117
 ----------
 2802690005
 ----------
```

*Note:* Only the answer is written automatically down by *Ūrdhva Tiryak Sūtra* (forwards or backwards).

II. *Division:*

(2) Express $\frac{1}{19}$ in its full recurring decimal shape (18 digits):

*By the current method:*

```
19)1.00(.0̇5263157894736842i
    95
    --
     50
     38
     ---
     120
     114
     ---
      60
      57
      --
       30
       19
       ---
       110
        95
        ---
        150
        133
        ---
         170
         152
         ---
          180
          171
          ---
            90
            76
            --
```

*The Sanskrit Sūtra (Formula) is:*

।। एकाधिकेन पूर्वेण ।।

*By the Vedic mental one-line method:*

(by the *Ekādhika Pūrva Sūtra*) (forwards or backwards), we merely write down the 18-digit answer:-

0̇52631578
94736842i

```
140              40
133              38
----             --
  70              20
  57              19
  ---             --
  130              1
  114
  ---
   160
   152
   ---
     80
     76
     --
     40
```

*Division continued:*

*Note:* $\frac{1}{49}$ gives 42 recurring decimal places in the answer but these too are written down mechanically in the same way (backwards or forwards). And the same is the case with all such divisions (whatever the number of digits may be):

3) *Divide* 7031985 by 823:

*By the current method:*

```
823)7031985(8544
    6584
    ----
     4479
     4115
     ----
      3648
      3292
      ----
       3565
       3292
       ----
        273
```

$\therefore Q = 8544$
$R = 273$

*By the mental Vedic one–line method:*

```
823)70319(85
      675
    --------
    8544(273
```

(4) *Divide .00034147 by 81425632 (to 6 decimal places)*

The current method is notoriously too long, tedious, cumbrous and clumsy and entails the expenditure of enormous time and toil. Only the Vedic mental one-line method is given here. The truth-loving student can work it out by the other method and compare the two for himself.

```
8 / 1425632).00034147
            )     3295
            ____________
            .0000419...
```

(5) *Find the reciprocal of 7246041 to eleven Decimal places:*
*By the Vedic mental one-line-method.*
(by the *Ūrdhva-Tiryak Sūtra*)

```
7 / 246041).000001000000
                  374610
            ______________
            .00000013800...
```

*N.B.*: The same method can be used for 200 or more places.

III. *Divisibility*

(6) *Find out whether 5293240096 is divisible by 139 :*

By the current method, nothing less than complete division will give a clue to the answer Yes or No.

But by the Vedic mental one-line method (by the *Ekādhika Pūrva Sūtra*), we can at once say:

```
for)    5   2   9   3    2   4    0   0   9   6 }
139)  139  89  36  131  29  131  19  51  93     } ∴ Yes
```

IV. *Square Root:*

(7) *Extract the square root of 738915489:*

*By the current method:*

```
  7̇3̇8̇91̇54̇8̇9̇(27183
  4
47) 338
    329
541) 991
     541
5428) 45054
      43424
54363) 163089
       163089
       ______
          0
```

*By the Vedic mental one-line method:*

```
4)738915489
   35513674
  __________
  27183.000  Ans.
```

(*By the Ūrdhva-Tiryak Sūtra*)

∴ *The square root is 27183*

(8) *Extract the square root of 19.706412814 to 6 decimal places:*
The current method is too cumbrous and may be tried by the student himself.

The Vedic mental one-line method (by *Ūrdhva-Tiryak Sūtra*) is as follows:

```
8)19.706412813
     .351010151713
  ----------------
   4.439190 . . .
  ----------------
```

V. *Cubing and Cube-Root:*

*The Sanskrit Sūtra (Formula) is :*

।।यावदूनं तावदूनीकृत्य वर्गं च योजयेत्।।

(9) *Find the cube of 9989.*

The current method is too cumbrous.

The Vedic mental one-line method by the *Yāvadūnam-Tāvadūnam Sūtra* is as follows:

$$9989^3 = 9967 / 0363 / 1331 = 9967 / 0362 / 8669$$

(10) *Extract the Cube-Root of 355045312441*:

The current method is too cumbrous.

The Vedic mental one-line method is as follows :

$$\sqrt[3]{355045312441} = 7\ .\ .1 = 7081$$

## Specimens from Algebra

I. *Sample Equations:*

(11) Solve : $\frac{3x+4}{6x+7} = \frac{x+1}{2x+3}$

*By the current method:*

$\therefore 6x^2 + 17x + 12 =$
$6x^2 + 13x + 7$

$\therefore 4x = -5$

$\therefore\ x = -1\frac{1}{4}$

*The Sanskrit Sūtra (Formula) is:*

।।शून्यं साम्यसमुच्चये।।

By the Vedic method by the *Sūnyam-Samuccaya Sūtra*

$\therefore 4x + 5 = 0 \therefore x = -1\frac{1}{4}$

(12) $\frac{4x+21}{x+5} + \frac{5x-69}{x-14} = \frac{3x-5}{x-2} + \frac{6x-41}{x-7}$

The current method is too cumbrous.

The Vedic method simply says : $2x - 9 = 0 \therefore\ x = 4\frac{1}{2}$

(13) $\left(\frac{x-5}{x-7}\right)^3 = \frac{x-3}{x-9}$

The current method is horribly cumbrous.

The Vedic method simply says : $4x-24=0$ $x=6$.

II. *Quadratic-Equations (and Calculus)*:

The same is the case here.

(14) $\frac{16x-3}{7x+7}=\frac{2x-15}{11x-25}\therefore x=1 \text{ or } 10/9$

(15) $\frac{3}{x+3}+\frac{4}{x+4}=\frac{2}{x+2}+\frac{5}{x+5}\therefore x=0 \text{ or } -7/2$

(16) $7x^2-11x-7=0$

By Vedic method (by *Calana-kalana Sūtra*) by Calculus Formula we say : $14x-11=\pm\sqrt{317}$.

*N.B.*: Every quadratic can thus be broken down into two binomial-factors. And the same principle can be utilised for cubic, biquadratic, pentic etc., expressions.

III. *Summation of Series*:

The current methods are horribly cumbrous. The Vedic mental one-line methods are very simple and easy.

(17) $\frac{1}{56}+\frac{1}{72}+\frac{1}{90}+\frac{1}{110}=4/77.$

(18) $\frac{3}{70}+\frac{9}{190}+\frac{27}{874}+\frac{99}{6670}=\frac{138}{1015}$

## Specimen From Geometry

(19) Pythagoras' Theorem is constantly required in all mathematical work, but the proof of it is ultra-notorious for its cumbrousness, clumsiness, etc. There are several Vedic proofs thereof (everyone of them much simpler than *Euclid's*). I give two of them below :

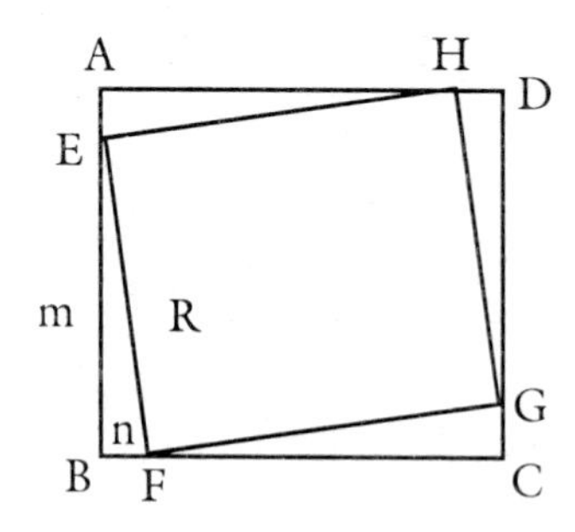


E,F, G and H are points on AB, BC, CD and DA such that AE=BF=CG=DH. Thus ABCD is split up into the square EFGH and 4 congruent triangles.

Their total area $= h^2+4x\frac{1}{2}\times mn$

$\rightarrow (h^2+4x\frac{1}{2}mn)$

$= h^2+2mn$

But the area of ABCD is $(m+n)^2$ $=m^2+2mn+n^2$

$\therefore h^2+2mn=m^2+2mn+n^2$

$\therefore h^2=m^2+n^2$. Q. E. D.

(20) *Second Proof*:

Draw BD $\perp$ to AC.

Then ABC, ADB and BDC are similar.

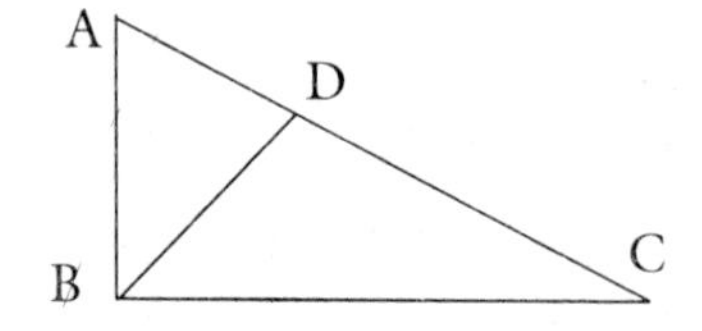

$\therefore \frac{ADB}{ABC} = \frac{AB^2}{AC^2}$ and $\frac{BDC}{ABC} = \frac{BC^2}{AC^2}$

$\therefore \frac{ADB + BDC}{ABC} = \frac{AB^2 + BC^2}{AC^2}$. But ADB + BDC = ABC

$\therefore AB^2 + BC^2 = AC^2$. Q.E.D.

*Note:* Apollonius Theorem, Ptolemy's Theorem, etc., are all similarly proved by very simple and easy methods.

### Specimens From Conics and Calculus

(21) *Equation of the straight line joining two points:*
For finding the equation of the straight line passing through two points (whose co-ordinates are given).
Say (9,17) and (7,–2).
*By the Current Method:*
Let the equation by y=mx+c.

$\therefore 9m + c = 17$; and $7m + c = -2$

Solving this simultaneous-equation in m and c.

We have 2m = 19; $\therefore m = 9\frac{1}{2} \therefore C = -68\frac{1}{2}$

Substituting these values, we have $y = 9\frac{1}{2}x - 68\frac{1}{2}$

$\therefore 2y = 19x - 137 \therefore 19x - 2y = 137$. But this method is cumbrous.

*Second method* using the formula $y - y' = \frac{y'' - y'}{x'' - x'}(x - x')$

is still more cumbrous (and confusing).
But the Vedic mental one-line method by the Sanskrit *Sūtra* (Formula), परावर्त्य योजयेत् (*Parāvartya-Sūtra*) enables us to write down the answer by a mere look at the given co-ordinates.

(22) *When does a general-equation represent two straight lines?*
Say, $12x^2+7xy-10y^2+13x+45y-35=0$

*By the Current Method*:
Prof. S.L. Loney devotes about 15 lines (section 119, Ex. 1 on page 97 of his "Elements of Co-ordinate Geometry") to his "model" solution of this problem as follows :
Here a=12, h=7/2, b= –10, g=13/2, f=45/2 and c=–35.

$\therefore abc + 2fgh - af^2 - bg^2 - ch^2$ turns out to be zero.

$$-12(-10)(-35) + \frac{2 \times 45}{2} \times \frac{13}{2} \times \frac{7}{2} - 12\left(\frac{45}{2}\right)^2 - (-10)\left(\frac{13}{2}\right)^2$$

$$-(-35)\left(\frac{7}{2}\right)^2 = 4200 + \frac{4095}{4} - 6075 + \frac{1690}{4} + \frac{1715}{4} - 1875 + \frac{7500}{4} = 0$$

The equation represents two straight lines.
Solving it for x, we have :

$$x^2 + \frac{7y+13}{12}x + \left(\frac{7y+13}{24}\right)^2 = \frac{10y^2-45y+35}{12} + \left(\frac{7y+13}{24}\right)^2$$

$$= \left(\frac{23y-43}{24}\right)^2$$

$$\therefore x + \frac{7y+13}{24} = \frac{23y-43}{24}$$

$$\therefore x = \frac{2y-7}{3} \text{ or } -\frac{5y+5}{4}$$

∴ The two straight lines are $3x - 2y - 7$ and $4x = -5y + 5$.

*By the Vedic method,* however, we at once apply the *Ādyamādyena Sūtra* and by merely looking at the quadratic write down the answer : Yes, and the straight lines are $3x-2y=-7$ and $4x+5y=5$.

(23) Dealing with the same principle and adopting the same procedure with regard to hyperbolas, conjugate hyperbolas and asymptotes, in articles 324 and 325 on pages 293 and 294 of his "Elements of Co-ordianate Geometry" Prof. S.L. Loney devotes 41 lines to the problem and says: As $3x^2-5xy-2y^2+5x-11y-8=0$ is the equation to the given hyperbola.

$$\therefore 3(-2)c + 2.\tfrac{5}{2}.\tfrac{11}{2}\left(-\tfrac{5}{2}\right) - 3\left(\tfrac{11}{2}\right)^2 - (-2)\left(\tfrac{5}{2}\right)^2 - C\left(-\tfrac{5}{2}\right)^2 = 0.$$

$$\therefore c = -12.$$

∴ The equation to the asymptotes is $3x^2 - 5xy - 2y^2 + 5x - 11y - 12 = 0$ and the equation to the conjugate-hyperbola is
$3x^2 - 5xy - 2y^2 + 5x + 15y - 16 = 0$

*By the Vedic method,* however, we use the same *Ādyam-ādyena Sūtra* and automatically write down the equation to the asymptotes and the equation to the conjugate-hyperbola.

The Vedic methods are so simple that their very simplicity is astounding, and, as Desmond Doig has aptly, remarked, *it is difficult for anyone to believe it until one actually sees it.*

It will be our aim in this and the succeeding volumes* to bring this long-hidden treasure-trove of mathematical knowledge within easy reach of everyone who wishes to obtain it and benefit by it.

*April, 2000*
New York (U.S.A.)

SWĀMĪ BHĀRATĪ KṚṢṆA TĪRTHA JĪ
JAGADGURU ŚAṄKARĀCĀRYA

* This is the only volume left by the author—EDITOR.

# 1

# Actual Applications of the Vedic Sūtras to Concrete Mathematical Problems

## A SPECTACULAR ILLUSTRATION

For the reasons just explained immediately hereinbefore let us take the question of the conversion of vulgar fractions into their equivalent decimal form.

### First Example

Case I Let us first deal with the case of a fraction say 1/19. 1/19 whose denominator ends in 9.

**By the Current Method**

```
19) 1.00(.0 5 2 6 3 1 5 7 8 }
     95 (9 4 7 3 6 8 4 2 1  }
    ---
     50        170
     38        152
    ---        ---
    120        180
    114        171
    ---        ---
      60        90     160
      57        76     152
      --        --     ---
      30       140      80
      19       133      76
      --       ---      --
     110        70      40
      95        57      38
     ---        --      --
     150       130      20
     133       114      19
     ---       ---      --
      170      160       1
```

**By the Vedic One-line Mental Method**

**A. First Method**

```
1/19 = .0̇ 5 2 6 3 1 5 7 8 }
        1 1     1 1 1 1   }
         9 4 7 3 6 8 4 2 1̇ }
          1   1 1         }
```

**B. Second Method**

```
1/19 = .0̇ 5 2 6 3 1 5 7 8 / 9 4 7 3 6 8 4 2 1̇
        1 1     1 1 1 1 /   1 1 1
```

This is the whole working. And the *modus operandi* is explained in the next few pages.

It is thus apparent that the 18-digit recurring-decimal answer requires 18 steps of working according to the current system but only *one* by the Vedic Method.

## Explanation

The relevant *Sūtra* reads: एकाधिकेन पूर्वेण (*Ekādhikena Pūrvena*) which, rendered into English, simply says: "By one more than the previous one". Its application and *modus operandi* are as follows:

(*i*) The last digit of the denominator in this case being 1 and the previous one being 1, "one more than the previous one" evidently means 2.

(*ii*) And the preposition "by" (in the *Sūtra*) indicates that the arithmetical operation prescribed is either multiplication or division. For, in the case of addition and subtraction, *to* and *from* respectively would have been the appropriate preposition to use. But "*by*" is the preposition actually found used in the *Sūtra*. The inference is therefore obvious that either multiplication or division must be enjoined. And, as both the meanings are perfectly correct and equally tenable according to grammar and literary usage and as there is no reason—in or from the text—for one of the meanings being accepted and the other one rejected, it further follows that *both* the processes are actually meant. And, as a matter of fact, each of them actually serves the purpose of the *Sūtra* and fits right into it as we shall presently show, in the immediately following explanation of the *modus operandi* which enables us to arrive at the right answer by either operation.

## A. The First Method

The first method is by means of multiplication by 2 (which is the "*Ekādhika Pūrva*", i.e. the number which is just one more than the penultimate digit in this case).

Here, for reasons which will become clear presently, we can know beforehand that the last digit of the answer is bound to be 1. For, the relevant rule hereon (which we shall explain and expound at a later stage) stipulates that the product of the last digit of the denominator and the last digit of the decimal equivalent of the fraction in question must invariably end in 9. Therefore, as the last digit of the denominator in this case is 9, it automatically follows that the last digit of the decimal equivalent is bound to be 1 (so that the product of the multiplicand and the multiplier concerned may end in 9).

We, therefore, start with 1 as the last (i.e. the right-hand-most) digit of the answer and proceed *leftward* continuously multiplying by 2 (which is the *Ekādhika Pūrva*, i.e. one more than the penultimate digit of the denominator in this case) until a repetition of the whole operation stares us in the face and intimates to us that we are dealing with a Recurring Decimal and may therefore put up the usual recurring marks (dots) and stop further multiplication-work.

Our *modus-operandi*-chart is thus as follows:

(*i*) We put down 1 as the right-hand most digit.

1

(*ii*) We multiply that last digit 1 by 2 and put the 2 down as the immediately preceding digit.

2 1

(*iii*) We multiply that 2 by 2 and put 4 down as the next previous digit.

4 2 1

(*iv*) We multiply that 4 by 2 and put it down, thus

8 4 2 1

(*v*) We multiply that 8 by 2 and get 16 as the product. But this has two digits. We therefore put the 6 down immediately to the left of the 8 and keep the 1 on hand to be carried over to the left at the next step (as we always do in all multiplication e.g. of 69 × 2 = 138 and so on).

6 8 4 2 1<br>1

(*vi*) We now multiply 6 by 2, get 12 as the product, add thereto the 1 (kept to be carried over from the right at the last step), get 13 as the consolidated product, put the 3 down and keep the 1 on hand for carrying over to the left at the next step.

3 6 8 4 2 1<br>1 1

(*vii*) We then multiply 3 by 2, add the one carried over from the right one, get 7 as the consolidated product. But, as this is a single-digit number with nothing to carry over to the left, we put it down as our next multiplicand.

7 3 6 8 4 2 1<br>1 1

(*viii-xviii*) We follow this procedure continually until we reach the 18th digit counting leftwards from the right, when we find that the whole decimal has begun to repeat itself. We therefore, put up the usual recurring marks (dots) on the first and the last digit of the answer (for betokening that the whole of it is a Recurring Decimal) and stop the multiplication there.

Our chart now reads as follows:

$$\frac{1}{19} = .\dot{0}\,5\,2\,6\,3\,1\,5\,7\,8/9\,4\,7\,3\,6\,8\,4\,2\,\dot{1}$$

1 1 1 1 1 $\dot{1}$ / 1 1 1

We thus find that this answer obtained by us with the aid of our Vedic one-line mental arithmetic is just exactly the same as we obtained by the current method with its 18 steps of Division-work.

In passing, we may also just mention that the current process not only takes 18 steps of working for getting the 18 digits of the answer not to talk of the time, the energy, paper, ink etc. consumed but also suffers under the additional and still more serious handicap that, at each step, a probable "trial" digit of the quotient has to be taken *on trial* for multiplying the divisor which is sometimes found to have played on us the scurvy trick of yielding a product larger than the dividend on hand and has thus—after trial—to be discarded in favour of another "trial" digit and so on. In the Vedic method just above propounded, however, there are no subtractions at all and no need for such trials, experiments etc., and no scope for any tricks, pranks and so on but only a straightforward multiplication of single-digit numbers; and the multiplier is not merely a simple one but also the same throughout each particular operation. All this lightens, facilitates and expedites the work and turns the study of mathematics from a burden and a bore into a thing of beauty and a joy for ever at any rate, as far as the children are concerned.

In this context, it must also be *transparently* clear that the long, tedious, cumbrous and clumsy methods of the current system tend to afford greater and greater scope for the children's making mistakes (in the course of all the long multiplications, subtractions etc. involved therein); and once one figure goes wrong, the rest of the work must inevitably turn out to be an utter waste of time, energy and so on and engender feelings of fear, hatred and disgust in the children's minds.

**B. The Second Method**

As already indicated, the second method is of division (instead of multiplication) by the self-same "*Ekādhika Pūrva*", namely 2. And, as division is the exact opposite of multiplication, it stands to reason that the operation of division should proceed, not from right to left (as in the case of multiplication as expounded hereinbefore) but in the exactly opposite direction, i.e. from left to right. And such is actually found to be the case. Its application and *modus operandi* are as follows :

(*i*) Dividing 1 (the first digit of the dividend) by 2, we see the quotient is zero and the remainder is 1. We, therefore, set 0 down as the first digit of the quotient and *prefix* the remainder 1 to that very digit of the quotient (as a sort of reverse-procedure to the carrying-to-the-left process used in multiplication) and thus obtain 10 as our next dividend. $\,_{1}0$

(*ii*) Dividing this 10 by 2, we get 5 as the second digit of the quotient; and, as there is no remainder to be prefixed thereto, we take up that digit 5 itself as our next dividend. . 0 5
1

(*iii*) So, the next quotient-digit is 2; and the remainder is 1. We, therefore, put 2 down as the third digit of the quotient and prefix the remainder 1 to that quotient-digit 2 and thus have 12 as our next dividend. . 0 5 2
1 1

(*iv*) This gives us 6 as quotient-digit and zero as remainder. So, we set 6 down as the fourth digit of the quotient; and as there is no remainder to be prefixed thereto, we take 6 itself as our next digit for division which gives the next quotient digit as 3. . 0 5 2 6 3 1
1 1 1

(*v*) That gives us 1 and 1 as quotient and remainder respectively. We therefore put 1 down as the 6th quotient-digit, prefix the 1 thereto and have 11 as our next dividend. 0 5 2 6 3 1 5
1 1 1 1

(*vi-xvii*) Carrying this process of straight, continuous division by 2, we get 2 as the 17th quotient-digit and 0 as remainder.

(*xviii*) Dividing this 2 by 2, we get 1 as 18th quotient digit and 0 as remainder. But this is exactly what we began with. This means that the decimal begins to repeat itself from here. So, we stop the mental-division process and put down the usual recurring symbols (dots) on the 1st and 18th digit to show that the whole of it is a circulating decimal.

$\dot{0}$ 5 2 6 3 1 5 7 8
1 1 1 1 1 1
9 4 7 3 6 8 4 2 $\dot{1}$
1 1 1

Note that, in the first method, i.e. of multiplication, each surplus digit is *carried* over to the *left* and that, in the second method, i.e. of division, each remainder is *prefixed* to the right, i.e. just immediately to the left of the next dividend digit.

## C. A Further Short-cut

This is not all. As a matter of fact, even this much or rather, this little work of mental multiplication or division is not really necessary. This will be self-evident from sheer observation.

Let us put down the first 9 digits of the answer in one horizontal row above and the other 9 digits in another horizontal row just below and observe the fun of it. We notice that each set of digits in

. $\dot{0}$ 5 2 6 3 1 5 7 8
9 4 7 3 6 8 4 2 $\dot{1}$
9 9 9 9 9 9 9 9 9

the upper row and the lower row totals 9. And this means that, when just half the work has been completed by either of the *Vedic one-line* methods, the other half need *not* be obtained by the same process but is mechanically available to us by subtracting from 9 each of the digits already obtained! *And this means a reduction of the work still further by* 50%.

Yes, but how should one know that the task is exactly half-finished so that one may stop the work of multiplication or division, as the case may be and proceed to reel off the remaining half of the answer by subtracting from 9 each of the digits already obtained? And the answer is—as we shall demonstrate later on—that, in either method, if and as soon as we reach the difference between the numerator and the denominator (i.e. $19-1=18$), we shall have completed *exactly half the work;* and, with this knowledge, we know exactly when and where we may stop the multiplication or division work and when and where we can begin reeling off the complements from 9 as the remaining digits of the answer!

Thus both in the multiplication method and in the division method, we reach 18 when we have completed half the work and can begin the mechanical-subtraction device for the other half.

Details of these principles and processes *and other allied matters,* we shall go into, in due course, at the proper place. In the meantime, the student will find it both interesting and profitable to know this rule and turn it into good account from time to time as the occasion may demand or justify.

SECOND EXAMPLE

Case 2 Let us now take another case of a similar type say, 1/29. 1/29 where too the denominator ends in 9.

BY THE CURRENT METHOD

```
29) 1.00 (.0̇ 3 4 4 8 2 7 5 8 6 2 0 6 8
     87     9 6 5 5 1 7 2 4 1 3 7 9 3 1
     ---
     130            180
     116            174
     ---            ---
      140            60            150
      116            58            145
      ---           ---            ---
       240           200             50          110
       232           174             29           87
       ---           ---            ---          ---
         80            260           210          230
         58            232           203          203
        ---            ---           ---          ---
```

```
220          280          70          270
203          261          58          261
 170          190         120           90
 145          174         116           87
  250          160          40          30
  232          145          29          29
   180          150         110          1
```

**By the Vedic One-line Mental Method**

**A. First Method**

```
1/29 = .0̇ 3 4 4 8 2 7 5 8 6 2 0 6 8 |
         1 1 1 2   2 1 2 1     2 2 2 |
       9 6 5 5 1 7 2 4 1 3 7 9 3 1̇   }
       1 1 1   2   1     1 2 2       |
```

**B. Second Method**

```
1/29 = .0̇ 3 4 4 8 2 7 5 8 6 2 0 6 8 |
         1 1 1 2   2 1 2 1     2 2 2 |
       9 6 5 5 1 7 2 4 1 3 7 9 3 1̇   }
       1 1 1   2   1     1 2 2       |
```

This is the whole working by both the processes. The procedures are explained hereunder:

**A. Explanation of The First Method**

Here too, the last digit of the denominator is 9; but the penultimate one is 2; and one more than that means 3. So, 3 is our *common*—i.e. *uniform multiplier* this time. And, following the same procedure as in the case of 1/19, we put down, 1 as the last (i.e. the right-hand-most) digit of the answer and carry on the multiplication continually leftward by 3 "carrying" the left-hand extra side-digit—if any—over to the left until the recurring decimal actually manifests itself as such. And we find that, by our mental one-line process, we get the same 28 digit-answer as we obtained by 28 steps of cumbrous and tedious working according to the current system, as shown on the left-hand side margin on the previous page.

Our *modus-operandi*-chart herein reads as follows:

```
1/29 = .0̇ 3 4 4 8 2 7 5 8 6 2 0 6 8 |
         1 1 1 2   2 1 2 1     2 2 2 |
       9 6 5 5 1 7 2 4 1 3 7 9 3 1̇   }
       1 1 1   2   1     1 2 2       |
```

### B. Explanation of the Second Method

The division-process to be adopted here is exactly the same as in the case of 1/19; but the divisor instead of the multiplier is uniformly 3 all through. And the chart reads as follows:

$\frac{1}{29}$ = . $\dot{0}$ 3 4 4 8 2 7 5 8 6 2 0 6 8
1 1 1 2 2 1 2 1 2 2 2
9 6 5 5 1 7 2 4 1 3 7 9 3 1
1 1 1 2 1 1 2 2

### C. The Complements from Nine

Here too, we find that the two halves are all complements of each other from 9. So, this fits in too.

$\therefore \frac{1}{29}$ = $\dot{0}$ 3 4 4 8 2 7 5 8 6 2 0 6 8
9 6 5 5 1 7 2 4 1 3 7 9 3 $\dot{1}$
9 9 9 9 9 9 9 9 9 9 9 9 9 9

### Third Example

Case 3 **By the Current System**

1/49

49)1.00 (.$\dot{0}$ 2 0 4 0 8 1 6 3 2 6 5 3 0 6 1 2 2 4 4 8
98 9 7 9 5 9 1 8 3 6 7 3 4 6 9 3 8 7 7 5 5 $\dot{1}$
200
196
400
392
80
49
310
294
160
147
130
98
320
294
260
245
150
147

120
98
220
196
240
196
440
392
480
441
390
343
470
441

410
392
180
147
330
294
360
343
170
147

190
147
430
392
380
343
370
343

| | | | |
|---|---|---|---|
| 300 | 290 | 230 | 270 |
| 294 | 245 | 196 | 245 |
| 60 | 450 | 340 | 250 |
| 49 | 441 | 294 | 245 |
| 110 | 90 | 460 | 50 |
| 98 | 49 | 441 | 49 |
| 120 | 410 | 190 | 1 |

### By the Vedic One-line Mental Method

Our multiplier or divisor as the case may be is now 5, i.e. one more than the penultimate digit. So, A. (By multiplication leftward from the right) by 5, we have—

$\frac{1}{49} = .\dot{0}$ 2 0 4 0 8 1 6 3 2 6 5 3 0 6 1 2 2 4 4 8
9 7 9 5 9 1 8 3 6 7 3 4 6 9 3 8 7 7 5 5 $\dot{1}$
3 4 2 4   4 1 3 3 1 2 3 4 1 4 3 3 2 2

or B. (By division rightward from the left) by 5;

$\frac{1}{49} = .\dot{0}$ 2 0 4 0 8 1 6 3 2 6 5 3 0 6 1 2 2 4 4 8
1   2   4   3 1 1 3 2 1   3   1 1 2 2 4 4
9 7 9 5 9 1 8 3 6 7 3 4 6 9 3 8 7 7 5 5 $\dot{1}$

*Note*:—At this point, in all the three processes, we find that we have reached 48 the difference between the numerator and the denominator. This means that half the work of multiplication or division, as the case may be has been completed and that we may therefore stop that process and may begin the easy and mechanical process of obtaining the remaining digits of the answer whose total number of digits is thus found to be 21+21=42. And yet, the remarkable thing is that the current system takes 42 steps of elaborate and cumbrous dividing with a series of multiplications and substractions and with the risk of the failure of one or more "trial digits" of the quotient and so on while a single, straight and continuous process—of multiplication or division—by a single multiplier or divisor is quite enough in the Vedic method.

The complements from nine are also there.

But this is not all. Our readers will doubtless be surprised to learn—but it is fact—that there are, in the Vedic system, still simpler and easier methods by which, without doing even the easy work explained hereinabove, we can put down digit after digit of the answer, right from the very start to the very end.

But, as these three examples of $\frac{1}{19}$, $\frac{1}{29}$, and $\frac{1}{49}$ have been dealt with and explained at this stage, not in the contemplated regular sequence but only by way of preliminary *demonstration* for the satisfaction of a certain, natural and understandable, nay, perfectly justifiable type of purely intellectual curiosity, we do not propose to go—here and now—into a further detailed and elaborate, comprehensive and exhaustive exposition of the other astounding processes found in the Vedic mathematical *Sūtras* on this particular subject. We shall hold them over to be dealt with, at their own appropriate place, in due course, in a later chapter.

ॐ तत् सत्

# 2

# Arithmetical Computations

## MULTIPLICATION

BY 'NIKHILAM' SŪTRA

We now pass on to a systematic exposition of certain salient, interesting, important and necessary formulae of the utmost value and utility in connection with arithmetical calculations beginning with the processes and methods described in the Vedic mathematical *Sūtras*.

At this point, it will not be out of place for us to repeat that there is a *general* formula which is simple and easy and can be applied to all cases; but there are also *special* cases—or rather, types of cases—which are simpler still and which are, therefore, first dealt with.

We may also draw the attention of all students and teachers of mathematics to the well-known and universal fact that, in respect of arithmetical multiplications, the usual present-day procedure everywhere in schools, colleges and universities is for the children in the primary classes to be told to cram up—or "get by heart"—the multiplication-tables up to 16 times 16, 20 × 20 and so on. But, according to the Vedic system, the multiplication tables are not really required above 5 × 5. And a school-going pupil who knows simple addition and subtraction (of single-digit numbers) and the multiplication-table up to five times five, can *improvise* all the necessary multiplication-tables for himself at any time and can himself do all the requisite multiplications involving bigger multiplicands and multipliers, with the aid of the relevant simple Vedic formulae which enable him to get at the required products, very easily and speedily—nay, practically, immediately. The *Sūtras* are very short; but, once one understands them and the *modus operandi* involved therein for their practical application, the whole thing becomes a sort of child's play and ceases to be a "problem".

1. Let us first take up a very easy and simple illustrative example, i.e. the multiplication of single-digit numbers above 5 and see how this can be done without previous knowledge of the higher multiplications of the multiplication-tables.

The *Sūtras* reads: निखिलं नवतश्चरमं दशतः (*Nikhilaṁ Navataścaramaṁ Daśataḥ*) which, literally translated, means: "all from 9 and the last from 10"! We shall give a detailed explanation, presently, of the meaning and applications of this cryptical-sounding formula. But just now, we state and explain the actual procedure, step by step.

Suppose we have to multiply 9 by 7.

(*i*) We should take, as *base* for our calculations, that power of 10 which is nearest to the numbers to be multiplied. In this case 10 itself is that power;

$$\begin{array}{r} (10) \\ 9-1 \\ \underline{7-3} \\ 6\,/\,3 \end{array}$$

(*ii*) Put the numbers 9 and 7 above and below on the left-hand side (as shown in the working alongside here on the right-hand side margin);

(*iii*) Subtract each of them form the base (10) and write down the remainders (1 and 3) on the right-hand side with a connecting *minus* sign (–) between them, to show that the numbers to be multiplied are both of them less than 10.

(*iv*) The product will have two parts, one on the left side and one on the right. A vertical dividing line may be drawn for the purpose of demarcation of the two parts.

(*v*) Now, the left-hand-side digit (of the answer) can be arrived at in one of 4 ways:

(*a*) Subtract the base 10 from the sum of the given numbers (9 and 7, i.e. 16). And put (16 – 10) i.e. 6, as the left-hand part of the answer. $9+7-10=6$

or (*b*) Subtract the sum of the two deficiencies (1 + 3 = 4) form the base (10). You get the same answer (6) again. $10-1-3=6$

or (*c*) Cross-subtract deficiency 3 on the second row from the original number 9 in the first row. And you find that you have got (9 – 3), i.e. 6 again. $9-3=6$

or (*d*) Cross-subtract in the converse way (i.e. 1 from 7). And you get 6 again as the left-hand side portion of the required answer. $7-1=6$

*Note*: This availability of the same result in several easy ways is a very common feature of the Vedic system and is of great advantage and help to the student as it enables him to test and verify the correctness of his answer, step by step.

(*vi*) Now, vertically multiply the two deficit figures (1 and 3). The product is 3. And this is the right-hand side portion of the answer.

(*vii*) Thus $9 \times 7 = 63$.

$$\begin{array}{r} (10)\ 9-1 \\ \underline{7-3} \\ 6/3 \end{array}$$

This method holds good in all cases and is, therefore, capable of infinite application. In fact, old historical traditions describe this cross-subtraction process as having been responsible for the acceptance of the × mark as the sign of multiplication.

$$\begin{array}{c} (10)\ 9-1 \\ \times \\ \underline{7-3} \\ 6/3 \end{array}$$

As further illustrations of the same rule, note the following examples:

| | | | | | | | |
|---|---|---|---|---|---|---|---|
| 9 – 1 | 9 – 1 | 9 – 1 | 9 – 1 | 8 – 2 | 8 – 2 | 8 – 2 | 7 – 3 |
| 9 – 1 | 8 – 2 | 6 – 4 | 5 – 5 | 8 – 2 | 7 – 3 | 6 – 4 | 7 – 3 |
| 8/1 | 7/2 | 5/4 | 4/5 | 6/4 | 5/6 | 4/8 | 4/9 |

This proves the correctness of the formula. The algebraical explanation for this is very simple:

$$(x-a)(x-b) = x(x-a-b) + ab$$

A slight difference, however, is noticeable when the vertical multiplication of the deficit digits (for obtaining the right-hand side portion of the answer) yields a product consisting of more than one digit. For example, if and when we have to multiply 6 by 7, and write it down as usual:

$$\begin{array}{c} 7-3 \\ \underline{6-4} \\ 3/_{1}2 \end{array}$$

we notice that the required vertical multiplication (of 3 and 4) gives us the product 12 (which consists of 2 digits; but, as our base is 10 and the right-hand-most digit is obviously of units, we are entitled only to one digit (on the right-hand side).

This difficulty, however, is easily surmounted with the usual multiplication rule that the surplus portion on the left should always be "carried" over to the left. Therefore, in the present case, we keep the 2 of the 12 on the right-hand side and "carry" the 1 over to the left and change the 3 into 4. We thus obtain 42 as the actual product of 7 and 6.

$$\begin{array}{l} 7-3 \\ 6-4 \\ \hline 3/_{1}2 = 4/2 \end{array}$$

A similar procedure will naturally be required in respect of other similar multiplications:

$$\begin{array}{llll} 8-2 & 7-3 & 6-4 & 6-4 \\ 5-5 & 5-5 & 6-4 & 5-5 \\ \hline 3/_{1}0 = 4/0 & 2/_{1}5 = 3/5 & 2/_{1}6 = 3/6 & 1/_{2}0 = 3/0 \end{array}$$

This rule of multiplication by means of cross-subtraction for the left-hand portion and of vertical multiplication for the right-hand half being an actual application of the absolute algebraic identity: $(x+a)(x+b) = x(x+a+b)+ab$, can be extended further without any limitation. Thus, as regards numbers of two digits each, we may notice the following examples:

*N.B.*: The base now required is 100.

$$\begin{array}{lllllll} 91-9 & 93-7 & 93-7 & 93-7 & 89-11 & 91-9 & 93-7 \\ 91-9 & 92-8 & 93-7 & 94-6 & 95-5 & 96-4 & 97-3 \\ \hline 82/81 & 85/56 & 86/49 & 87/42 & 84/55 & 87/36 & 90/21 \end{array}$$

$$\begin{array}{lllllll} 92-8 & 88-12 & 78-22 & 88-12 & 56-44 & 67-33 & 25-75 \\ 98-2 & 98-2 & 97-3 & 96-4 & 98-2 & 97-3 & 99-1 \\ \hline 90/16 & 86/24 & 75/66 & 84/48 & 54/88 & 64/99 & 24/75 \end{array}$$

*Note 1*: In all these cases, note that both the cross-subtractions always give the same remainder (for the left-hand-side portion of the answer).

*Note 2*: Here too, note that the vertical multiplication for the right-hand side portion of the product may, in some cases, yield a more-than-two-digit product; but, with 100 as our base, we can have only two digits on the right-hand side. We should therefore adopt the same method as before (i.e. keep the permissible two digits on the right-hand side and "carry" the surplus or extra digit over to the left) as in the case of ordinary addition, compound addition etc. Thus:

$$\begin{array}{lll} 88-12 & 88-12 & 25-75 \\ 88-12 & 91-9 & 98-2 \\ \hline 76/_{1}44 = 77/44 & 79/_{1}08 = 80/08 & 23/_{1}50 = 24/50 \end{array}$$

*Note*: Also, how the meaning of the *Sūtra* comes out in all the examples just above dealt with and tells us how to write down immediately the deficit figures on the right-hand side. The rule is that all the other digits of the given original numbers are to be subtracted from 9 but the last, i.e. the right hand-most one should be deducted from 10. Thus, if 63 be the given number, the deficit from the base is 37; and so on. This process helps us in the work of ready on-sight subtraction and enables us to put the deficiency down immediately.

A new point has now to be taken into consideration, i.e. that, just as the process of vertical multiplication may yield a *larger* number of digits in the product than is permissible and this contingency has been provided for, so, it may similarly yield a product consisting of a *smaller* number of digits than we are entitled to. What is the remedy? Well, this contingency too has been provided for. And the remedy is—as in the case of decimal multiplications—merely the filling up of all such vacancies with zeroes. Thus,

| 99 – 1 | 98 – 2 | 96 – 4 | 97 – 3 |
|---|---|---|---|
| 97 – 3 | 99 – 1 | 98 – 2 | 97 – 3 |
| 96/03 | 97/02 | 94/08 | 94/09 |

With these 3 procedures for meeting the 3 possible contingencies in question, i.e. of normal, abnormal and sub-normal number of digits in the vertical-multiplication-products and with the aid of the subtraction-rule, i.e. of all the digits from 9 and the last one from 10, for writing down the amount of the deficiency from the base, we can extend this multiplication-rule to numbers consisting of a larger number of digits, thus:

| 888 – 112 | 879 – 121 | 697 – 303 | 598 – 402 |
|---|---|---|---|
| 998 – 002 | 999 – 001 | 997 – 003 | 998 – 002 |
| 886/224 | 878/121 | 694/909 | 596/804 |

| 988 – 012 | 888 – 112 | 112 – 888 |
|---|---|---|
| 988 – 012 | 991 – 009 | 998 – 002 |
| 976/144 | $879/_{1}008 = 880/008$ | $110/_{1}776 = 111/776$ |

| 988 – 012 | 998 – 002 | 9997 – 0003 |
|---|---|---|
| 998 – 002 | 997 – 003 | 9997 – 0003 |
| 986/024 | 995/006 | 9994/0009 |

| 99979 – 00021 | 999999997 – 000000003 |
|---|---|
| 99999 – 00001 | 999999997 – 000000003 |
| 99978/00021 | 999999994/000000009 |

Yes, but, in all these cases, the multiplicand and the multiplier are just a little below a certain power of ten (taken as the base). What about numbers which are above it?

And the answer is that the same procedure will hold good there too, except that, instead of *cross-subtracting*, we shall have to *cross-add.* And all the other rules regarding digit-surplus, digit-deficit etc., will be exactly the same as before. Thus,

$$\begin{array}{lllllll} 12+2 & 13+3 & 11+1 & 16+6 & 18+8 & 108+8 & 111+11 \\ 11+1 & 12+2 & 15+5 & 11+1 & 11+1 & 108+8 & 109+9 \\ \hline 13/2 & 15/6 & 16/5 & 17/6 & 19/8 & 116/64 & 120/99 \end{array}$$

$$\begin{array}{lll} 16+6 & 17+7 & 18+8 \\ 12+2 & 12+2 & 12+2 \\ \hline 18/2 = 19/2 & 19/_{1}4 = 20/4 & 20/_{1}6 = 21/6 \end{array}$$

$$\begin{array}{llll} 18+8 & 19+9 & 1005+5 & 1016+16 \\ 18+8 & 19+9 & 1009+9 & 1006+6 \\ \hline 26/_{6}4 = 32/4 & 28/_{8}1 = 36/1 & 1014/045 & 1022/096 \end{array}$$

In passing, the algebraical principle involved may be explained as follows:

$$(x+a)(x+b) = x(x+a+b)+ab$$

Yes, but if one of the numbers is *above* and the other is *below* a power of 10 (the base taken), what then?

The answer is that the plus and the minus will, on multiplication, behave as they always do and produce a minus-product and that the right-hand portion obtained by vertical multiplication will therefore have to be *subtracted.* A vinculum may be used for making this clear. Thus,

$$\begin{array}{llll} 12+2 & 108+8 & 107+7 & 102+2 \\ 8-2 & 97-3 & 93-7 & 98-2 \\ \hline 10/\bar{4} = 96 & 105/\bar{2}\bar{4} = 104/76 & 100/\bar{4}\bar{9} = 99/51 & 100/\bar{0}\bar{4} = 99/96 \end{array}$$

$$\begin{array}{ll} 1026+26 & 1033+33 \\ 997-\ \ 3 & 997-\ \ 3 \\ \hline 1023/\bar{0}\bar{7}\bar{8} = 1022/922 & 1030/\bar{0}\bar{9}\bar{9} = 1029/901 \end{array}$$

$$\begin{array}{l} 10006+6 \\ 9999-1 \\ \hline 10005/000\bar{6} = 10004/9994 \end{array}$$

*Note* : Note that even the subtraction of the vinculum-portion may be easily done with the aid of the *Sūtra* under discussion, i.e. all from 9 and the last from 10.

## Multiples and Sub-multiples

Yes, but, in all these cases, we find both the multiplicand and the multiplier, or at least one of them, very near the base taken in each case; and this gives us a small multiplier and thus renders the multiplication very easy. What about the multiplication of two numbers, neither of which is near a convenient base?

The needed solution for this purpose is furnished by a small '*Upasūtra*' (or sub-formula) which is so-called because of its practically axiomatic character.

This sub-*sūtra* consists of only one word आनुरूप्येण (*Ānurūpyeṇa*) which simply means "Proportionately". In actual application, it connotes that, in all cases where there is a rational ratio-wise relationship, the ratio should be taken into account and should lead to a proportionate multiplication or division as the case may be.

In other words, when neither the multiplicand nor the multiplier is sufficiently near a convenient power of 10 which can suitably serve as a base, we can take a convenient multiple or sub-multiple of a suitable base, as our "working base", perform the necessary operation with its aid and then multiply or divide the result proportionately, i.e. in the same proportion as the original base may bear to the working base actually used by us. A concrete illustration will make the *modus operandi* clear.

Suppose we have to multiply 41 by 41. Both these numbers are so far away from the base 100 that by our adopting that as our actual base, we shall get 59 and 59 as the deficiency from the base. And thus the consequent vertical multiplication of 59 by 59 would prove too cumbrous a process to be permissible under the Vedic system and will be positively inadmissible.

We therefore, accept 100 merely as a theoretical base and take sub-multiple 50 which is conveniently near 41 and 41 as our working basis, work the sum up accordingly and then do the proportionate multiplication or division, for getting the correct answer.

Our chart will then take this shape:

$$\frac{100}{2} = 50$$

(*i*) We take 50 as our working base.

(*ii*) By cross-subtraction, we get 32 on the left-hand side.

$$\begin{array}{r} 41 - 9 \\ 41 - 9 \\ \hline 2)\,32/81 \\ \hline 16/81 \\ \hline \end{array}$$

(*iii*) As 50 is a half of 100, we therefore divide 32 by 2 and put 16 down as the real left-hand side portion of the answer.

(*iv*) The right-hand side portion 81 remains *unaffected.*

(*v*) The answer therefore is 1681.

or, *secondly*, instead of taking 100 as our theoretical base and its half 50 as our working base and dividing 32 by 2, we may take 10, as our theoretical base and its multiple 50 as our working base and ultimately multiply 32 by 5 and get 160 for the left-hand side. And as 10 was our theoretical base and we are therefore entitled to only one digit on the right hand side, we retain 1 of the 81 on the right hand side, "carry" the 8 of the 81 over to the left, add it to the 160 already there and thus obtain 168 as our left-hand-side portion of the answer. The product of 41 and 41 is thus found to be 1681 (the same as we got by the first method).

$$\begin{array}{c} 10 \times 5 = 50 \\ \hline 41 - 9 \\ 41 - 9 \\ \hline 32/_{8}1 \\ \times 5/ \\ \hline 160/_{8}1 = 1681 \\ \hline \end{array}$$

or, *thirdly*, instead of taking 100 or 10 as our theoretical base and 50 a sub-multiple or multiple thereof as our working base, we may take 10 and 40 as the bases respectively and work at the multiplication as shown on the margin here. And we find that the product is 1681 the same as we obtained by the first and the second methods.

$$\begin{array}{c} 10 \times 4 = 40 \\ 41 + 1 \\ 41 + 1 \\ \hline 42/1 \\ \times 4/ \\ \hline 168/1 \\ \hline \end{array}$$

Thus, as we get the same answer 1681 by all the three methods, we have option to decide—according to our own convenience—what theoretical base and what working base we shall select for ourselves.

As regards the principle underlying and the reason behind the vertical-multiplication operation on the right-hand-side remaining unaffected and not having to be multiplied or divided "proportionately" a very simple illustration will suffice to make this clear.

Suppose we have to divide 65 successively by 2, 4, 8, 16, 32 and 64 which bear a certain internal ratio or ratios among themselves. We may write down our table of answers as follows:

$$\frac{65}{2} = 32\frac{1}{2}; \ \frac{65}{4} = 16\frac{1}{4}; \ \frac{65}{8} = 8\frac{1}{8}; \ \frac{65}{16} = 4\frac{1}{16};$$

*and*

$$\frac{65}{32} = 2\frac{1}{32}; \text{ and } \frac{65}{64} = 1\frac{1}{64} \text{ R \textit{is constant.}}$$

We notice that, as the denominator, i.e. the divisor goes on *increasing* in a certain ratio, the quotient goes on *decreasing*, proportionately; but the *remainder remains constant*. And this is why it is rightly called the *remainder* (शिष्यते शेषसंज्ञः ।।)

The following additional examples will serve to illustrate the principle and process of आनुरूप्येण, i.e. the selecting of a multiple or sub-multiple as our working base and doing the multiplication work in this way.

(1) $49 \times 49$

Working Base $100/2 = 50$

$$\begin{array}{r} 49-1 \\ 49-1 \\ \hline 2)\,48/01 \\ \hline 24/01 \end{array}$$

*Or* (2) $49 \times 49$

Working Base $10 \times 5 = 50$

$$\begin{array}{r} 49-1 \\ 49-1 \\ \hline 48/1 \\ \times 5/ \\ \hline 240/1 \end{array}$$

(3) $46 \times 46$

Working Base $100/2 = 50$

$$\begin{array}{r} 46-4 \\ 46-4 \\ \hline 2)\,42/16 \\ \hline 21/16 \end{array}$$

*Or* (4) $46 \times 46$

Working Base $10 \times 5 = 50$

$$\begin{array}{l} 46-4 \\ 46-4 \\ \hline 42/{}_{1}6 \\ \times 5/ \\ \hline 210/{}_{1}6 = 211/6 \end{array}$$

(5) $46 \times 44$

Working Base $10 \times 5 = 50$

$$\begin{array}{r} 46-4 \\ 44-6 \\ \hline 40/{}_{2}4 \\ \times 5/ \\ \hline 200{}_{2}4 \\ =202/4 \end{array}$$

Or (6) $46 \times 44$

Working Base $100/2 = 50$

$$\begin{array}{r} 46-4 \\ 44-6 \\ \hline 2)\,40/24 \\ \hline 20/24 \end{array}$$

(7) $59 \times 59$

Working Base $10 \times 6 = 60$

Or (8) $59 \times 59$

Working Base $10 \times 5 = 50$

$$\begin{array}{l} 59-1 \\ \underline{59-1} \\ 58/1 \\ \underline{\times 6/} \\ 348/1 \end{array}$$

$$\begin{array}{l} 59+9 \\ \underline{59+9} \\ 68/_{8}1 \\ \underline{\times 5/} \\ 348/1 \end{array}$$

*Or* (9) $59 \times 59$

Working Base $100/2 = 50$

$$\begin{array}{r} 59+9 \\ \underline{59+9} \\ \underline{2)68/81} \\ 34/81 \end{array}$$

(10) $23 \times 23$

Working Base $10 \times 2 = 20$

$$\begin{array}{l} 23+3 \\ \underline{23+3} \\ 26/9 \\ \underline{\times 2/} \\ 52/9 \end{array}$$

(11) $54 \times 46$

Working Base $10 \times 5 = 50$

$$\begin{array}{l} 54+4 \\ \underline{46-4} \\ 50/_{\bar{1}}\bar{6} \\ \underline{\times 5} \\ \underline{250/_{\bar{1}}\bar{6}} \\ = 248/4 \end{array}$$

*Or* (12) $54 \times 46$

Working Base $100/2 = 50$

$$\begin{array}{r} 54+4 \\ \underline{46-4} \\ \underline{2)50/\bar{1}\bar{6}} \\ \underline{25/\bar{1}\bar{6}} \\ = 24/84 \end{array}$$

(13) $19 \times 19$

Working Base $10 \times 2 = 20$

$$\begin{array}{l} 19-1 \\ \underline{19-1} \\ 18/1 \\ \underline{\times 2/} \\ = 36/1 \end{array}$$

*Or* (14) $19 \times 19$

Working Base $10 \times 1 = 10$

$$\begin{array}{l} 19+9 \\ \underline{19+9} \\ 28/_{8}1 \\ \underline{+8/} \\ = 36/1 \end{array}$$

(15) $62 \times 48$

Working Base $10 \times 4 = 40$

$$\begin{array}{r} 62+22 \\ \underline{48+\ \ 8} \\ 70/_{17}6 \\ \underline{\times 4/} \\ \underline{280/_{17}6} \\ = 297/6 \end{array}$$

*Or* (16) $62 \times 48$

Working Base $10 \times 6 = 60$

$$\begin{array}{l} 62+\ \ 2 \\ \underline{48-12} \\ 50/_{2}\bar{4} \\ \underline{\times 6/} \\ 300/_{2}\bar{4} \\ = 297/6 \end{array}$$

*Or* (17) $62 \times 48$

Working Base $10 \times 5 = 50$

$62 + 12$

$\underline{48 - \ \ 2}$

$60/{}_{2}\bar{4}$

$\underline{\times 5/}$

$300/{}_{2}\bar{4}$

$= 297/6$

(18) $62 \times 48$

Working Base $100/2 = 50$

$62 + 12$

$\underline{48 - \ \ 2}$

$\underline{2)60/{}_{2}\bar{4}}$

$30/{}_{2}\bar{4}$

$= 29/76$

(19) $23 \times 21$

Working Base $10 \times 3 = 30$

$23 - 7$

$\underline{21 - 9}$

$14/{}_{6}3$

$\underline{\times 3/}$

$42/{}_{6}3$

$= 48/3$

*Or* (20) $23 \times 21$

Working Base $10 \times 2 = 20$

$23 + 3$

$\underline{21 + 1}$

$24/3$

$\underline{\times 2/}$

$= 48/3$

(21) $249 \times 245$

Working Base $1000/4 = 250$

$249 - 1$

$\underline{245 - 5}$

$\underline{4)244 / 005}$

$= 61 / 005$

(22) $48 \times 49$

Working Base $10 \times 5 = 50$

$48 - 2$

$\underline{49 - 1}$

$47/2$

$\underline{\times 5/}$

$= 235/2$

*Or* (23) $48 \times 49$

Working Base $100/2 = 50$

$48 - 2$

$\underline{49 - 1}$

$\underline{2)47/02}$

$23\frac{1}{2}/02 = 23/52$

*Note* : Here 47 being odd, its division by 2 gives us a fractional quotient $23\frac{1}{2}$ and that, just as half a rupee or half a pound or half a dollar is taken over to the right-hand-side (as 8 annas or 10 shillings or 50 cents), so the half here (in the $23\frac{1}{2}$) is taken over to the right-hand-side (as 50). So, the answer is 23/52.

(24) $249 \times 246$

Working Base $1000/4 = 250$

(25) $229 \times 230$

Working Base $1000/4 = 250$

$$\begin{array}{r} 249 - 1 \\ 246 - 4 \\ \hline 4)\ 245 / 004 \\ \hline 61\frac{1}{4} / 004 \\ = 61 / 254 \end{array} \qquad \begin{array}{r} 229 - 21 \\ 230 - 20 \\ \hline 4)\ 209 / 420 \\ \hline 52\frac{1}{4} / 420 \\ = 52 / 670 \end{array}$$

*Note*: In the above two cases, the $\frac{1}{4}$ on the left hand side is carried over to the right hand as 250 as $\frac{1}{4}$ of 1000.

The following additional worked out examples will serve to further elucidate the principle and process of multiplication according to the Vedic *Sūtra* (*'Nikhilaṁ'*) and facilitate the student's practice and application thereof:

(1) $87965 \times 99998$

$$\begin{array}{r} 87965 - 12035 \\ 99998 - \quad\quad 2 \\ \hline = 87963 / 24070 \end{array}$$

(2) $48 \times 97$

$$\begin{array}{r} 48 - 52 \\ 97 - \ 3 \\ \hline 45 /_{1}56 \\ = 46/56 \end{array}$$

(3) $72 \times 95$

$$\begin{array}{r} 72 - 28 \\ 95 - \ 5 \\ \hline 67 /_{1}40 \\ \hline = 68/40 \end{array}$$

(4) $889 \times 9998$

$$\begin{array}{r} 0889 - 9111 \\ 9998 - \quad\ 2 \\ \hline 887 /_{1}8222 \\ \hline = 888/8222 \end{array}$$

(5) $77 \times 9988$

$$\begin{array}{r} 0077 - 9923 \\ 9988 - 0012 \\ \hline 65 /_{11}9076 \\ \hline = 76/9076 \end{array}$$

(6) $299 \times 299$

Working Base

(W.B.) $100 \times 3 = 300$

$$\begin{array}{r} 299 - 1 \\ 299 - 1 \\ \hline 298 / 01 \\ \times 3 / \quad \\ \hline = 894 / 01 \end{array}$$

(7) $687 \times 699$

W.B. $100 \times 7 = 700$

$$\begin{array}{r} 687 - 13 \\ 699 - \ 1 \\ \hline 686 / 13 \\ \times 7 / \quad \\ \hline = 4802 / 13 \end{array}$$

(8) $128 \times 672$

W.B. $100 \times 7 = 700$

$$\begin{array}{r} 128 - 572 \\ 672 - \ 28 \\ \hline 100 /_{160}16 \\ \times 7 / \quad \\ \hline 700 /_{160}16 \\ \hline = 860/16 \end{array}$$

(9) $231 \times 582$

W.B. $100 \times 6 = 600$

$$\begin{array}{r} 231 - 369 \\ 582 - \phantom{0}18 \\ \hline 213/_{66}42 \\ \times 6/\phantom{_{66}42} \\ \hline 1278/_{66}42 \\ \hline =1344/42 \end{array}$$

(10) $362 \times 785$

W.B. $100 \times 8 = 800$

$$\begin{array}{r} 362 - 438 \\ 785 - \phantom{0}15 \\ \hline 347/_{65}70 \\ \times 8/\phantom{_{65}70} \\ \hline 2776/_{65}70 \\ \hline = 2841/70 \end{array}$$

(11) $532 \times 528$

W.B. $100 \times 5 = 500$

$$\begin{array}{r} 532 + 32 \\ 528 + 28 \\ \hline 560/_{8}96 \\ \times 5/\phantom{_{8}96} \\ \hline 2800/_{8}96 \\ \hline = 2808/96 \end{array}$$

(12) $532 \times 528$

W.B. $1000/2 = 500$

$$\begin{array}{r} 532 + 32 \\ 528 + 28 \\ \hline 2)560/896 \\ \hline = 280/896 \end{array}$$

(13) $532 \times 472$

W.B. $100 \times 5 = 500$

$$\begin{array}{r} 532 + 32 \\ 472 - 28 \\ \hline 504/_{\bar{8}}\bar{9}\bar{6} \\ \times 5/\phantom{_{\bar{8}}\bar{9}\bar{6}} \\ \hline 2520/_{\bar{8}}\bar{9}\bar{6} \\ \hline = 2511/04 \end{array}$$

(14) $532 \times 472$

W.B. $1000/2 = 500$

$$\begin{array}{r} 532 + 32 \\ 472 - 28 \\ \hline 2)504/_{\bar{8}}\bar{9}\bar{6} \\ \hline 252/_{\bar{8}}\bar{9}\bar{6} \\ \hline = 251/104 \end{array}$$

(15) $235 \times 247$

W.B. $1000/4 = 250$

$$\begin{array}{r} 235 - 15 \\ 247 - \phantom{0}3 \\ \hline 4)232/045 \\ \hline = \phantom{0}58/045 \end{array}$$

(16) $3998 \times 4998$

W.B. $10000/2 = 5000$

$$\begin{array}{r} 3998 - 1002 \\ 4998 - \phantom{000}2 \\ \hline 2)3996/2004 \\ \hline = 1998/2004 \end{array}$$

(17) $19 \times 499$

W.B. $100 \times 5 = 500$

$$\begin{array}{r} 19 - 481 \\ 499 - \phantom{48}1 \\ \hline 18/_{4}81 \\ \times 5\phantom{/_{4}81} \\ \hline = 94/81 \end{array}$$

*Or* (18) $19 \times 499$

W.B. $1000/2 = 500$

$$\begin{array}{r} 19 - 481 \\ 499 - \phantom{48}1 \\ \hline 2)18/481 \\ \hline = 9/481 \end{array}$$

(19) 635 × 502
W.B. 1000/2 = 500

$$\begin{array}{r} 635 + 135 \\ 502 + \phantom{13}2 \\ \hline 2)637/270 \\ \hline 318\frac{1}{2}/270 \\ \hline = 318/770 \end{array}$$

(20) 18 × 45
W.B. 100/2 = 50

$$\begin{array}{r} 18 - 32 \\ 45 - \phantom{3}5 \\ \hline 2)13/{}_{1}60 \\ \hline 6\frac{1}{2}/{}_{1}60 \\ \hline 8/10 \end{array}$$

(21) 389 × 516
W.B. 1000/2 = 500

$$\begin{array}{r} 389 - 111 \\ 516 + \phantom{1}16 \\ \hline 2)405/\bar{1}\overline{776} \\ \hline 202\frac{1}{2}/\bar{1}\overline{776} \\ \hline 202/\bar{1}\overline{276} \\ = 200/724 \end{array}$$

*Note*: Most of these examples are quite easy, in fact much easier, by the उर्ध्वतिर्यग्भ्याम् (*Ūrdhva-Tiryagbhyām*) *Sūtra* which is to be expounded in the next chapter. They have been included here, merely for demonstrating that they too can be solved by the '*Nikhilaṁ*' *Sūtra* expounded in this chapter.

But before we actually take up the '*Ūrdhva-Tiryak*' formula and explain its *modus operandi* for multiplication, we shall just now explain a few corollaries which arise out of the '*Nikhilaṁ*' *Sūtra* which is the subject-matter of this chapter.

### The First Corollary

The first corollary naturally arising out of the '*Nikhilaṁ*' *Sūtra* reads as यावदूनं तावदूनीकृत्य वर्गं च योजयेत् ॥ which means : "whatever the extent of its deficiency, lessen it still further to that very extent; and also set up the square of that deficiency".

This evidently deals with the squaring of numbers. A few elementary examples will suffice to make its meaning and application clear:

Suppose we have to find the square of 9. The following will be the successive stages in our mental working:

(*i*) We should take up the nearest power of 10, i.e. 10 itself as our base.

(*ii*) As 9 is 1 less than 10, we should decrease it still further by 1 and set 8 down as our left-side portion of the answer.

8/

(*iii*) And, on the right hand, we put down the square of that deficiency $1^2$ 8/1

(*iv*) Thus $9^2 = 81$

$$\begin{array}{r} 9-1 \\ 9-1 \\ \hline 8/1 \end{array}$$

Now, let us take up the case of $8^2$. As 8 is 2 less than 10, we lessen it still further by 2 and get 8–2, i.e. 6 for the left-hand and putting $2^2=4$ on the right-hand side, we say $8^2=64$

6/
6/4

$$\begin{array}{r} 8-2 \\ 8-2 \\ \hline 6/4 \end{array}$$

In exactly the same manner, we say

$$7^2 = (7-3)/3^2 = 4/9$$

$$6^2 = (6-4) \text{ and } 4^2 = 2/_1 6 = 3/6$$

$$5^2 = (5-5) \text{ and } 5^2 - 0/_2 5 = 25; \text{ and so on}$$

$$\begin{array}{r} 7-3 \\ 7-3 \\ \hline 4/9 \end{array}$$

Yes, but what about numbers *above* 10 ? We work exactly as before; but, instead of *reducing* still further by the deficit, we *increase* the number still further by the *surplus* and say:

$$11^2 = (11+1)/1^2 = 12/1$$

$$\begin{array}{r} 11+1 \\ 11+1 \\ \hline 12/1 \end{array}$$

$$12^2 = (12+2)/2^2 = 14/4$$

$$13^2 = (13+3)/3^2 = 16/9$$

$$14^2 = (14+4)/4^2 = 18/_1 6 = 19/6$$

$$15^2 = (15+5)/5^2 = 20/_2 5 = 225$$

$$19^2 = (19+9)/9^2 = 28/_8 1 = 361; \text{ and so on.}$$

$$\begin{array}{r} 19+9 \\ 19+9 \\ \hline 28/81 = 361 \end{array}$$

And then, extending the same rule to numbers of two or more digits, we proceed further and say:

$91^2 = 82/81$; $92^2 = 84/64$; $93^2 = 86/49$;

$94^2 = 88/36$; $95^2 = 90/25$; $96^2 = 92/16$;

$97^2 = 94/09$; $98^2 = 96/04$; $99^2 = 98/01$;

$108^2 = 116/64$; $103^2 = 106/09$;

$989^2 = 978/121$; $988^2 = 976/144$; $993^2 = 986/049$;

$89^2 = 78/_121 = 79/21$; $88^2 = 76/_1 44 = 77/44$;

$9989^2 = 9978/0121$; $9984^2 = 9968/0256$; $9993^2 = 9986/0049$.

*The Algebraical Explanation* for this is as follows:

$(a \pm b)^2 = a^2 \pm 2ab + b^2$

$\therefore 97^2 = (100-3)^2 = 10000 - 600 + 9 = 94/09$;

$92^2 = (100-8)^2 = 10000 - 1600 + 64 = 84/64$;

$108^2 = (100+8)^2 = 10000 + 1600 + 64 = 116/64$; and so on

*A Second Algebraical Explanation is as follows*:

$a^2 - b^2 = (a+b)(a-b)$

$\therefore a^2 = (a+b)(a-b) + b^2$

So, if we have to obtain the square of any number a, we can add any number b to it, subtract the same number b from it and multiply the two and finally add the square of that number b (on the right hand side). Thus, if 97 has to be squared, we should select such a number b as will, by addition or by subtraction, give us a number ending in a zero (or zeros) and thereby lighten the multi-multiplication work. In the present case, if our b be 3, $a+b$ will become 100 and $a-b$ will become 94. Their product is 9400; and $b^2 = 9 \therefore 97^2 = 94/09$. This proves the corollary.

Similarly, $92^2 = (92+8)(92-8) + 64 = 84/64$;

$93^2 = (93+7)(93-7) + 49 = 86/49$;

$988^2 = (988+12)(988-12) + 144 = 976/144$;

$108^2 = (108+8)(108-8) + 64 = 116/64$; and so on.

*The Third Algebraical Explanation* is based on the *Nikhilaṁ Sūtra* and has been indicated already.

$$\begin{array}{r} 91-9 \\ 91-9 \\ \hline 82/81 \end{array}$$

The following additional sample-examples will further serve to enlighten the student on this corollary:

(1) $19^2$
$$\begin{array}{l} 19+9 \\ 19+9 \\ \hline 28/_8 1 \\ \hline =36/1 \end{array}$$

*Or* (2) $19^2$
$$\begin{array}{l} 19-1 \\ 19-1 \\ \hline 18/1 \\ \times 2 \\ \hline =36/1 \end{array}$$

(3) $29^2$
$$\begin{array}{l} 29+9 \\ 29+9 \\ \hline 38/_8 1 \\ \times 2 \\ \hline =84/1 \end{array}$$

*Or* (4) $29^2$
$$\begin{array}{l} 29-1 \\ 29-1 \\ \hline 28/1 \\ \times 3 \\ \hline =84/1 \end{array}$$

(5) $49^2$
$$\begin{array}{l} 49-1 \\ 49-1 \\ \hline 48/1 \\ \times 5/ \\ \hline =240/1 \end{array}$$

*Or* (6) $49^2$
$$\begin{array}{l} 49-1 \\ 49-1 \\ \hline 2)48/01 \\ \hline =24/01 \end{array}$$

(7) $59^2$

$$\begin{array}{l} 59+9 \\ 59+9 \\ \hline 68\,/_{8}1 \\ \times 5 \\ \hline 340\,/_{8}1 = 348/1 \end{array}$$

Or (8) $59^2$

$$\begin{array}{l} 59+9 \\ 59+9 \\ \hline 2)68/81 \\ \hline = 34/81 \end{array}$$

(9) $41^2$

$$\begin{array}{l} 41+1 \\ 41+1 \\ \hline 42/1 \\ \times 4 \\ \hline = 168/1 \end{array}$$

Or (10) $41^2$

$$\begin{array}{l} 41-9 \\ 41-9 \\ \hline 2)32/81 \\ \hline = 16/81 \end{array}$$

(11) $989^2$

$$\begin{array}{l} 989-11 \\ 989-11 \\ \hline = 978/121 \end{array}$$

(12) $775^2$

$$\begin{array}{l} \text{W.B. } 100 \times 8 = 800 \\ 775-25 \\ 775-25 \\ \hline 750\,/_{6}25 \\ \times 8 \\ \hline = 6006/25 \end{array}$$

*Note* : All the cases dealt with hereinabove are doubtless of numbers just a little below or just a little above a power of ten or of a multiple or sub-multiple thereof. This corollary is specially suited for the squaring of such numbers. Seemingly more complex and "difficult" cases will be taken up in the next chapter relating to the *Ūrdhva-Tiryak Sūtra*; and still most "difficult" will be explained in a still later chapter dealing with the squaring, cubing etc., of bigger numbers.

### The Second Corollary

The second corollary is applicable only to a special case under the first corollary, i.e. the squaring of numbers ending in 5 and other cognate numbers. Its wording is exactly the same as that of the *Sūtra* which we used at the outset for the conversion of vulgar fractions into their recurring decimal equivalents, i.e. एकाधिकेन पूर्वेण. The *Sūtra* now takes a totally different meaning and, in fact, relates to a wholly different set-up and context.

Its literal meaning is the same as before (i.e. "by one more than the previous one"); but it now relates to the squaring of numbers ending in 5, e.g. say, 15. Here, the last digit is 5; and the "previous" one is 1. So, one more than that is 2. Now, the *Sūtra* in this context tells us to multiply the previous digit 1 by one more than itself, i.e. by 2. So the left-hand side digit is 1×2; and the right-hand side is the vertical-multiplication-product, i.e. 25 as usual.

$$\begin{array}{l} 1/5 \\ \hline 2/25 \\ \hline \end{array}$$

Thus $15^2 = 1 \times 2/25 = 2/25$

Similarly, $25^2 = 2 \times 3/25 = 6/25$;

$35^2 = 3 \times 4/25 = 12/25$;

$55^2 = 5 \times 6/25 = 30/25;$

$65^2 = 6 \times 7/25 = 42/25;$

$75^2 = 7 \times 8/25 = 56/25;$

$85^2 = 8 \times 9/25 = 72/25;$

$95^2 = 9 \times 10/25 = 90/25;$

$105^2 = 10 \times 11/25 = 110/25;$

$115^2 = 11 \times 12/25 = 132/25;$

$125^2 = 156/25; 135^2 = 182/25; 145^2 = 210/25;$

$155^2 = 240/25; 165^2 = 272/25; 175^2 = 306/25;$

$185^2 = 342/25; 195^2 = 380/25;$ and so on.

*The Algebraical Explanation* is quite simple and follows straightaway from the *Nikhilaṁ Sūtra* and still more so from the *Ūrdhva-Tiryak* formula to be explained in the next chapter (q.v.).

A sub-corollary to this corollary (relating to the squaring of numbers ending in 5) reads : अन्त्ययोर्दशकेऽपि (*Antyayor-daśake'pi*) and tells us that the above rule is applicable not only to the squaring of a number ending in 5 but also to the multiplication of two numbers whose last digits together total 10 and whose previous part is exactly the same.

For example, if the numbers to be multiplied are not 25 and 25, but, say 27 and 23 whose last digits, i.e. 7 and 3 together total 10 and whose previous part is the same namely 2, even then the same rule will apply, i.e. that the 2 should be multiplied by 3 the next higher number. Thus we have 6 as our left-hand part of the answer; and the right-hand one is, by vertical multiplication as usual.

$7 \times 3 = 21$. And so $27 \times 23 = 6/21$

$$\begin{array}{r} 27 \\ 23 \\ \hline =6/21 \\ \hline \end{array}$$

We can proceed further on the same lines and say:

$96 \times 94 = 90/24; 97 \times 93 = 90/21; 98 \times 92 = 90/16;$
$99 \times 91 = 90/09; 37 \times 33 = 12/21; 79 \times 71 = 56/09;$
$87 \times 83 = 72/21; 114 \times 116 = 132/24;$ and so on

This sub-corollary too is based on the same *Nikhilaṁ Sūtra*; and harder examples thereof will more appropriately come under the *Ūrdhva-Tiryak* formula of the next chapter or the still later chapter on more difficult squarings and cubings.

At this point, however, it may just be pointed out that the above rule is capable of further application and come in handy, for the multiplication of numbers whose last digits in sets of 2, 3 and so on together total 100, 1000 etc. For example—

$$\left.\begin{array}{l} 191 \times 109 = 20 / 819 \\ 793 \times 707 = 560 / 651 \\ 884 \times 816 = 720 / 1\,344 = 721 / 344 \end{array}\right\}$$

*N.B.*—Note the added zero at the end of the left-hand side of the answer.

### The Third Corollary

Then comes a Third Corollary to the *Nikhilaṁ Sūtra*, which relates to a very special type of multiplication and which is not frequently in requisition elsewhere but is often required in mathematical astronomy etc. The wording of the sub-*sūtra* (corollary) एकन्यूनेन पूर्वेण (*Ekanyūnena Pūrveṇa*) sounds as if it were the converse of the *Ekādhika Sūtra*. It actually is; and it relates to and provides for multiplications wherein the multiplier-digits consist entirely of nines. It comes up under three different headings as follows:

### The First Case

The annexed table of products produced by the single-digit multiplier 9 gives us the necessary clue to an understanding of the *Sūtra*:

| | | |
|---|---|---|
| 9 × 2 = | 1 | 8 |
| 9 × 3 = | 2 | 7 |
| 9 × 4 = | 3 | 6 |
| 9 × 5 = | 4 | 5 |
| 9 × 6 = | 5 | 4 |
| 9 × 7 = | 6 | 3 |
| 9 × 8 = | 7 | 2 |
| 9 × 9 = | 8 | 1 |
| 9 × 10 = | 9 | 0 |

We observe that the left-hand-side is invariably one less than the multiplicand and that the right-side-digit is merely the complement of the left-hand-side digit from 9. And this tells us what to do to get both the portions of the product.

The word '*Pūrva*' in this context has another technico-terminological usage and simply means the "multiplicand" while the word 'Apara' signifies the multiplier.

The meaning of the sub-corollary thus fits in smoothly into its context, i.e. that the multiplicand has to be decreased by 1; and as for the right-hand-side, that is mechanically available by the subtraction of the left-hand-side from 9 which is practically a direct application of the *Nikhilaṁ Sūtra*.

As regards multiplicands and multipliers of 2 digits each, we have the following table of products:

| | | |
|---|---|---|
| 11 × 99 = 10 | 89 | =(11–1)/99–(11–1)=1089 |
| 12 × 99 = 11 | 88 | |
| 13 × 99 = 12 | 87 | |
| 14 × 99 = 13 | 86 | |
| 15 × 99 = 14 | 85 | |
| 16 × 99 = 15 | 84 | |
| 17 × 99 = 16 | 83 | |
| 18 × 99 = 17 | 82 | |
| 19 × 99 = 18 | 81 | |
| 20 × 99 = 19 | 80 | |

And this table shows that the rule holds good here too. And by similar continued observation, we find that it is uniformly applicable to all cases, where the multiplicand and the multiplier consist of the same number of digits. In fact, it is a simple application of the *Nikhilaṁ Sūtra* and is bound to apply.

| | | |
|---|---|---|
| 7 – 3 | 77 – 23 | 979 – 021 |
| 9 – 1 | 99 – 1 | 999 – 1 |
| 6 / 3 | 76 / 23 | 978/021 |

We are thus able to apply the rule to all such cases and say, for example:

| | | |
|---|---|---|
| 777 | 9879 | 1203579 |
| 999 | 9999 | 9999999 |
| 776 / 223 | 9878 / 0121 | 1203578 / 8796421 |

| | |
|---|---|
| 9765431 | 1 2 3 4 5 6 7 8 0 9 |
| 9999999 | 9 9 9 9 9 9 9 9 9 9 |
| 9765430 / 0234569 | 1234567808 / 8765432191 |

Such multiplications involving multipliers of this special type come up in advanced astronomy etc.; and this sub-formula *Ekanyūnena Pūrveṇa* is of immense utility there.

### The Second Case

The second case falling under this category is one wherein the multiplicand consists of a smaller number of digits than the multiplier. This, however, is easy enough to handle; and all that is necessary is to fill the blank (on the left) with the required number of zeroes and proceed exactly as before and then leave the zeroes out. Thus—

| | | | |
|---|---|---|---|
| 7 | 79 | 798 | 79 |
| 99 | 999 | 99 999 | 9999999 |
| 06/93 | 078/921 | 00797/99202 | 0000078/9999921 |

THE THIRD CASE

(To be omitted during a first reading.)

The third case coming under the heading is one where the multiplier contains a smaller number of digits than the multiplicand. Careful observation and study of the relevant table of products gives us the necessary clue and helps us to understand the correct application of the *Sūtra* to this kind of examples.

| *Column 1* | | *Column 2* | | *Column 3* |
|---|---|---|---|---|
| 11×9=9 | 9 | 21×9=18 | 9 | 37×9=33/3 |
| 12×9=10 | 8 | 22×9=19 | 8 | 46×9=41/4 |
| 13×9=11 | 7 | 23×9=20 | 7 | 55×9=49/5 |
| 14×9=12 | 6 | 24×9=21 | 6 | 64×9=57/6 |
| 15×9=13 | 5 | 25×9=22 | 5 | 73×9=65/7 |
| 16×9=14 | 4 | 26×9=23 | 4 | 82×9=73/8 |
| 17×9=15 | 3 | 27×9=24 | 3 | 91×9=81/9 |
| 18×9=16 | 2 | 28×9=25 | 2 | and so on. |
| 19×9=17 | 1 | 29×9=26 | 1 | |
| 20×9=18 | 0 | 30×9=27 | 0 | |

We note here that, in the first column of products where the multiplicand starts with 1 as its first digit the left-hand-side part of the product is uniformly 2 less than the multiplicand; that, in the second column where the multiplicand begins with 2, the left-hand-side part of the product is exactly 3 less; and that, in the third column of miscellaneous first-digits the difference between the multiplicand and the left-hand portion of the product is invariably one more than the excess portion to the extreme left of the multiplicand.

The procedure applicable in this case is therefore evidently as follows:

(*i*) Divide the multiplicand off by a vertical line—into a right-hand portion consisting of as many digits as the multiplier; and subtract from the multiplicand one more than the whole excess portion on the left. This gives us the left-hand-side portion of the product.

*or* take the *Ekanyūna* and subtract therefrom the previous, i.e. the excess portion on the left; and

(*ii*) Subtract the right-hand-side part of the multiplicand by the *Nikhilaṁ* rule. This will give you the right-hand-side of the product.

The following examples will make the process clear:

(1) 43×9
```
 4 : 3 :
   :-5 : 3
 ---------
 3 : 8 : 7
```

(2) 63×9
```
 6 : 3 :
   :-7 : 3
 ---------
 5 : 6 : 7
```

(3) 122×9
```
 12 : 2 :
 -1 : 3 : 2
 ----------
 10 : 9 : 8
```

(4) $112 \times 99$

$$\begin{array}{rcrcr} 1 & : & 12 & : & \\ - & : & 2 & : & 12 \\ \hline 1 & : & 10 & : & 88 \end{array}$$

(5) $11119 \times 99$

$$\begin{array}{rcrcr} 111 & : & 19 & : & \\ -1 & : & 12 & : & 19 \\ \hline 110 & : & 07 & : & 81 \end{array}$$

(6) $4599 \times 99$

$$\begin{array}{rcrcr} 45 & : & 99 & : & \\ & : & -46 & : & 99 \\ \hline 45 & : & 53 & : & 01 \end{array}$$

(7) $15639 \times 99$

$$\begin{array}{rcrcr} 156 & : & 39 & : & \\ -1 & : & 57 & : & 39 \\ \hline 154 & : & 82 & : & 61 \end{array}$$

(8) $25999 \times 999$

$$\begin{array}{rcrcr} 25 & : & 999 & : & \\ & : & -26 & : & 999 \\ \hline 25 & : & 973 & : & 001 \end{array}$$

(9) $777999 \times 9999$

$$\begin{array}{rcrcr} 77 & : & 7999 & : & \\ & : & -78 & : & 7999 \\ \hline 77 & : & 7921 & : & 2001 \end{array}$$

(10) $111011 \times 99$

$$\begin{array}{rcrcr} 1110 & : & 11 & : & \\ -11 & : & 11 & : & 11 \\ \hline 1099 & : & 00 & : & 89 \end{array}$$

(11) $1000001 \times 99999$

$$\begin{array}{rcrcr} 10 & : & 00001 & : & \\ & : & -11 & : & 00001 \\ \hline 9 & : & 99990 & : & 99999 \end{array}$$

# 3
# Multiplication

## By Ūrdhva-Tiryak Sūtra

Having dealt in fairly sufficient detail with the application of the *Nikhilaṁ Sūtra* to special cases of multiplication, we now proceed to deal with the ऊर्ध्वतिर्यग्भ्याम् (*Ūrdhva-Tiryagbhyām*)*Sūtra* which is the General Formula applicable to all cases of multiplication and will also be found very useful, later on, in the division of a large number by another large number.

The formula itself is very short and terse, consisting of only one compound word and means "vertically and cross-wise". The applications of this brief and terse *Sūtra* are manifold (as will be seen again and again, later on). First we take it up in its most elementary application namely, to Multiplication in general.

A simple example will suffice to clarify the *modus operandi* thereof. Suppose we have to multiply 12 by 13.

$$\begin{array}{l} 12 \\ \underline{13} \\ \underline{1:3+2:6=156} \end{array}$$

(*i*) We multiply the left-hand-most digit 1 of the multiplicand *vertically* by the left-hand-most digit 1 of the multiplier, get their product 1 and set it down as the left-hand-most part of the answer;

(*ii*) We then multiply 1 and 3, and 1 and 2 *cross-wise*, add the two, get 5 as the sum and set it down as the middle part of the answer; and

(*iii*) We multiply 2 and 3 vertically, get 6 as their product and put it down as the last the right-hand-most part of the answer. Thus $12 \times 13 = 156$.

A few other examples may also be tested and will be found to be correct:

(1)
$$\begin{array}{r} 12 \\ 11 \\ \hline 1:1+2:2 \\ =132 \end{array}$$

(2)
$$\begin{array}{r} 16 \\ 11 \\ \hline 1:1+6:6 \\ =176 \end{array}$$

(3)
$$\begin{array}{r} 21 \\ 14 \\ \hline 2:8+1:4 \\ =294 \end{array}$$

(4)
$$\begin{array}{r} 23 \\ 21 \\ \hline 4:2+6:3 \\ =483 \end{array}$$

(5)
$$\begin{array}{r} 41 \\ 41 \\ \hline 16:4+4:1 \\ =1681 \end{array}$$

*Note*: When one of the results contains more than 1 digit, the right-hand-most digit thereof is to be put down there and the preceding, i.e. left-hand-side digit or digits should be carried over to the left and placed under the previous digit or digits of the upper row until sufficient practice has been achieved for this operation to be performed mentally. The digits carried over may be shown in the working (as illustrated below):

(1)
$$\begin{array}{r} 15 \\ 15 \\ \hline 105 \\ 12 \\ \hline 225 \end{array}$$

(2)
$$\begin{array}{r} 25 \\ 25 \\ \hline 405 \\ 22 \\ \hline 625 \end{array}$$

(3)
$$\begin{array}{r} 32 \\ 32 \\ \hline 924 \\ 1 \\ \hline 1024 \end{array}$$

(4)
$$\begin{array}{r} 35 \\ 35 \\ \hline 905 \\ 32 \\ \hline 1225 \end{array}$$

(5)
$$\begin{array}{r} 37 \\ 33 \\ \hline 901 \\ 32 \\ \hline 1221 \end{array}$$

(6)
$$\begin{array}{r} 49 \\ 49 \\ \hline 1621 \\ 78 \\ \hline 2401 \end{array}$$

The Algebraical principle involved is as follows:

Suppose we have to multiply $(ax+b)$ by $(cx+d)$. The product is $acx^2 + x(ad+bc) + bd$. In other words, the first term, i.e. the coefficient of $x^2$ is got by vertical multiplication of *a* and *c*; the middle term, i.e. the coefficient of x is obtained by the cross-wise multiplication of *a* and *d* and of *b* and *c* and the addition of the two products; and the independent term is arrived at by vertical multiplication of the absolute terms. And, as all arithmetical numbers are merely algebraic expressions in x (with $x = 10$), the algebraic principle explained above is readily applicable to arithmetical numbers too. Now, if our multiplicand and multiplier be of 3 digits each, it merely means that we are multiplying

$(ax^2 + bx + c)$ by $(dx^2 + ex + f)$ (where $x=10$):

$$\begin{array}{l} ax^2 + bx + c \\ dx^2 + ex + f \\ \hline adx^4 + x^3(ae+bd) + x^2(af+be+cd) + x(bf+ce) + cf \end{array}$$

We observe here the following facts:

(*i*) that the coefficient of $x^4$ is got by the vertical multiplication of the first digit from the left side;

(*ii*) that the coefficient of $x^3$ is got by the cross-wise multiplication of the first two digits and by the addition of the two products;

(*iii*) that the coefficient of $x^2$ is obtained by the multiplication of the first digit of the multiplicand by the last digit of the multiplier, of the middle one by the middle one and of the last one by the first one and by the addition of all the three products;

(*iv*) that the coefficient of x is obtained by the cross-wise multiplication of the second digit by the third one and conversely by the addition of the two products; and

(*v*) that the independent term results from the vertical multiplication of the last digit by the last digit.

We thus follow a process of ascent and of descent going forward with the digits on the upper row and coming rearward with the digits on the lower row. If and when this principle of ordinary Algebraic multiplication is properly understood and carefully applied to the Arithmetical multiplication on hand (where x stands for 10), the *Ūrdhva Tiryak Sūtra* may be deemed to have been successfully *mastered* in actual practice.

A few illustrations will serve to explain this *Ūrdhva-Tiryak* process of vertical and cross-wise multiplications:

| (1) | (2) | (3) |
|---|---|---|
| 111 | 108 | 109 |
| 111 | 108 | 111 |
| 12321 | 10 60 4 | 11 099 |
| | 1 6 | 1 |
| | 11 66 4 | 12 099 |

| (4) | (5) | (6) |
|---|---|---|
| 116 | 116 | 582 |
| 114 | 116 | 231 |
| 12 1 0 4 | 12 32 6 | 10 1 3 42 |
| 1 1 2 | 1 13 | 3 3 1 |
| 13 2 2 4 | 13 45 6 | 13 4 4 42 |

| (7) | (8) | (9) |
|---|---|---|
| 532 | 785 | 321 |
| 472 | 362 | 52 |
| 20 7 9 04 | 21 6 7 6 0 | 0 5 692 |
| 4 3 2 | 6 7 4 1 | 1 1 |
| 25 1 1 04 | 28 4 1 7 0 | 1 6 692 |

(10)
$$\begin{array}{r} 795 \\ 362 \\ \hline 21\ 9\ 3\ 8\ 0 \\ 6\ 8\ 4\ 1\phantom{\ 0} \\ \hline 28\ 7\ 7\ 9\ 0 \end{array}$$

(11)
$$\begin{array}{r} 1021 \\ 2103 \\ \hline 2147163 \end{array}$$

(12)
$$\begin{array}{r} 621 \\ 547 \\ \hline 30\ 4\ 5\ 87 \\ 3\ 5\ 1\phantom{\ 87} \\ \hline 33\ 9\ 6\ 87 \end{array}$$

(13)
$$\begin{array}{r} 6471 \\ 6212 \\ \hline 36\ 6\ 6\ 6\ 752 \\ 3\ 5\ 3\ 1\ 1\phantom{\ 752} \\ \hline 40\ 1\ 9\ 7\ 852 \end{array}$$

(14)
$$\begin{array}{r} 8\ 7\ 2\ 6\ 5 \\ 3\ 2\ 1\ 1\ 7 \\ \hline 2\ 4\ 7\ 8\ 7\ 2\ 7\ 5\ 7\ 5 \\ 3\ 2\ 3\ 9\ 6\ 2\ 4\ 3\phantom{\ 5} \\ \hline 2\ 8\ 0\ 2\ 6\ 9\ 0\ 0\ 0\ 5 \end{array}$$

*Note*: It need hardly be mentioned that we can carry out this *Ūrdhva-Tiryak* process of multiplication from left to right or from right to left as we prefer. The only difference will be that, in the former case, two-line multiplication will be necessary at least mentally while, in the latter case, one-line multiplication will suffice, but careful practice is necessary.

Owing to their relevance to this context, a few Algebraic examples of the *Ūrdhva-Tiryak* type are being given.

(1)
$$\begin{array}{l} a+b \\ a+9b \\ \hline a^2+10ab+9b^2 \end{array}$$

(2)
$$\begin{array}{l} a+3b \\ 5a+7b \\ \hline 5a^2+22ab+21b^2 \end{array}$$

(3)
$$\begin{array}{l} 3x^2+5x+7 \\ 4x^2+7x+6 \\ \hline 12x^4+41x^3+81x^2+79x+42 \end{array}$$

(4)
$$\begin{array}{l} x^5+3x^4+5x^3+3x^2+x+1 \\ 7x^5+5x^4+3x^3+x^2+3x+5 \\ \hline 7x^{10}+26x^9+53x^8+56x^7+43x^6+40x^5+41x^4+38x^3+ \\ 19x^2+8x+5 \end{array}$$

*Note*: If and when a power of x is absent, it should be given a zero coefficient; and the work should be carried on exactly as before. For example, for $(7x^3+5x+1)(3x^3+x^2+3)$, we work as follows:

$$\begin{array}{l} 7x^3+0+5x+1 \\ 3x^3+x^2+0+3 \\ \hline 21x^6+7x^5+15x^4+29x^3+x^2+15x+3 \end{array}$$

### The Use of the Vinculum

It may, in general, be stated that multiplications by digits higher than

5 may some times be facilitated by the use of the vinculum. The following example will illustrate this:

$$\begin{array}{r} 576 \\ 214 \\ \hline 109944 \\ 1332 \\ \hline 123264 \end{array} \qquad \textit{Or} \quad (2) \quad \begin{array}{r} 6\bar{2}\bar{4} \\ 214 \\ \hline 1224\bar{2}\bar{6} \\ 1\bar{1}\bar{1} \\ \hline 123264 \end{array}$$

(1) is the first working shown above.

But the vinculum process is one which the student must very carefully practise, before he resorts to it and relies on it.

### Miscellaneous Examples

There being so many methods of multiplication one of them the *Ūrdhva-Tiryak* being perfectly general and therefore applicable to all cases and the others the *Nikhilaṁ*, the *Yāvadūnam* etc. being of use in certain special cases only, it is for the student to think of and weigh all the possible alternative processes available, make up his mind as to the simplest method in each particular case and apply the formula prescribed therefor.

We now conclude this chapter with a number of miscellaneous examples and with our own "running comments" thereon giving the students the necessary experience for making the best possible selection from amongst the various alternative methods in question:

(1) $73 \times 37$

(*i*) By *Ūrdhva-Tiryak* rule,

$$\begin{array}{r} 73 \\ 37 \\ \hline 2181 \\ 52 \\ \hline =2701 \end{array}$$

or (*ii*) By the same method but with the use of the vinculum

$$\begin{array}{r} 1\ \bar{3}\ 3 \\ 0\ 4\ \bar{3} \\ \hline 0\ 4\ \bar{5}\ 1\ \bar{9} \\ \bar{1}\ 2\ \ \  \\ \hline =2\ 7\ 0\ 1 \end{array}$$

*Evidently, the former is better.*

(2) $94 \times 81$

(*i*) By *Ūrdhva-Tiryak*,

$$\begin{array}{l} 94 \\ 81 \\ \hline 7214 = 7614 \\ \ \ 4 \\ \hline \end{array} \qquad \textit{Or}\ (ii) \quad \begin{array}{r} 1\bar{1}4 \\ 1\bar{2}1 \\ \hline 1\bar{3}7\bar{9}4 = 7614 \end{array}$$

(*ii*) By *ibid* (with the use of the Vinculum).
Evidently the former is better; but

(*iii*) The *Nikhilaṁ* Method is still better:

$$\begin{array}{r} 81-19 \\ 94-\ 6 \\ \hline 75/_1 14 = 7614 \end{array}$$

(3) $123 \times 89$

(*i*)
$$\begin{array}{r} 123 \\ 089 \\ \hline 08527 \\ 242 \\ \hline =10947 \end{array}$$

*Or* (*ii*)
$$\begin{array}{r} 123 \\ 1\bar{1}\bar{1} \\ \hline 110\bar{5}\bar{3} \\ =10947 \end{array}$$

*Or* (*iii*)
$$\begin{array}{r} 123+23 \\ 89-11 \\ \hline 112/\overline{253} \\ \hline 110/\overline{53} = 109/47 \end{array}$$

(4) $652 \times 43$

(*i*)
$$\begin{array}{r} 652 \\ 043 \\ \hline 04836 \\ 232 \\ \hline 28036 \end{array}$$

(*ii*) The Vinculum method is manifestly cumbrous in this case and need not be worked out.

$$\begin{array}{r} (1\bar{3}\bar{5}2 \\ \times 0043) \\ \hline \end{array}$$

(*iii*) The *Nikhilaṁ* method may be used and will be quite easy; but we will have to take a multiple of 43 which will bring it very near 1000. Such a multiple is 43 × 23 = 989; and we can work with it and finally divide the whole thing out by 23. This gives us the same answer (28/036).

$$\begin{array}{r} 652-348 \\ 989-011 \\ \hline 641/_3 828 \\ 23)644828 \\ \hline 28036 \end{array}$$

Therefore, the *Ūrdhva* (general) process is obviously the best in this case.

(5) $123 \times 112$

(*i*)
$$\begin{array}{r} 123 \\ 112 \\ \hline 13776 \\ \hline =13776 \end{array}$$

(*ii*) As all the digits are within 5, the Vinculum method is manifestly out of place.

(Nikhilaṁ)

(*iii*)
$$\begin{array}{r} 123+23 \\ 112+12 \\ \hline 135/_2 76 \\ =137/76 \\ \hline \end{array}$$

Both the first and the third methods seem equally good.

(6) $99 \times 99$

(*i*)
$$\begin{array}{r} 99 \\ 99 \\ \hline 8121 \\ 168 \\ \hline =9801 \end{array}$$

(*ii*)
$$\begin{array}{r} 1\bar{0}\bar{1} \\ 1\bar{0}\bar{1} \\ \hline 10\bar{2}01 \\ =9801 \end{array}$$

(*iii*)
$$\begin{array}{r} 99-1 \\ 99-1 \\ \hline =98/01 \end{array}$$
appropriate

(*iv*) The *Yāvadūnam* method is also quite appropriate and easy
$99^2 = 98/01$

(7)
$$\begin{array}{r} 246 \\ 131 \\ \hline 20026 \\ 122 \\ \hline =32226 \end{array}$$

(8)
$$\begin{array}{r} 222 \\ 143 \\ \hline 20646 \\ 111 \\ \hline =31746 \end{array}$$

(9)
$$\begin{array}{r} 642 \\ 131 \\ \hline 62002 \\ 221 \\ \hline =84102 \end{array}$$

(10)
$$\begin{array}{r} 321 \\ 213 \\ \hline 67373 \\ 1 \\ \hline =68373 \end{array}$$

In all these 4 cases Nos. 7—10, the General formula fits in at once.

(11) $889 \times 898$

(*i*)
$$\begin{array}{r} 889 \\ 898 \\ \hline 646852 \\ 13047 \\ 21 \\ \hline =798322 \end{array}$$

*Or* (*ii*)
$$\begin{array}{r} 1\bar{1}\bar{1}\bar{1} \\ 1\bar{1}0\bar{2} \\ \hline 1\bar{2}0\bar{2}322 \\ =798322 \end{array}$$

*Or* (*iii*)
$$\begin{array}{r} 889-111 \\ 898-102 \\ \hline 787/_{11}322 \\ =798/322 \end{array} \quad \Bigg| \quad \begin{array}{r} {}^{*}111+11 \\ 102+\ 2 \\ \hline 113/22 \\ \hline \end{array}$$

*Note* : Here in (*iii*) *Nikhilaṁ* method, the vertical multiplication of 111 and 102 is also performed in the same manner (as shown in the *marked margin).

(12) (*i*)
$$\begin{array}{r} 576 \\ \times 328 \\ \hline 151288 \\ 3764 \\ \hline =188928 \end{array}$$

*Or* (*ii*) Vinculum method inappropriate

*Or* (*iii*)
$$\begin{array}{r} 576-424 \\ 984-\ 16 \\ \hline 3)560/_{6}784 \\ \hline =188/928 \end{array}$$

*N.B.* 984 being $3 \times 328$, we have made use of it and then divided out by 3.

(13) $817 \times 332$

(*i*)
$$\begin{array}{r} 817 \\ 332 \\ \hline 247034 \\ 2421 \\ \hline =271244 \end{array}$$

*Or* (*ii*) Vinculum method may also be used.

*Or* (*iii*) $\therefore 332 \times 3 = 996$
$$\begin{array}{r} \therefore 817-183 \\ 996-4 \\ \hline 3)813/732 \\ \hline =271/244 \end{array}$$

(14) $989 \times 989$

(*i*)
$$\begin{array}{r} 989 \\ 989 \\ \hline 814641 \\ 14248 \\ 21 \\ \hline =978121 \end{array}$$

*Or* (*ii*) Vinculum method also useful
$$\begin{array}{r} 10\bar{1}\bar{1} \\ 10\bar{1}\bar{1} \\ \hline 10\bar{2}\bar{2}121 \\ =978/121 \end{array}$$

*Or* (*iii*)
$$\begin{array}{r} 989-11 \\ 989-11 \\ \hline =978/121 \end{array}$$

*Or* (*iv*) *Yāvadūnam* $989^2 = 978/121$. This is the best.

(15) $8989 \times 8898$

(*i*)
8989
8898
64681652
1308147
2221
= 79984122

*Or* (*ii*)
$1\bar{1}0\bar{1}\bar{1}$
$1\bar{1}\bar{1}0\bar{2}$
$1\bar{2}00\bar{2}4122$
$= 79984122$
$^{*}1011 + 11$
$1102 + 102$
$1113/_1122$
$= 1114/122$

*Or* (*iii*)
8989 – 1011
8898 – 1102
7887 / 4122 *
111 /
7998/4122

(16) $213 \times 213$

(*i*)
213
213
44369
1
= 45369

*Or* (*ii*) Vinculum method not suitable.

*Or* (*iii*)
213 + 13
213 + 13
$226/_169$
× 2/
$452/_169$
= 453 / 69

*N.B.* The digits being small, the general formula is always best.

## PRACTICAL APPLICATION IN COMPOUND MULTIPLICATION

### A. Square Measure, Cubic Measure

This is not a separate subject, all by itself. But it is often of practical interest and importance, even to lay people and deserves our attention on that score. We therefore deal with it briefly.

#### Areas of Rectangles

Suppose we have to know the area of a rectangular piece of land whose length and breadth are 7'8" and 5'11" respectively.

According to the conventional method, we put both these measurements into uniform shape either as inches or as vulgar fractions of feet—preferably the latter and say:

$$\text{Area} = \frac{92}{12} \times \frac{71}{12} = \frac{6532}{144} = \frac{1633}{36}$$

36) 1633 (45 sq. ft.
144
193
180
13
× 144
36) 1872 (52 sq. in.
180
72
72

∴ Area = 45 sq. ft. 52. sq. in.

In the Vedic method, however, we make use of the Algebraical multiplication and the *Ādyam Sūtra* and say:

$$\text{Area} = \left.\begin{array}{r} 5x + 11 \\ \times\ 7x + 8 \end{array}\right\} = 35x^2 + 117x + 88$$

Splitting the middle term (by dividing by 12), we get 9 and 9 as Q and R.

$$\begin{aligned} \therefore E &= 35x^2 + (9 \times 12 + 9)x + 88 \\ &= 44x^2 + (9 \times 12) + 88 \\ &= 44 \text{ sq. ft.} + 196 \text{ sq. in.} \\ &= 45 \text{ sq. ft.} + 52 \text{ sq. in.} \end{aligned}$$

And the whole work can be done *mentally*:

(2) Similarly $\left.\begin{array}{r} 3'\ 7'' \\ \times 5'10'' \end{array}\right\} \begin{array}{l} = 15x^2 + 65x + 70 \\ = 20x^2 + (5 \times 12) + 70 \\ = 20 \text{ sq. ft.} + 130 \text{ sq. in.} \end{array}$

and (3) $\left.\begin{array}{r} 7x + 11 \\ \times 5x + 8 \end{array}\right\} \begin{array}{l} = 35x^2 + 111x + 88 \\ = 44 \text{ sq. ft.} + 124 \text{ sq. in.} \end{array}$

## Volumes of Parallelepipeds

We can extend the same method to sums relating to 3 dimensions also. Suppose we have to find the volume of a parallele-piped whose dimensions are 3'7", 5'10" and 7'2".
By the customary method, we will say:

$$C.C. = \frac{43}{12} \times \frac{70}{12} \times \frac{86}{12}$$

(with all the big multiplications and divisions involved). But, by the Vedic process, we have:

$$\left.\begin{array}{r} 3x + 7 \\ 5x + 10 \end{array}\right\} \begin{array}{l} = 20x^2 + 10x + 10 \\ \qquad 7x + 2 \\ \hline 140x^3 + 110x^2 + 90x + 20 \\ = 149x^3 + 9x^2 + 7x + 8 \\ = 149 \text{ cub. ft and } 1388 \text{ cub. in.} \end{array}$$

Thus, even in these small computations, the customary method seems to have a natural or ingrained bias in favour of needlessly big multiplications, divisions, vulgar fractions etc., for their own sake. The Vedic *Sūtras*, however, help us to avoid these and make the work a pleasure and not an infliction.

## PRACTICE AND PROPORTION IN COMPOUND MULTIPLICATION

The same procedure under the ऊर्ध्वतिर्यक् (*Ūrdhva-Tiryak Sūtra*) is readily applicable to most questions which come under the headings "Simple Practice" and "Compound Practice", wherein "ALIQUOT" parts are taken and many steps of working are resorted to under the current system but wherein according to the Vedic method, all of it is mental Arithmetic.

For example, suppose the question is:

"In a certain investment, each rupee invested brings Rupees two and five annas to the investor. How much will an outlay of Rs. 4 and annas nine therein yield?"

THE FIRST CONVENTIONAL METHOD
(*By Means of Aliquot Parts*)

| | Rs. As. Ps. |
|---|---|
| For One Rupee | 2 – 5 – 0 |
| For 4 Rupees | 9 – 4 – 0 |
| 8 As. = $\frac{1}{2}$ of Re.1 | 1 – 2 – 6 |
| 1 a = $\frac{1}{8}$ of 8 As. | 0 – 2 – $3\frac{3}{4}$ |
| Total for Rs. 4 and annas 9 | 10 – 8 – $9\frac{3}{4}$ |

SECOND CURRENT METHOD
(*By Simple Proportion*)

$$\text{Rs. } 2-5-0=\frac{37}{16};$$

$$\text{and Rs. } 4-9-0=\text{Rs.}\frac{73}{16}$$

$$\because \text{ On Re. 1, the yield is Rs. } \frac{37}{16}$$

$$\therefore \text{ On Rs. } \frac{73}{16}, \text{ the yield is Rs. } \frac{37}{16}\times\frac{73}{16}=\text{Rs. }\frac{2701}{256}$$

$$256)\ 2701\ (10-8-9\frac{3}{4}$$

```
256) 2701 (10 – 8 – 9 3/4
     256
      141
     ×16
```

$$\begin{array}{r l}
256)\ 2045\ (8 \\
\underline{2304} \\
208 \\
\underline{\times 12} \\
256)\ 2496\ (9 \\
\underline{2304} \\
\frac{192}{256} = 3/4
\end{array}$$

BY THE VEDIC ONE-LINE METHOD

$$\begin{array}{l}
2x+5 \\
\underline{4x+9} \\
8x^2/38x/45
\end{array}$$

Splitting the middle term (or by simple division from right to left):

$$10x^2 + 6x + 2\tfrac{13}{16}$$

$$=\text{Rs. } 10 \text{ and } 8\tfrac{13}{16} \text{ annas}$$

A few more instances may be taken :

(1) Rs. 2/5 × Rs. 2/5

$$\begin{array}{l}
2-5 \\
\underline{2-5} \\
4/20/25 = \text{Rs. } 5/5\frac{9}{16} \text{ annas}
\end{array}$$

(2) Rs. 4/9 × Rs. 4/9

$$\begin{array}{l}
4-9 \\
\underline{4-9} \\
16/72/81 = \text{Rs. } 20/13\frac{1}{16} \text{ annas}
\end{array}$$

(3) Rs. 16/9 × Rs. 16/9

$$\begin{array}{l}
16-9 \\
\underline{16-9} \\
256/288/81 = \text{Rs. } 274/\text{annas } 5\frac{1}{16}
\end{array}$$

(4) Rs. 4/13 × Rs. 4/13

(*i*) By the current 'Practice' method

| | Rs. – As. |
|---|---|
| For Re. 1 | 4 – 13 |
| For Re. 4 | 19 – 4 |

$8 \text{ annas} = \frac{1}{2} \text{of Re. } 1 \quad 2-6\frac{1}{2}$

$4 \text{ annas} = \frac{1}{2} \text{of 8 As.} \quad 1-3\frac{1}{4}$

$1 \text{ a} = \frac{1}{4} \text{of 4 annas} \quad 0-4\frac{13}{16}$

$\text{Total } 23-2\frac{9}{16}$

(*ii*) By the current 'Proportion' method

$\text{Rs. } 4/13 = \text{Rs. } \frac{77}{16}$

$\therefore \frac{77}{16} \times \frac{77}{16} = \frac{5929}{256}$

$$\begin{array}{r} 256)\ 5929\ (23-2\frac{9}{16} \\ 512 \\ \hline 809 \\ 768 \\ \hline 41 \\ \times 16 \\ \hline 256)\ 656\ (2 \\ 512 \\ \hline 144 \end{array}$$

$\frac{144}{256} = 9/16$

(*iii*) By the one-line Vedic method

$$\begin{array}{l} 4-13 \\ 4-13 \\ \hline 16/104/169 \end{array} = \text{Rs. } 23/2\frac{9}{16} \text{ annas}$$

*N.B.*: Questions relating to paving, carpeting, ornamenting etc., which are under the current system usually dealt with by the 'Practice' method or by the 'Proportion' process can all be readily answered by the *Ūrdhva Tiryak* method.

For example, suppose the question is:

At the rate of 7 annas 9 pies per foot, what will be the cost for 8 yards 1 foot 3 inches?

$$\begin{array}{l} 25-3 \\ \ \ 7-9 \\ \hline 175/246/27 \end{array}$$

$= 195 \text{ annas } 8\frac{1}{4} \text{ pies}$

$= \text{Rs. } 12/3/8\frac{1}{4}$

# 4

# Division

### By the Nikhilaṁ Method

Having dealt with Multiplication at fairly considerable length, we now go on to Division; and there we start with the *Nikhilaṁ* method which is a special one.

Suppose we have to divide a number of dividends (of two digits each) successively by the same Divisor 9 we make a chart therefor as follows:

(1) $9)\ \begin{array}{r} 1/2 \\ /1 \\ \hline 1/3 \end{array}$ (2) $9)\ \begin{array}{r} 2/1 \\ /2 \\ \hline 2/3 \end{array}$ (3) $9)\ \begin{array}{r} 3/3 \\ /3 \\ \hline 3/6 \end{array}$

(4) $9)\ \begin{array}{r} 4/0 \\ /4 \\ \hline 4/4 \end{array}$ (5) $9)\ \begin{array}{r} 5/2 \\ /5 \\ \hline 5/7 \end{array}$ (6) $9)\ \begin{array}{r} 6/1 \\ /6 \\ \hline 6/7 \end{array}$

(7) $9)\ \begin{array}{r} 7/0 \\ /7 \\ \hline 7/7 \end{array}$ (8) $9)\ \begin{array}{r} 8/0 \\ /8 \\ \hline 8/8 \end{array}$

Let us first split each dividend into a left-hand part for the quotient and a right-hand part for the remainder and divide them by a vertical line.

In all these particular cases, we observe that the first digit of the dividend becomes the quotient and the sum of the two digits becomes the remainder. This means that we can mechanically take the first digit down for the quotient-column and that, by adding the quotient to the second digit, we can get the remainder.

Next, we take as dividends, another set of bigger numbers of 3 digits each and make a chart of them as follows:

```
(1)  9) 10/3      (2)  9) 11/3      (3)  9) 12/4
         1/1                1/2                1/3
        ----               ----               ----
        11/4               12/5               13/7

(4)  9) 16/0      (5)  9) 21/1      (6)  9) 31/1
         1/7                2/3                3/4
        ----               ----               ----
        17/7               23/4               34/5
```

In these cases, we note that the remainder and the sum of the digits are still the same and that, by taking the first digit of the dividend down mechanically and adding it to the second digit of the dividend, we get the second digit of the quotient and that by adding it to the third digit of the dividend, we obtain the remainder.

And then, by extending this procedure to still bigger numbers (consisting of still more digits) we are able to get the quotient and remainder correctly. For example,

```
(1)  9) 1203/1    (2)  9) 1230/1    (3)  9) 120021/2
         133/6              136/6             13335/6
        ------             ------            --------
        1336/7             1366/7            133356/8
```

And, thereafter, we take a few more cases as follows:

```
(1)  9) 1/8       (2)  9) 22/5      (3)  9) 13/6
         /1                 2/4                1/4
        ---                ----               -----
        1/9                24/9               14/10

(4)  9) 23/7
         2/5
        -----
        25/12
```

But in all these cases, we find that the remainder is the same as or greater than the divisor. As this is not permissible, we redivide the remainder by 9, carry the quotient over to the quotient column and retain the final remainder in the remainder column, as follows:

```
(1)  9) 1/8       (2)  9) 22/5      (3)  9) 13/6
         /1                 2/4                1/4
        ---                ----               -----
        1/9                24/9               14/10
        ---                ----               -----
        2/0                25/0               15/ 1

(4)  9) 23/7
         2/5
        -----
        25/12
        -----
        26/ 3
```

We also notice that, when the remainder is greater than the divisor, we can do the consequent final division by the same method, as follows:

(1) 9) 13 / 6
1 / 4
14 / 1 / 0
/ 1
1 / 1
15 / 1

(2) 9) 23 / 7
2 / 5
25 / 1 / 2
/ 1
1 / 3
26 / 3

(3) 9) 101164 / 9
11239 / 13
1123913 / 2 / 2
/ 2
2 / 4
112405 / 4

We next take up the next lower numbers 8, 7 etc., as our divisors and note the results, as follows:

(1) 8) 2 / 3
/ 4
2 / 7

(2) 7) 1 / 2
/ 3
1 / 5

(3) 6) 1 / 1
/ 4
1 / 5

Here we observe that, on taking the first digit of the dividend down mechanically, we do not get the remainder by adding that digit of the quotient to the second digit of the dividend but have to add to it twice, thrice or 4 times the quotient-digit already taken down. In other words, we have to multiply the quotient digit by 2 in the case of 8, by 3 in the case of 7, by 4 in the case of 6 and so on. And this again means that we have to multiply the quotient-digit by the divisor's complement from 10.

And this suggests that the *Nikhilaṁ* rule about the subtraction of all from 9 and of the last from 10 is at work; and, to make sure of it, we try with bigger divisions, as follows:

(1) 89) 1 / 11
11   11
1 / 22

(2) 73) 1 / 11
27   27
1 / 38

(3) 888) 1 / 234
112   / 112
1 / 346

(4) 8888) 1 / 2345
1112   / 1112
1 / 3457

(5) 7999) 1 / 2345
2001   / 2001
1 / 4346

(6) 8897) 1 / 2345
1103   / 1103
1 / 3448

(7) 8897) 1 / 1203
1103   / 1103
1 / 2306

(8) 7989) 1 / 0102
2011   / 2011
1 / 2113

(9) 899997) 1 / 010101
100003   / 100003
1 / 110104

(10)

89) 11 / 11
11 1 / 1
/ 22
12 / 43

(11)

89) 100 / 13
11 11 /
1 / 1
/ 22
112/45

(12)

888) 12 / 345
112 1 / 12
/ 336
13 / 801

(13)

8997) 21 / 0012
1003 2 / 006
/ 3009
23 / 3081

(14)

8998) 30/ 0000
1002 3 / 006
/ 3006
33 / 3066

(15)

8888) 10 /1020
1112 1 / 112
/ 1112
11 / 3252

(16)

8987) 20 / 0165
1013 2 / 026
/ 2026
22 / 2451

(17)

899) 10 /102
101 1 / 01
/ 101
11 / 213

(18)

89998) 20 / 02002
10002 2 / 0004
/ 20004
22 / 22046

(19)

89997) 10 / 10101
10003 1 / 0003
/ 10003
11 / 20134

(20)

89997) 12/34567
10003 1 / 0003
/ 30009
13 / 64606

(21)

98987) 10 / 30007
01013 0 / 1013
/ 00000
10 / 40137

(22)

99979) 111/11111
00021 00 / 021
0 / 0021
/ 00021
111 / 13442

(23)

88 110 / 01
12 12 /
2 / 4
/ 48
124 / 89
= 125 / 1

In all the above examples, we have deliberately taken as divisors, numbers containing big digits. The reason therefore is as follows:

(*i*) It is in such division that the student finds his chief difficulty, because he has to multiply long big numbers by the "trial" digit of the quotient at every step and subtract that result from each dividend at each step; but, in our method of the *Nikhilaṁ* formula, the bigger the digits, the smaller will be the required complement from 9 or 10 as the case may be; and the multiplication-task is lightened thereby.

(*ii*) There is no subtraction to be done at all!

(*iii*) And, even as regards the multiplication, we have no multiplication of numbers by numbers as such but only *of* a single digit *by* a single digit, with the pleasant consequence that, at no stage, is a student called upon to multiply more than 9 by more than 9. In other words $9 \times 9 = 81$, is the utmost multiplication he has to perform.

A single sample example will suffice to prove this:

$$
(24) \quad \begin{array}{r|l}
9819) & 2 \quad 01\;37 \\
\overline{0181} & \quad\;\; 02_{1}62 \\
\hline
 & 2 \quad 04\;99
\end{array}
$$

*Note*: In this case, the product of 8 and 2 is written down in its proper place, as 16 with no "carrying" over to the left and so on.

Thus, in our "division" process by the *Nikhilaṁ* formula, we perform only small single-digit multiplications; we do *no* subtraction and *no* division at all; and yet we readily obtain the required quotient and the required remainder. In fact, we have accomplished our division-work in full, without actually doing any division at all!

As for divisors consisting of small digits, another simple formula will serve our purpose and is to be dealt with in the next chapter. Just at present we deal only with big divisors and explain how simple and easy such difficult divisions can be made with the aid of the *Nikhilaṁ Sūtra*.

And herein, we take up a few more illustrative examples relating to the cases already referred to wherein the remainder exceeds the divisor and explain the process, by which this difficulty can be easily surmounted by further application of the same *Nikhilaṁ* method:

$$
(25) \quad \begin{array}{r|r}
88) & 1 \quad 98 \\
\overline{12} & 12 \\
\hline
 & 1 \quad 110
\end{array}
$$

The remainder here 110 being greater than the divisor 88 we have to divide 110 by 88 and get the quotient and the final remainder and carry the former over and add it to the quotient already obtained. Thus, we say:

$$\begin{array}{l|ll}
88) & 1 & 10 \\
\overline{12} & & 12 \\
\hline
 & 1 & 22 \\
\hline
\end{array}$$

so, we add the newly obtained 1 to the previously obtained 1 and put down 2 as the quotient and 22 as the remainder.

This double process can be combined into one as follows:

$$\begin{array}{l|lr}
88) & 1 & 98 \\
\overline{12} & & 12 \\
\hline
 & 1 & 1/10 \\
 & & /12 \\
 & 1 & 1/22 \\
\hline
 & 2 & /22 \\
\hline
\end{array}$$

A few more illustrations will serve to help the student in practising this method:

(26)
$$\begin{array}{l|lr}
89997) & 12 & 94567 \\
\overline{10003} & 1 & 0003 \\
 & & 30009 \\
\hline
 & 13\ 1 & 24606 \\
 & & 10003 \\
\hline
 & 14 & 34609 \\
\hline
\end{array}$$

(27)
$$\begin{array}{l|lr}
97) & 1 & 98 \\
\overline{03} & & 03 \\
\hline
 & 1\ 1 & 01 \\
 & & 03 \\
\hline
 & 2 & 04 \\
\hline
\end{array}$$

(28)
$$\begin{array}{l|ll}
99979) & 111 & 99171 \\
\overline{00021} & 00 & 021 \\
 & 0 & 0021 \\
 & & 00021 \\
\hline
 & 111\ 1 & 01502 \\
 & & 00021 \\
\hline
 & 112 & 01523 \\
\hline
\end{array}$$

Thus, even the whole lengthy operation of division of 11199171 by 99979 involves *no* division and *no* subtraction and consists of a few multiplications of single digits by single digits and a little addition of an equally easy character.

Yes, this is all good enough so far as it goes; but it provides only for a particular type namely, of divisions involving large-digit numbers. Can it help us in other divisions, i.e. those which involve small-digit divisors?

The answer is a candidly emphatic and unequivocal No. An actual sample specimen will prove this:

Suppose we have to divide 1011 by 23. By the *Nikhilaṁ* method, the working will be as follows:

| | | | | | |
|---|---|---|---|---|---|
| (29) | 23) | 10 | | 1 | 1 |
| | 77 | 7 | | 7 | |
| | | | 4 | 9 4 | 9 |
| | | 17 | 6 | 2 | 0 |
| | | | | 42 | 42 |
| | | 23 | 4 | 8 | 2 |
| | | | | 28 | 28 |
| | | 27 | 3 | 9 | 0 |
| | | | | 21 | 21 |
| | | 30 | 3 | 2 | 1 |
| | | | | 21 | 21 |
| | | 33 | 2 | 5 | 2 |
| | | | | 14 | 14 |
| | | 35 | 2 | 0 | 6 |
| | | | | 14 | 14 |
| | | 37 | 1 | 6 | 0 |
| | | | | 7 | 7 |
| | | 38 | 1 | 3 | 7 |
| | | | | 7 | 7 |
| | | 39 | 1 | 1 | 4 |
| | | 4 | | −9 | 2 (4 times the divisor) |
| | | 43 | | 2 | 2 |

This is manifestly not only too long and cumbrous but much more so than the current system which, in this particular case, is indisputably shorter and easier.

In such a case, we can use a multiple of the divisor and finally multiply again by the *Ānurūpya* rule. Thus,

| | | | |
|---|---|---|---|
| (30) | $23 \times 4 = 92$) | 10 | 11 |
| | 08 | 0 | 8 |
| | | 10 | 91 |
| | | ×4 | |
| | | 40 | − 69 |
| | | 43 | 22 |

But even this is too long and cumbrous; and this is a suitable case for the application of the परावर्त्य (*Parāvartya*) method. This we proceed to explain in the next chapter.

# 5
# Division

### By the Parāvartya Method

We have thus found that, although admirably suited for application in the special or particular cases wherein the divisor-digits are big ones, yet the *Nikhilaṁ* method does not help us in other cases namely, those wherein the divisor consists of small digits. The last example with 23 as divisor at the end of the last chapter has made this perfectly clear. Hence the need for a formula which will cover other cases. And this is found provided for in the *Parāvartya Sūtra*, which is a special-case formula, which reads "*Parāvartya Yojayet*" and which means "Transpose and apply".

The well-known rule relating to transposition enjoins invariable change of sign with every change of side. Thus + becomes – and conversely; and × becomes ÷ and conversely. In the current system, this law is known but only in its application to the transposition of terms from left to right and conversely and from numerator to denominator and conversely in connection with the solution of equations, the proving of Identities etc., and also with regard to the Remainder Theorem, Horner's process of Synthetic Division etc. According to the Vedic system, however, it has a number of applications, one of which is discussed in the present chapter.

At this point, we may make a reference to the Remainder Theorem and Horner's process and then pass on to the other most interesting applications of the *Parāvartya Sūtra*.

### The Remainder Theorem

We may begin this part of the exposition with a simple proof of the Remainder Theorem, as follows:

If E, D, Q and R be the dividend, the divisor, the quotient and the remainder in a case of division and if the divisor is $(x - p)$, we may put this relationship down algebraically as follows:

$E = D.\ Q + R$ i.e. $E = Q\ (x - p) + R$.
And if we put $x = p$, $x - p$ becomes zero; and the identity takes the shape, $E = R$. In other words, the given expression E itself (with p substituted for x) will be the remainder.

Thus, the given expression E (i.e. the dividend itself) (with p substituted for x) automatically becomes the remainder. And p is automatically available by putting $x - p = 0$, i.e. by merely reversing the sign of $-p$ which is the absolute term in the binomial divisor. In general terms, this means that, if E be $ax^n + bx^{n-1} + cx^{n-2} + dx^{n-3}$ etc., and if D be $x-p$, the remainder is $ap^n + bp^{n-1} + cp^{n-2} + dp^{n-3}$ and so on (i.e. E with p substituted for x). This is the Remainder Theorem.

Horner's process of Synthetic Division carries this still further and tells us the quotient too. It is, however, only a very small part of the *Parāvartya* formula which goes much farther and is capable of numerous applications in other directions also.

Now, suppose we have to divide

$(12x^2 - 8x - 32)$ by $(x - 2)$.

$$\begin{array}{c|l} x-2 & 12x^2 - 8x - 32 \\ 2 & \qquad\quad 24 + 32 \\ \hline & 12x + 16 \quad 0 \end{array}$$

We put $x - 2$ (the divisor) down on the left (as usual); just below it, we put down the $-2$ with its sign changed; and we do the multiplication work exactly as we did in the previous chapter.

A few more algebraic examples may also be taken:

(1) Divide $7x^2 + 5x + 3$ by $x - 1$

$$\begin{array}{c|l} x-1 & 7x^2 + 5x \ + 3 \\ 1 & \qquad\ 7 \quad +12 \\ \hline & 7x + 12 \quad 15 \end{array}$$

$\therefore$ The quotient is $7x + 12$; and the remainder is 15.

(2)
$$\begin{array}{c|l} x+1 & 7x^2 + 5x \ + 3 \\ -1 & \qquad -7 \ + 2 \\ \hline & 7x - \ 2 \ + 5 \end{array}$$

(3)
$$\begin{array}{c|l} x-2 & x^3 + 7x^2 + 6x \ + \ 5 \\ 2 & \qquad\ 2 \quad + 18 \ \ + 48 \\ \hline & x^2 + 9x + 24 \quad 53 \end{array}$$

(4)
$$\begin{array}{c|l} x-3 & x^3 - x^2 + 7x \ + 3 \\ 3 & \qquad 3 \ + 6 \quad 39 \\ \hline & x^2 + 2x + 13 \quad 42 \end{array}$$

(5)
$$\begin{array}{c|l} x-5 & x^3 - 3x^2 + 10x \ \ - 7 \\ 5 & \qquad\ 5 \ \ + 10 \quad 100 \\ \hline & x^2 + 2x \ + 20 \ \ + 93 \end{array}$$

At this stage, the student should practise the whole process as a mental exercise in respect of binomial divisors at any rate. For example, with regard to the division of $(12x^2 - 8x - 32)$ by the binomial $(x - 2)$, one should be able to say:

$$\frac{12x^2 - 8x - 32}{x - 2} = 12x + 16; \text{ and } R = 0$$

The procedure is as follows:

(*i*) $\frac{12x^2}{x}$ gives 12 as the first coefficient in the quotient; and we put it down;

(*ii*) multiply 12 by $-2$, reverse the sign and add to the next coefficient on the top (numerator). Thus $12x - 2 = -24$ reversed, it is 24. Add $-8$ and obtain 16 as the next coefficient of the quotient.

(*iii*) multiply 16 by $-2$; reverse the sign and add to the next coefficient on top. Thus $16x - 2 = -32$; reversed, it is 32; add $-32$ and obtain 0 as the remainder.

Similarly, (1) $\frac{7x^2 + 5x + 3}{x - 1}$ $\therefore Q = 7x + 12;$ and $R = 15$

(2) $\frac{7x^2 + 5x + 3}{x + 1}$ $\therefore Q = 7x - 2;$ and $R = 5$

(3) $\frac{x^3 + 7x^2 + 6x + 5}{x - 2}$ $\therefore Q = x^2 + 9x + 24;$ and $R = 53$

(4) $\frac{x^3 - x^2 + 7x + 3}{x - 3}$ $\therefore Q = x^2 + 2x + 13;$ and $R = 42$

and (5) $\frac{x^3 - 3x^2 + 10x - 7}{x - 5}$ $\therefore Q = x^2 + 2x + 20;$ and $R = 93$

This direct and straight application of the *Parāvartya Sūtra* should be so well practised as to become very simple mental arithmetic. And the student should be able to say at once:

(6) $\frac{x^3 + 7x^2 + 9x + 11}{x - 2}$ $\therefore Q = x^2 + 9x + 27;$ and $R = 65;$

and (7) $\frac{x^4 - 3x^3 + 7x^2 + 5x + 7}{x - 4}$ $\therefore Q = x^3 + x^2 + 11x + 49;$ and $R = 203.$

Extending this process to the case of divisors containing three terms, we should follow the same method, but should also take care to reverse the signs of the coefficient in all the other terms except the first:

(1)
$$\begin{array}{l|lllll}
x^2 - x - 1 & x^4 - x^3 + x^2 & & +3x+5 \\
1+1 & \quad 1+1 & & \\
 & \qquad 0 & +0 & \\
 & & 2 & +2 \\
\hline
 & x^2+0+2 & & +5x+7
\end{array}$$
$\therefore Q = x^2 + 2$; and $R = 5x + 7$

(2)
$$\begin{array}{l|lllll}
x^2 - 2x - 9 & 6x^4 + 13x^3 + 39x^2 & & +37x & +45 \\
2+9 & \quad 12 \quad +54 & & & \\
 & \qquad 50 & & +225 & \\
 & & & 286 & +1287 \\
\hline
 & 6x^2 + 25x + 143 & & +548x & +1332
\end{array}$$

$\therefore Q = 6x^2 + 25x + 143$; and $R = 548x + 1332$

(3)
$$\begin{array}{l|lllll}
x^2 + 1 & 2x^4 - 3x^3 + 0 & & -3x - 2 \\
0-1 & \quad 0-2 & & \\
 & \qquad 0 & +3 & \\
 & & 0 & +2 \\
\hline
 & 2 \quad -3 \quad -2 & 0 & +0
\end{array}$$

*Note*: The zero $x^2$ and the zero x carefully.

$\therefore Q = 2x^2 - 3x - 2$ and $R = 0$

(4)
$$\begin{array}{l|llll}
x^2 - 2x + 1 & x^3 - 3x^2 & +3x - 1 \\
2-1 & \quad 2 & -1 \\
 & & -2 \quad +1 \\
\hline
 & 1 \quad -1 & 0 \quad +0
\end{array}$$
$\therefore Q = x - 1$; and $R = 0$

(5)
$$\begin{array}{l|llll}
x^2 + 2x + 4 & 2x^3 + 9x^2 & +18x + 20 \\
-2-4 & \quad -4 & -8 \\
 & & -10 \quad -20 \\
\hline
 & 2 \quad +5 & 0 \quad +0
\end{array}$$
$\therefore Q = 2x + 5$; and $R = 0$

(6)
$$\begin{array}{l|llll}
x^3 - x^2 + 2x - 3 & x^5 + 0 + x^3 & -7x^2 & +0+9 \\
1-2+3 & \quad 1-2 & +3 & \\
 & \qquad 1 & -2 & +3 \\
 & & 0 & +0+0 \\
\hline
 & 1+1+0 & -6 & +3+9
\end{array}$$

*Note*: the zero $x^4$ and the zero x carefully.

$\therefore Q = x^2 + x$; and $R = -6x^2 + 3x + 9$

(7)
$$
\begin{array}{l|lllll}
x^2-x+1 & x^4 & +0 & +x^2 & +0 & +1 \\
1-1 & & 1 & -1 & & \\
 & & & 1 & -1 & \\
 & & & & 1 & -1 \\
\hline
 & 1 & +1 & +1 & 0 & +0
\end{array}
$$

*Note*: the zero $x^3$ and zero x carefully.

$\therefore Q = x^2 + x + 1$; and $R = 0$

(8)
$$
\begin{array}{l|llllll}
x^3-2x^2+1 & -2x^5 & -7x^4 & +2x^3 & +18x^2 & -3x & -8 \\
2+0-1 & & -4 & +0 & +2 & & \\
 & & & -22 & +0 & +11 & \\
 & & & & -40 & +0 & +20 \\
\hline
 & -2 & -11 & -20 & -20 & +8 & +12
\end{array}
$$

*Note*: the zero x in the divisor carefully.

$\therefore Q = -2x^2 - 11x - 20$; and $R = -20x^2 + 8x + 12$

In all the above cases, the first coefficient in the divisor happened to be 1; and therefore there was no risk of fractional coefficients coming in. But what about the cases wherein, the first coefficient not being unity, fractions will have to be reckoned with?

The answer is that all the work may be done as before, with a simple addition to the effect that every coefficient in the answer must be divided by the first coefficient of the divisor.

Thus,
$$
\begin{array}{l|llll}
2x-4 & -4x^3 & -7x^2 & +9x & -12 \\
4 & & -8 & -30 & -42 \\
\hline
 & -2 & -15/2 & -21/2 & -54
\end{array}
$$

$\therefore Q = -2x^2 - 7\frac{1}{2}x - 10\frac{1}{2}$; and $R = -54$

This, however, means a halving of each coefficient (at every step); and this is not only more cumbrous but also likely to lead to confusion, reduplication etc.

The better method therefore would be to divide the divisor itself at the very outset by its first coefficient, complete the working and divide it all again, once for all at the end. Thus:

$$
\begin{array}{l|llll}
2x-4 & -4x^3 & -7x^2 & +9x & -12 \\
x-2 & & -8 & -30 & -42 \\
2 & & & & \\
\hline
2) & -4 & -15 & -21 & -54 \\
\hline
 & -2 & -7\frac{1}{2} & -21/2 & -54
\end{array}
$$

*N.B.* : Note that the R always remains constant.

Two more illustrative examples may be taken:

(1)
$$
\begin{array}{l|ll}
3)\ 3x-7 & 3x^2-x & -5 \\
x-2\frac{1}{3} & 7 & +14 \\
2\frac{1}{3} & & \\
\hline
 & 3)\ 3x+6 & 9 \\
\hline
 & x+2 & 9
\end{array}
$$

$\therefore Q = x + 2$; and $R = 9$

(2)

$$\begin{array}{l|llll|ll} 2)\ 2x^2-3x+1 & 2x^5 & -9x^4 & +5x^3 & +16x^2 & -16x & +36 \\ \hline x^2-1\frac{1}{2}x+\frac{1}{2} & & 3 & -1 & & & \\ \hline 1\frac{1}{2}\ -\frac{1}{2} & & & -9 & +3 & & \\ & & & & -7\frac{1}{2} & +2\frac{1}{2} & \\ \hline & 2) & 2 & -6 & -5 & +11\frac{1}{2} & 17\frac{1}{4} \quad -5\frac{3}{4} \\ \hline & & 1 & -3 & -2\frac{1}{2} & +5\frac{3}{4} & 15/4+30\frac{1}{4} \end{array}$$

$\therefore Q = x^3 - 3x^2 - 2\frac{1}{2}x + 5\frac{3}{4}$; and $R = 3\frac{3}{4}x + 30\frac{1}{4}$

*N.B.*: Note that R is constant in every case.

## Arithmetical Applications (Miscellaneous)

We shall now take up a number of arithmetical applications and get a clue as to the utility and jurisdiction of the *Nikhilaṁ* formula and why and where we have to apply the *Parāvartya Sūtra.*

(1) Divide 1234 by 112

$$\begin{array}{r|rrr} 112 & 1 & & 234 \\ \overline{888} & & & 888 \\ \hline & 1 & 1 & 122 \\ \hline & 2 & 1 & 122 \\ & & & 888 \\ \hline \end{array}$$

$$\begin{array}{r|r} 3 & 010 \\ & 888 \\ \hline 3+8 & 898 \\ & -896 \\ \hline 11 & 2 \\ \hline \end{array}$$

But this is too cumbrous. The *Parāvartya* formula will be more suitable. Thus,

$$\begin{array}{r|rrrr} 112 & 1 & 2 & +3 & 4 \\ \overline{-1-2} & & -1 & -2 & \\ & & & -1 & -2 \\ \hline & 1 & 1 & 0 & 2 \end{array}$$

$\therefore Q = 11$; and $R = 2$

This is so much simpler.

(2) Divide 1241 by 112.

(i)

$$\begin{array}{c|ll|l}
\frac{112}{888} & 1 & & 241 \\
 & & & 888 \\ \hline
 & 1 & 1 & 129 \\
 & & & .888 \\ \hline
 & 2 & 1 & 017 \\
 & & & 888 \\ \hline
 & 3 & & 905 \\
 & 8 & & -896 \\ \hline
 & 11 & 9 & \\ \hline
\end{array}
\qquad \frac{112}{-1-2}$$

(ii) This too is too long. Therefore use *Parāvartya.*

$$\begin{array}{l}
1\ 2\ 4\ 1 \\
-1-2 \\
\quad -1-2 \\ \hline
11\ 09 \\ \hline
\end{array}
\qquad \therefore Q = 11; \text{ and } R = 9$$

(3) Divide 1234 by 160

(i) *Nikhilaṁ* method is manifestly unsuitable. We should therefore use the *Parāvartya* formula.

(ii)

$$\begin{array}{c|cc|cc}
\frac{160}{-6+0} & 1 & 2 & 3 & 4 \\
 & & -6 & 0 & \\
 & & & 240 & \\ \hline
 & & 6 & 274 & \\
 & & 1 & -160 & \\ \hline
 & & 7 & 114 & \\ \hline
\end{array}$$

But this is a case where *Vilokanenaiva*, i.e. by simple inspection or observation, we can put the answer down.

(4)

$$\begin{array}{c|cccccc}
\frac{11203}{-1-2-0-3} & 2 & 3 & 9 & 4 & 7 & 9 \\
 & & -2 & -4 & -0 & -6 & \\
 & & & -1 & -2 & -0 & -3 \\ \hline
 & 2 & 1 & 4 & 2 & 1 & 6 \\ \hline
\end{array}$$

(5)

$$\begin{array}{c|ccccc}
\frac{112}{-1-2} & 1 & 3 & 0 & 4 & 5 \\
 & & -1 & -2 & -4 & \\
 & & & -2 & 4+8 & \\ \hline
 & 12 & 4 & & 53 & \\
 & & = 116 / 53 & & & \\ \hline
\end{array}$$

In all these cases where the digits in the divisor are small the *Nikhilaṁ* method is generally unsuitable; and the *Parāvartya* one is always to be preferred.

(6) Divide 13456 by 1123

$$\begin{array}{c|c|ccc}
\frac{1123}{-1-2-3} & 13 & 4 & 5 & 6 \\
 & -1 & -2 & -3 & \\
 & & -2 & -4 & -6 \\ \hline
 & 12 & 0 & -2 & +0 \\ \hline
\end{array}$$

Here, as the remainder portion is a negative quantity, we should follow the device used in subtractions of larger numbers from smaller ones in coinage etc.

| Rs. | as. | ps. | | £ | s. | d. |
|---|---|---|---|---|---|---|
| 7 | 5 | 3 | | 7 | 5 | 3 |
| | 9 | 9 | | | 9 | 9 |
| 6 | 11 | 6 | | 6 | 15 | 6 |

In other words, take 1 over from the quotient column to the remainder column, i.e. take 1123 over to the right side, subtract 20 therefrom and say, Q = 11 and R = 1103

(7) Divide 13905 by 113 (similar)

| 113 | 1 3 9 | 0 5 |
|---|---|---|
| –1 – 3 | – 1 – 3 | |
| | – 2 | – 6 |
| | | – 4 – 12 |
| | 1 2 4 | – 107 |
| | 1 2 3 | 06 |

*N.B.*: Always remember that just as one Rupee = 16 annas, One Pound = 20 shillings and one Dollar = 100 cents and so on, so one taken over from the quotient to the remainder—column stands, in concrete value, for the divisor.

(8)

| 1012 | 11 | 1 1 1 |
|---|---|---|
| 0 – 1 – 2 | 0 | – 1 – 2 |
| | | 0 – 1 – 2 |
| | 11 | 0 – 2 – 1 |
| | –1 | +10 1 2 |
| | 10 | 9 9 1 |

(9)

| 1133 | 1 2 | 3 4 9 |
|---|---|---|
| –1 – 3 – 3 | – 1 | – 3 – 3 |
| | | – 1 – 3 – 3 |
| | 11 | – 1 – 2 + 6 = –114 |
| | –1 | + 1133 |
| | 10 | 1019 |

(10) Divide 13999 by 112

| 112 | 1 3 9 | 9 9 |
|---|---|---|
| –1 – 2 | – 1 – 2 | |
| | – 2 – 4 | |
| | – 5 | – 10 |
| | 1 2 5 | 0 – 1 |
| | | – 1 + 11 2 |
| | 1 2 4 | 11 1 |

∴ Q = 124 and R = 111

(11)
$$\begin{array}{r|l}
1132 & 11\quad 3\ 2\ 9 \\
-1-3-2 & -1\quad -3-2 \\
 & \qquad 0\ 0\ 0 \\ \hline
 & 10\quad 0\ 0\ 9
\end{array}$$

Also by *Vilokanam*, i.e. mere observation.

(12)
$$\begin{array}{r|l}
82 & 1\ 0\ 3 \\
\overline{1-2+2} & \quad 2-2 \\
2-2 & \overline{1\quad 21}
\end{array}$$

Also by *Vilokanam*, i.e. by mere observation.

(13) (*i*)
$$\begin{array}{r|l}
819 & 2\quad 3\ \ 4\ \ 1 \\
181 & \quad\ 2\ 16\ 2 \\ \hline
 & 2\quad 703
\end{array}$$

This is by the *Nikhilaṁ* method.

But 18 can be counted as $10+8$ or as $20-2$. So, put 181 down as $2-2+1$. We can thus avoid multiplication by big digits, i.e. by more than five.

(*ii*)
$$\begin{array}{r|l}
819 & 2\ 3\quad 4\ \ 1 \\
181 & \quad 4\ -4+2 \\
2-2+1 & 2\qquad 703
\end{array}$$

(14) Divide 39999 by 9819 Or (*ii*) by Vinculum and *Parāvartya*

(*i*)
$$\begin{array}{r|l}
9819 & 3\ 9\ \ 9\ 9\ 9 \\
0181 & \quad 3\ \ 324\ 3 \\ \hline
 & 3\ 1\ \ 0542 \\
 & \qquad 0181 \\ \hline
 & 4\qquad 0723
\end{array}$$

$$\begin{array}{r|l}
9819 & 3\ 9\ 9\ 9\ 9 \\
10\bar{2}\bar{2}\bar{1} & \quad 0\ 6-6\ 3 \\ \hline
0+2-2+1 & 3\ 1\ 0542 \\
 & \quad 02-2+1 \\ \hline
 & 4\quad 0723
\end{array}$$

(15) Divide 1111 by 839

(*i*)
$$\begin{array}{r|l}
839 & 1\quad 111 \\
161 & \quad\ 161 \\ \hline
 & 1\quad 272
\end{array}$$

Or (*ii*) But $839 = 1\bar{2}4\bar{1}$; and $161 = 24\bar{1}$

$$\therefore \frac{1\bar{2}4\bar{1}}{2-4+1} \qquad \begin{array}{l}
1\ 1\ 1\ 1 \\
\ \ 2-4\ 1 \\ \hline
1\ 3\ \bar{3}\ 2 \\ \hline
=1\ 272
\end{array}$$

(*iii*)
$$\begin{array}{r|l}
839 & 1\quad 111 \\
161 & \quad\ \ 2\bar{4}1 \\
2\bar{4}1 & 1\quad 3\bar{3}2 \\
 & 1\quad 272
\end{array}$$

(16)
$$\begin{array}{r|l}
818 & 5\ 0\quad 1\quad 2 \\
182 & \quad\ 5\ \ 40\ \ 10 \\ \hline
 & 5\qquad 922 \\
 & =6\qquad 104
\end{array}$$

by simple subtraction of the divisor as in the case of 16 annas, 20 shillings, 100 cents etc.

(17)

| 988 | 13 | 0 | 4 | 5 |
|---|---|---|---|---|
| 012 | 0 | 1 | 2 | |
| | | 0 | 3 | 6 |
| | 13 | 2 | 0 | 1 |

(18)

| 858 | 7 | 1 | 1 | 1 |
|---|---|---|---|---|
| 142 | | 7 | 28 | 14 |
| | 7 | 1 | 105 | |
| | | | 142 | |
| | 8 | | 247 | |

(19) (*i*)

| 828 | 43 | 9 | 9 | 9 |
|---|---|---|---|---|
| 172 | 4 | 28 | 8 | |
| | | 7 | 49 | 14 |
| | 47 | 50 | 8 | 3 |
| | | 5 | 35 | 10 |
| | 52 | 9 | 4 | 3 |
| | | −8 | −2 | −8 |
| | 53 | 1 | 1 | 5 |

*Or* (*ii*)

| 828 | 43 | 9 | 9 9 |
|---|---|---|---|
| 172 | | 8 − 12 | 8 |
| 2 − 3 + 2 | | | 22 − 33 + 22 |
| | 51 | | 19 − 16 + 31 |
| | 51 | 1 | 771 |
| | | | 2 − 32 |
| | 52 | | 943 |
| | | | −828 |
| | 53 | | 115 |

(20) Divide 1771 by 828

(*i*)

| 828 | 1 | 771 |
|---|---|---|
| 172 | | 172 |
| | 1 | 943 |
| | | −828 |
| | 2 | 115 |

*Or* (*ii*)

| 828 | 1 | 7 | 7 | 1 |
|---|---|---|---|---|
| 172 | | 2 | −3 | +2 |
| 2 − 3 + 2 | 1 | 9 | 4 | 3 |
| | | −8 | 2 | 8 |
| | 2 | 1 | 1 | 5 |

(21) Divide 2671 by 828

(*i*)

| 828 | 2 | 6 | 7 | 1 |
|---|---|---|---|---|
| 172 | | 2 | 14 | 4 |
| | 2 | 1 | 015 | |
| | | | 172 | |
| | 3 | | 187 | |

*Or* (*ii*)

| 828 | 2 | 6 | 7 | 1 |
|---|---|---|---|---|
| 172 | | 4 | −6 | +4 |
| 2 − 3 + 2 | 2 1 | 0 | 1 | 5 |
| | | 2 | −3 | +2 |
| | 3 | 1 | 8 | 7 |

*Or* (*iii*) Subtract 828 straight off in both cases from 1015.

(22) Divide 39893 by 829

(*i*)

| 829 | 39 | 8 | 9 | 3 |
|---|---|---|---|---|
| 171 | 3 | 21 | 3 | |
| | | 12 | 84 | 12 |
| | 42 | 5 0 | 7 | 5 |
| | | | 5 3 | 5 5 |
| | 47 | | 9 3 | 0 |
| | | | −8 2 | 9 |
| | 48 | | 1 0 | 1 |

*Or* (*ii*)

| 829 | 39 | 8 | 9 | 3 |
|---|---|---|---|---|
| 171 | | 6 | −9 | +3 |
| 2 − 3 + 1 | | 30 | −45 | +15 |
| | 45 | 25 | 8 | 8 |
| | | 4 | −6 | +2 |
| | 47 | 9 | 3 | 0 |
| | | −8 | 2 | 9 |
| | 48 | 1 | 0 | 1 |

(23) Divide 21011 by 799

799 | 21 0 1 1
201 | 4 0 2
| 10 0 5
| 25 1 036
| 201
| 26 237

(24) Divide 13045 by 988

988 | 13 0 4 5
012 | 0 1 2
| 0 3 6
| 13 2 0 1

(25) Divide 21999 by 8819

(*i*) 8819 | 2 1 9 9 9
1181 | 2+4−4+2
1+2−2+1 | 2 4 3 6 1

*Or* (*ii*) 8819 | 2 1 9 9 9
1181 | 2 21 6 2
| 2 4 3 6 1

Even this is too cumbrous. *Ānurūpya* and *Parāvartya* will be more suitable.

(26) Divide 1356 by 182

(*i*) 182 | 13 5 6
−8 −2 | −8 −2
| 40 10
| 5 4 4 6
| −32 −8
| −28 −2
| 9 +364
| −2
| 7 82

(*ii*) 182 | 13 5 6
2) 2 −2 +2 | 1 −1
1 −1 +1 | 4 −4
1 −1 | 2)14 8 2
| 7 8 2

*N.B*.R is constant.

(27) 882 | 3 1 2 8
118 | 3+6−6
1+2−2 | 3 4 8 2

(28) Divide 4009 by 882

(*i*) 882 | 4 0 0 9
1−1−2+2 | 4+8−8
1+2−2 | 4 4 8 1

(*ii*) 882 | 4 0 0 9
118 | 4+8−8
1+2−2 | 4 4 8 1

*Note*: In both these methods, the working is exactly the same.

(29) 2) 224 | 2 6 9 9
112 | −2 −4
−1 −2 | −4 −8
2 | 2)24 1 1
| 12 R 11

R is constant

(30) 2) 223 | 1 6 9 9
$111\frac{1}{2}$ | $-1 -1\frac{1}{2}$
$-1 -1\frac{1}{2}$ | $-5 -7\frac{1}{2}$
| $2)\ 15\quad 2\frac{1}{2} + 1\frac{1}{2}$
| $7\frac{1}{2}\quad 2\frac{1}{2}\quad 1\frac{1}{2}$
| $7\ 111\frac{1}{2}$
| 7 138

(31)

$$\begin{array}{r|r}
2)\ 222 & 1\ 2\ 3\ 4 \\
\overline{1}\overline{1}\overline{1} & -1-1 \\
\hline
-1-1 & -1-1 \\
\hline
 & 2)\ 11\ 1\ 3 \\
\hline
 & 5\ \ 111 \\
\hline
 & 5\ \ 124
\end{array}$$

32) Divide 7685 by 672

$$\begin{array}{r|rrrr}
672 & 7 & 6 & 8 & 5 \\
\overline{3}28 & & 21+21 & & -14 \\
\hline
3+3-2 & 7 & 2 & 9\ \ 8 & 1 \\
 & & & 6+6 & -4 \\
\hline
 & 9 & & 1\ \ 6 & +3\ \ 7 \\
 & & & 3 & +3-2 \\
\hline
 & 10 & & 9 & 6\ \ 5 \\
 & & & -6 & 7\ \ 2 \\
\hline
 & 11 & & 2 & 9\ \ 3
\end{array}$$

This work can be curtailed—or at least rendered a bit easier—by the *Ānurūpyeṇa Sūtra*. We can take 168 which is one-fourth of 672 or 84 which is one-eighth of it or, better still, 112 which is one-sixth thereof; and work it out with that divisor and finally divide the quotient *proportionately*.

The division with 112 as divisor works out as follows:

$$\begin{array}{l|l}
\because 672 = 6 \times 112 & 7\ 6\quad 8\ 5 \\
\therefore\ \ 112 & -7\ -14 \\
\hline
\quad -1-2 & \qquad\quad 1+2 \\
\hline
 & 7-1\quad -5+7 \\
\hline
 & 6) = 69\ -5+7 \\
\hline
 & \quad 11\tfrac{1}{2} - 5 + 7 \\
 & = 11 \quad 336 - 50 + 7 = 293
\end{array}$$

It will thus be seen that, in all such cases, a fairly easy method is for us to take the nearest multiple or sub-multiple to a power of 10 as our temporary divisor, use the *Nikhilaṁ* or the *Parāvartya* process and then multiply or divide the quotient *proportionately*. A few more examples are given below, in illustration hereof:

(1) Divide 1400 by 199

(*i*)

$$\begin{array}{c|l}
199 & 14\ \ 0\ \ 0 \\
\hline
2)20-1 & 0\ \ \tfrac{1}{2} \\
\hline
1+0-\tfrac{1}{2} & \quad\ 0+2 \\
\hline
0+\tfrac{1}{2} & 2)\ 14\tfrac{1}{2}+2 \\
\hline
 & \quad 7\tfrac{1}{2}+2 = 7/7
\end{array}$$

*Or* (*ii*) Since $5 \times 199 = 995$

$$\begin{array}{c|r}
\therefore\ 995 & 1\ 4\ 0\ 0 \\
\hline
005 & 0\ 0\ 5 \\
\hline
 & 1\ \ 4\ 0\ 5 \\
 & \times 5 - 3\ \ 9\ 8 \\
\hline
 & 7\ \ 7
\end{array}$$

(2) Divide 1699 by 223

$\because 4 \times 223 = 892$

$$\therefore \begin{array}{c} 892 \\ \hline 1-1-1+2 \\ \hline 1+1-2 \end{array} \qquad \begin{array}{llll} 1 & 6 & 9 & 9 \\ & 1+ & 1- & 2 \\ \hline 1 & 8 & 0 & 7 \\ \times 4 & -6 & 6 & 9 \\ \hline 4 & 1 & 3 & 8 \\ +3 & & & \\ \hline 7 & 1 & 3 & 8 \\ \hline \end{array}$$

(3) Divide 1334 by 439

$\because 2 \times 439 = 878$

$$\therefore \frac{878}{122} \qquad \begin{array}{llll} 1 & 3 & 3 & 4 \\ & 1 & 2 & 2 \\ \hline 1 & 4 & 5 & 6 \\ \times 2 & -4 & 3 & 9 \\ \hline 3 & 17 & & \\ \hline \end{array}$$

(4) Divide 1234 by 511

$\because 2 \times 511 = 1022$

$$\therefore \frac{1022}{0-2-2} \qquad \begin{array}{llll} 1 & 2 & 3 & 4 \\ & 0- & 2- & 2 \\ \hline 1 & 2 & 1 & 2 \\ \times 2 & & & \\ \hline 2 & 2 & 1 & 2 \\ \hline \end{array}$$

(5) Divide 1177 by 516

$\because 2 \times 516 = 1032$

$$\therefore \frac{1032}{0-3-2} \qquad \begin{array}{llll} 1 & 1 & 7 & 7 \\ & 0- & 3- & 2 \\ \hline 1 & 1 & 4 & 5 \\ \times 2 & & & \\ \hline 2 & 1 & 4 & 5 \\ \hline \end{array}$$

*Note*: The remainder is constant in all the cases.

# 6
# Argumental Division

### BY SIMPLE ARGUMENT PER THE ŪRDHVA TIRYAK SŪTRA

In addition to the *Nikhilaṁ* method and the *Parāvartya* method which are of use only in certain special cases there is a third method of division which is one of simple *argumentation* based on the '*Ūrdhva Tiryak*' *Sūtra* and practically amounts to a converse thereof.

The following examples will explain and illustrate it:

(1) Suppose we have to divide $(x^2+2x+1)$ by $(x+1)$, we make a chart, as in the case of an ordinary multiplication (by the '*Ūrdhva Tiryak*' process) and got down the dividend and the divisor. Then the argumentation is as follows:

$$\begin{array}{l} x+1 \\ \underline{x+1} \\ \underline{x^2+2x=1} \end{array}$$

(*i*) $x^2$ and x being the first terms of the dividend and the divisor or the product and the multiplicand respectively, the first term of the quotient or the multiplier *must* be x.

(*ii*) As for the coefficient of x in the product, it must come up as the sum of the cross-wise-multiplication-products of these. We have already got x by the cross-multiplication of x in the upper row and 1 in the lower row; but the coefficient of x in the product is 2. The other x *must* therefore be the product of x in the lower row and the absolute term in the upper row.
$\therefore$ The latter is 1. And thus is quotient is $x+1$.

(2) Divide $(12x^2-8x-32)$ by $(x-2)$.

$$\frac{12x^2-8x-32}{x-2}=12x+16$$

(*i*) $12x^2$ divided by x gives us 12x.

(*ii*) The twelve multiplied by $-2$ gives us $-24$; but Q = 12x the actual coefficient of x in the product or the dividend is $-8$

∴ We must get the remaining 16x by multiplying the x of the divisor by 16. ∴ The absolute term in the divisor must be 16. ∴ $Q = 12x + 16$. And as $-2 \times 16 = -32$, ∴ $R = 0$.

(3) Divide $(x^3 + 7x^2 + 6x + 5)$ by $(x - 2)$

(*i*) $x^3$ divided by x gives us $x^2$ which is therefore the first term of the quotient.

$$\frac{x^3 + 7x^2 + 6x + 5}{x - 2}$$

$$\therefore Q = x^2 + \ldots$$

(*ii*) $x^2 \times -2 = -2x^2$; but we have $7x^2$ in the dividend. This means that we have to get $9x^2$ more. This must result from the multiplication of x by 9x. Hence the second term of the divisor must be 9x.

$$\frac{x^3 + 7x^2 + 6x + 5}{x - 2} \quad \therefore Q = x^2 + 9x + \ldots$$

(*iii*) As for the third term, we already have $-2 \times 9x = -18x$. But we have 6x in the dividend. We must therefore get an additional 24x. This can only come in by the multiplication of x by 24. ∴ This is the third term of the quotient.

$\therefore Q = x^2 + 9x + 24$

(*iv*) Now this last term of the quotient multiplied by $-2$ gives us −48. But the absolute term in the dividend is 5. We have therefore to get an additional 53 from somewhere. But there is no further term left in the dividend. This means that the 53 will remain as the remainder. ∴ $Q = x^2 + 9x + 24$; and $R = 53$.

*Note*: All the work explained in detail above can be easily performed by means of the '*Parāvartya*' *Sūtra* as already explained in the '*Parāvartya*' chapter, in connection with *Mental* division by Binomial divisors.

The procedure is very simple; and the following examples will throw further light thereon and give the necessary practice to the student:

(1) $\dfrac{x^3 + 7x^2 + 9x + 11}{x - 2} \quad \therefore Q = x^2 + 9x + 27$; and $R = 65$

(2) $\dfrac{x^3 - x^2 - 7x + 3}{x - 3} \quad \therefore Q = x^2 + 2x - 1$; and $R = 0$

(3) $\dfrac{x^4 - 3x^3 + 7x^2 + 5x + 7}{x - 4} \quad \therefore Q = x^3 + x^2 + 11x + 49$; and $R = 203$

(4) $\dfrac{-4x^3 + 9x^2 + 9x - 12}{2x - 4} \quad \therefore Q = -2x^2 + \frac{1}{2}x + 5\frac{1}{2}$; and $R = 10$

(5) $\dfrac{3x^2 - x - 5}{3x - 7}$ $\therefore Q = x + 2$; and $R = 9$

(6) $\dfrac{16x^2 + 8x + 1}{4x + 1}$ $\therefore Q = 4x + 1$; and $R = 0$

(7) $\dfrac{x^4 - 4x^2 + 12x - 9}{x^2 - 2x + 3}$ *N.B.* : Put zero coefficients for absent powers.

$\therefore Q = x^2 + 2x - 3$; and $R = 0$

(8) $\dfrac{x^3 + 2x^2 + 3x + 5}{x^2 - x - 1}$ $\therefore Q = x + 3$; and $R = 7x + 8$

(9) $\dfrac{x^4 + 4x^3 + 6x^2 + 4x + 1}{x^2 + 2x + 1}$ $\therefore Q = x^2 + 2x + 1$; and $R = 0$

(10) $\dfrac{x^4 + 2x^3 + 3x^2 + 2x + 1}{x^2 + x + 1}$ $\therefore Q = x^2 + x + 1$; and $R = 0$

(11) $\dfrac{x^4 - x^3 + 3x^2 + 5x + 5}{x^2 - x - 1}$ $\therefore Q = x^2 + 4$; and $R = 9x + 9$

(12) $\dfrac{6x^4 + 13x^3 + 39x^2 + 37x + 45}{x^2 - 2x - 9}$ $\therefore Q = 6x^2 + 25x + 143$; and $R = 548x + 1332$

(13) $\dfrac{12x^4 - 3x^3 - 3x - 12}{x^2 + 1}$ $\therefore Q = 12x^2 - 3x - 12$; and $R = 0$

(14) $\dfrac{12x^4 + 41x^3 + 81x^2 + 79x + 42}{3x^2 + 5x + 7}$ $\therefore Q = 4x^2 + 7x + 6$; and $R = 0$

(15) $\dfrac{x^4 - 4x^2 + 12x - 9}{x^2 + 2x - 3}$ $\therefore Q = x^2 - 2x + 3$ and $R = 0$

(16) $\dfrac{12x^4 - 3x^3 - 3x - 12}{12x^2 - 3x - 12} = x^2 + 1$; and $R = 0$

(17) $\dfrac{12x^4 + 41x^3 + 81x^2 + 79x + 42}{4x^2 + 7x + 6} = 3x^2 + 5x + 7$; and $R = 0$

(18) $\dfrac{x^4 - 4x^2 + 12x - 9}{x^2 - 2x + 3} = x^2 + 2x - 3$; and $R = 0$

(19) $\dfrac{2x^3 + 9x^2 + 18x + 20}{2x + 5} = x^2 + 2x + 4$; and $R = 0$

(20) $\dfrac{2x^3 + 9x^2 + 18x + 20}{x^2 + 2x + 4} = 2x + 5$; and $R = 0$

(21) $\dfrac{6x^4 + 13x^3 + 39x^2 + 37x + 45}{3x^2 + 2x + 9} = 2x^2 + 3x + 5$; and R = 0

(22) $\dfrac{6x^4 + 13x^3 + 39x^2 + 37x + 45}{2x^2 + 3x + 5} = 3x^2 + 2x + 9$; and R = 0

(23) $\dfrac{16x^4 + 36x^2 + 81}{4x^2 + 6x + 9} = 4x^2 - 6x + 9$; and R = 0

(24) $\dfrac{16x^4 + 36x^2 + 81}{4x^2 - 6x + 9} = 4x^2 + 6x + 9$; and R = 0

(25) $\dfrac{16x^4 + 36x^2 + 16x + 9}{4x^2 + 2x + 1} = 4x^2 - 2x + 9$; and R = 0

(26) $\dfrac{16x^4 + 36x^2 + 16x + 9}{4x^2 - 2x + 9} = 4x^2 + 2x + 1$; and R = 0

(27) $\dfrac{x^5 + x^3 - 7x^2 + 9}{x^3 - 2x^2 + 2x - 3} = x^2 + 2x + 3$; and $R = -2x^2 + 18$

(28) $\dfrac{16x^4 + 36x^2 + 6x + 86}{4x^2 + 6x + 9} = 4x^2 - 6x + 9$; and $R = 6x + 5$

(29) $\dfrac{-2x^5 - 7x^4 + 2x^3 + 18x^2 - 3x - 8}{x^3 - 2x^2 + 1} = -2x^2 - 11x - 20$;
and $R = -20x^2 + 8x + 12$

(30) $\dfrac{x^4 + 3x^3 - 16x^2 + 3x + 1}{x^2 + 6x + 1} = x^2 - 3x + 1$; and R = 0

(31) $\dfrac{x^4 + 3x^3 - 16x^2 + 3x + 1}{x^2 - 3x + 1} = x^2 + 6x + 1$; and R = 0

(32) $\dfrac{2x^5 - 9x^4 + 5x^3 + 16x^2 - 16x + 36}{2x^2 - 3x + 1} = x^3 - 3x^2 - 2\frac{1}{2}x + 5\frac{3}{4}$;
and $R = 3\frac{3}{4}x + 30\frac{1}{4}$

(33) $\dfrac{21x^6 + 7x^5 + 15x^4 + 29x^3 + x^2 + 15x + 3}{7x^3 + 5x + 1} = 3x^3 + x^2 + 3$

(34) $\dfrac{21x^6 + 7x^5 + 15x^4 + 29x^3 + x^2 + 15x + 3}{3x^3 + x^2 + 3} = 7x^3 + 5x + 1$

(35) $\dfrac{7x^{10} + 26x^9 + 53x^8 + 56x^7 + 43x^6 + 40x^5 + 41x^4 + 38x^3 + 19x^2 + 8x + 5}{x^5 + 3x^4 + 5x^3 + 3x^2 + x + 1}$

$= 7x^5 + 5x^4 + 3x^3 + x^2 + 3x + 5$

(36) $\frac{\text{(Same dividend as above)}}{7x^5 + 5x^4 + 3x^3 + x^2 + 3x + 5} = x^5 + 3x^4 + 5x^3 + 3x^2 + x + 1$

## LINKING NOTE

### RECAPITULATION AND CONCLUSION OF (ELEMENTARY) DIVISION SECTION

In chapters Four, Five and Six relating to division, we have dealt with a large number and variety of instructive examples and we now feel justified in postulating the following conclusions:

(1) The three methods expounded and explained are, no doubt, free from the big handicap which the current system labours under, namely, (*i*) the multiplication of large numbers (the Divisors) by "trial digits" of the quotient at every step with the chance of the product being found too big for the dividend and so on, (*ii*) the subtraction of large numbers from large numbers, (*iii*) the length, cumbrousness, clumsiness etc., of the whole procedure, (*iv*) the consequent liability of the student to get disgusted with and become sick of it all, (*v*) the resultant risk of errors being committed and so on;

(2) And yet, although *comparatively* superior to the process now in vogue everywhere, yet, they too suffer, in some cases, from these disadvantages. At any rate, they do not, in such cases, conform to the Vedic system's ideal of "Short and Sweet";

(3) And, besides, all three of them are suitable only for some *special* and particular type or types of cases; and none of them is suitable for general application to *all* cases:

(*i*) The '*Nikhilaṁ*' method is generally unsuitable for Algebraic divisions; and almost invariably, the '*Parāvartya*' process suits them better;

(*ii*) And, even as regards Arithmetical computations, the '*Nikhilaṁ*' method is serviceable only when the divisor-digits are large numbers, i.e. 6, 7, 8 or 9 and not at all helpful when the divisor digits are small ones, i.e. 1, 2, 3, 4 and 5; and it is only the '*Parāvartya*' method that can be applied in the latter kind of cases!

(*iii*) Even when a convenient multiple (or sub-multiple) is made use of, there is room for a choice having to be made—by the pupil—as to whether the '*Nikhilaṁ*' method or the '*Parāvartya*' one should be preferred;

(*iv*) And there is no exception-less criterion by which the student can make the requisite final choice between the two alternative methods;

(*v*) And, as, for the third method, i.e. by the reversed '*Ūrdhva-Tiryak*' *Sūtra*, the Algebraic utility thereof is plain enough; but it is difficult in respect of Arithmetical calculations to say when, where and why it should be resorted to as against the other two methods.

All these considerations arising from our detailed-comparative study of a large number of examples add up, in effect, to the simple conclusion that none of these methods can be of general utility in *all* cases, that the selection of the most suitable method in each particular case may owing to want of uniformity be confusing to the student and that this element of uncertainty is bound to cause confusion. And the question therefore naturally—nay, unavoidably arises as to whether the Vedic *Sūtras* can give us a General Formula applicable to *all* cases.

And the answer is—Yes, most certainly Yes! There *is* a splendid and beautiful and very easy method which conforms with the Vedic ideal of ideal simplicity all-round and which in fact gives us what we have been describing as "*Vedic one-line mental answers*"!

This astounding method we shall, however, expound in a later chapter under the caption 'Straight-Division'—which is one of the crowning beauties of the Vedic mathematics *Sūtras*. (Chapter Twenty-seven. q.v.).

# 7

# I. Factorisation

## Factorisation of Simple Quadratics

Factorisation comes in naturally at this point, as a form of what we have called "Reversed multiplication" and as a particular application of division. There is a lot of good material in the Vedic *Sūtras* on this subject too, which is *new* to the modern mathematical world but which comes in at a very early stage in our Vedic Mathematics.

We do not, however, propose to go into a detailed and exhaustive exposition of the subject but shall content ourselves with a few simple sample examples which will serve to throw light thereon and especially on the *Sūtraic* technique by which a *Sūtra* consisting of only one or two simple words, makes comprehensive provision for explaining and elucidating a procedure whereby a so-called "difficult" mathematical problem which, in the other system puzzles the students' brains ceases to do so nay, is actually laughed at by them as being worth rejoicing over and not worrying over!

For instance, let us take the question of factorisation of a quadratic expression into its component binomial factors. When the coefficient of $x^2$ is 1, it is easy enough, even according to the current system wherein you are asked to think out and find two numbers whose algebraic total is the middle coefficient and whose product is the absolute term. For example, let the quadratic expression in question be $x^2 + 7x + 10$; we mentally do the multiplication of the two factors $(x+2)$ and $(x+5)$ whose product is $x^2 + 7x + 10$; and by a mental process of reversing thereof, we think of 2 and 5 whose sum is 7 and whose product is 10; and we

$$\begin{array}{r} x+2 \\ x+5 \\ \hline x^2+7x+10 \\ \hline \end{array}$$

thus factorise $(x^2 + 7x + 10)$ into $(x+2)$ and $(x+5)$. And the actual working thereof is as follows:

$$x^2+7x+10$$
$$=x^2+2x+5x+10$$
$$=x(x+2)+5(x+2)$$
$$=(x+2)(x+5)$$

The procedure is, no doubt, mathematically correct; but the process is needlessly long and cumbrous. However, as the mental process actually employed is as explained above, there is no great harm done.

In respect, however, of quadratic expressions whose first coefficient is not unity, e.g. $2x^2+5x+2$, the students do not follow the mental process in question but helplessly depend on the 4-step method shown above and work it out as follows:

$$2x^2+5x+2$$
$$=2x^2+4x+x+2$$
$$=2x(x+2)+1(x+2)$$
$$=(x+2)(2x+1)$$

As the pupils are never taught to apply the mental process which can give us this result immediately, it means a real harm. The Vedic system, however, prevents this kind of harm, with the aid of two small sub-*Sūtras* which say (*i*) आनुरूप्येण (*Ānurūpyeṇa*) and (*ii*) आद्यमाद्येनान्त्यमन्त्येन (*Ādyamādyenāntyamantyena*) and which mean 'proportionately' and 'the first by the first and the last by the last'.

The former has been explained already in connection with the use of multiples and sub-multiples, in multiplication and division; but, alongside of the latter sub-*Sūtra*, it acquires a new and beautiful double application and significance and works out as follows:

(*i*) Split the middle coefficient into two such parts that the ratio of the first coefficient to that first part is the same as the ratio of that second part to the last coefficient. Thus, in the quadratic $2x^2+5x+2$, the middle term 5 is split into two such parts 4 and 1 that the ratio of the first coefficient, to the first part of the middle coefficient, i.e. $2:4$ and the ratio of the second part to the last coefficient, i.e. $1:2$ are the same. Now, this ratio, i.e. $x+2$ is one factor.

(*ii*) And the second factor is obtained by dividing the first coefficient of the quadratic by the first coefficient of the factor already found and the last coefficient of the quadratic by the last coefficient of that factor. In other words the second Binomial factor is obtained thus: $\frac{2x^2}{x}+\frac{2}{2}=2x+1$

Thus we say : $2x^2 + 5x + 2 = (x + 2)(2x + 1)$.

*Note* : The middle coefficient [which we split up above into $(4 + 1)$] may also be split up into $1 + 4$, that the ratio in that case is $2x + 1$ and that the other Binomial factor according to the above-explained method is $x + 2$. Thus, the change of sequence in the splitting up of the middle term makes no difference to the factors themselves !

This sub-*Sūtra* has actually been used already in the chapters on division; and it will come up again and again, later on, i.e. in Co-ordinate Geometry etc., in connection with straight lines, Hyperbolas, Conjugate Hyperbolas, Asymptotes etc.

But, just now, we make use of it in connection with the factorisation of quadratics into their Binomial factors. The following additional examples will be found useful:

(1) $2x^2 + 5x - 3 = (x + 3)(2x - 1)$

(2) $2x^2 + 7x + 5 = (x + 1)(2x + 5)$

(3) $2x^2 + 9x + 10 = (x + 2)(2x + 5)$

(4) $2x^2 - 5x - 3 = (x - 3)(2x + 1)$

(5) $3x^2 + x - 14 = (x - 2)(3x + 7)$

(6) $3x^2 + 13x - 30 = (x + 6)(3x - 5)$

(7) $3x^2 - 7x + 2 = (x - 2)(3x - 1)$

(8) $4x^2 + 12x + 5 = (2x + 1)(2x + 5)$

(9) $6x^2 + 11x + 3 = (2x + 3)(3x + 1)$

(10) $6x^2 + 11x - 10 = (2x + 5)(3x - 2)$

(11) $6x^2 + 13x + 6 = (2x + 3)(3x + 2)$

(12) $6x^2 - 13x - 19 = (x + 1)(6x - 19)$

(13) $6x^2 + 37x + 6 = (x + 6)(6x + 1)$

(14) $7x^2 - 6x - 1 = (x - 1)(7x + 1)$

(15) $8x^2 - 22x + 5 = (2x - 5)(4x - 1)$

(16) $9x^2 - 15x + 4 = (3x - 1)(3x - 4)$

(17) $12x^2 + 13x - 4 = (3x + 4)(4x - 1)$

(18) $12x^2 - 23xy + 10y^2 = (3x - 2y)(4x - 5y)$

(19) $15x^2 - 14xy - 8y^2 = (3x - 4y)(5x + 2y)$

An additional sub-*Sūtra* is of immense utility in this context, for the purpose of verifying the correctness of our answers in multiplications, divisions and factorisations. It reads: गुणितसमुच्चयः समुच्चयगुणितः and means: "*The product of the sum of the coefficients in the factors is equal to the sum of the coefficients in the product*".

In symbols, we may put this principle down thus:

$S_c$ of the product=Product of the $S_c$ (in the factors).

For example, $(x+7)(x+9)=(x^2+16x+63)$;

and we observe that $(1+7)(1+9)=1+16+63=80$

Similarly, in the case of Cubics, Bi-quadratics etc., the same rule holds good. For example:

$(x+1)(x+2)(x+3)=x^3+6x^2+11x+6$;

and we observe that $2\times3\times4=1+6+11+6=24$

Thus, if and when some factors are known, this rule helps us to fill in the gaps.

It will be found useful in the factorisation of cubics, biquadratics etc., and will be discussed in that context and in some other such contexts later on.

# 8

# II. Factorisation

### Factorisation of "Harder" Quadratics

There is a class of quadratic expressions known as Homogeneous Expressions of the second degree, wherein several letters x, y, z etc. figure and which are generally fought shy of by students and teachers as being too "difficult" but which can be very easily tackled by means of the *Ādyamādyena Sūtra* just explained and another sub-*Sūtra* which consists of only one compound word, which reads लोपनस्थापनाभ्यां and means: "by Alternate Elimination and Retention".

Suppose we have to factorise the Homogeneous quadratic $2x^2+6y^2+3z^2+7xy+11yz+7zx$. This is obviously a case in which the ratios of the coefficients of the various powers of the various letters are difficult to find out; and the reluctance of students to go into a troublesome thing like this, is quite understandable.

The '*Lopana-Sthāpana*' sub-*Sūtra*, however, removes the whole difficulty and makes the factorisation of a quadratic of this type as easy and simple as that of the ordinary quadratic already explained. The procedure is as follows:

Suppose we have to factorise the following long quadratic:

$2x^2+6y^2+3z^2+7xy+11yz+7zx$

(*i*) We first eliminate z by putting z = 0 and retain only x and y and factorise the resulting ordinary quadratic in x and y with the *Ādyam Sūtra*;

(*ii*) We then similarly eliminate y and retain only x and z and factorise the simple quadratic in x and z;

(*iii*) With these two sets of factors before us, we fill in the gaps caused by our own deliberate elimination of z and y respectively. And that gives us the real factors of the given long expression. The procedure is an argumentative one and is as follows:

If $z = 0$, then E (the given expression) $= 2x^2 + 7xy + 6y^2$

$= (x + 2y)(2x + 3y)$

Similarly, if $y = 0$, then $E = 2x^2 + 7xz + 3z^2 = (x + 3z)(2x + z)$

$\therefore$ Filling in the gaps which we ourselves had created by leaving out z and y, we say : $E = (x + 2y + 3z)(2x + 3y + z)$

The following additional examples will be found useful:

(1) $3x^2 + y^2 - 2z^2 - 4xy - yz - zx$

$E = (x - y)(3x - y)$ and also $(x - z)(3x + 2z)$

$\therefore E = (x - y - z)(3x - y + 2z)$

(2) $3x^2 + xy - 2y^2 + 19xz + 28z^2 + 9xw - 30w^2 - yz + 19wy + 46zw$

By eliminating two letters at a time, we get:

$E = (x + y)(3x - 2y)$, $(x + 4z)(3x + 7z)$ and also

$(x - 2w)(3x + 15w)$

$\therefore E = (x + y + 4z - 2w)(3x - 2y + 7z + 15w)$

(3) $2x^2 + 2y^2 + 5xy + 2x - 5y - 12 = (x + 3)(2x - 4)$ and

also $(2y + 3)(y - 4)$

$\therefore E = (x + 2y + 3)(2x + y - 4)$

(4) $3x^2 + 8xy + 4y^2 + 4y - 3 = (x - 1)(3x + 3)$ and also

$(2y - 1)(2y + 3)$

$\therefore E = (x + 2y - 1)(3x + 2y + 3)$

(5) $6x^2 - 8y^2 - 6z^2 + 2xy + 16yz + 5xz$

$= (2x - 2y)(3x + 4y)$ and also $(2x + 3z)(3x - 2z)$

$\therefore E = (2x - 2y + 3z)(3x + 4y - 2z)$

*Note*: We could have eliminated x also and retained only y and z and factorised the resultant simple quadratic. That would not, however, have given us any additional material but would have only confirmed and verified the answer we had already obtained. Thus, when 3 letters x, y and z are there, only two eliminations will generally suffice. The following exceptions to this rule should be noted:

(1) $x^2 + xy - 2y^2 + 2xz - 5yz - 3z^2$

$= (x - y)(x + 2y)$ and $(x - z)(x + 3z)$

But x is to be found in all the terms; and there is no means for deciding the proper combinations.

In this case, therefore, x too may be eliminated; and y and z retained. By so doing, we have:

$E = -2y^2 - 5yz - 3z^2 = (-y - z)(2y + 3z)$

$\therefore E = (x - y - z)(x + 2y + 3z)$

or, avoid the $x^2$ which gives the same coefficient and take only $y^2$ or $z^2$. And then, the confusion caused by the oneness of the coefficient in all the 4 factors is avoided; and we get,

$E = (x - y - z)(x + 2y + 3z)$ (as before)

(2) $x^2 + 2y^2 + 3xy + 2xz + 3yz + z^2$

(*i*) By eliminating z, y and x one after another,

we have $E = (x + y + z)(x + 2y + z)$

or (*ii*) By y or z both times, we get the same answer.

(3) $x^2 + 3y^2 + 2z^2 + 4xy + 3xz + 7yz$

Both the methods yield the same result:

$E = (x + y + 2z)(x + 3y + z)$

(4) $3x^2 + 7xy + 2y^2 + 11xz + 7yz + 6z^2 + 14x + 8y + 14z + 8$.

Here too, we can eliminate two letters at a time and thus keep only one letter and the independent term, each time.

Thus, $E = 3x^2 + 14x + 8 = (x + 4)(3x + 2)$;

$2y^2 + 8y + 8 = (2y + 4)(y + 2)$; and also

$6z^2 + 14z + 8 = (3z + 4)(2z + 2)$

$\therefore E = (x + 2y + 3z + 4)(3x + y + 2z + 2)$

*Note*: This "*Lopana-Sthāpana*" method of alternate elimination and retention will be found highly useful, later on in H.C.F., in Solid Geometry and in Co-ordinate Geometry of the straight line, the Hyperbola, the Conjugate Hyperbola, the Asymptotes etc.

# 9

# III. Factorisation of Cubics etc.

### By Simple Argumentation

We have already seen how, when a polynomial is divided by a binomial, a trinomial etc., the remainder can be found by means of the remainder theorem and how both the quotient and the remainder can be easily found by one or other method of division explained already.

From this it follows that, if, in this process, the remainder is found to be zero, it means that the given dividend is divisible by the given divisor, i.e. the divisor is a factor of the dividend.

And this means that, if, by some such method, we are able to find out a certain factor of a given expression, the remaining factor or the product of all the remaining factors can be obtained by simple division of the expression in question by the factor already found out by some method of division. (In this context, the student need hardly be reminded that, in all *Algebraic* divisions, the '*Parāvartya*' method is always to be preferred to the '*Nikhilaṁ*' method).

Applying this principle to the case of a cubic, we may say that, if, by the remainder theorem or otherwise, we know one binomial factor of a cubic, simple division by that factor will suffice to enable us to find out the quadratic which is the product of the remaining two binomial factors. And these two can be obtained by the '*Ādyamādyena*' method of factorisation already explained.

A simpler and easier device for performing this operation will be to write down the first and the last terms by the '*Ādyamādyena*' method and the middle term with the aid of the *Guṇita-Samuccaya* rule, i.e. the principle—already explained with regard to the $S_c$ of the product being the same as the product of the $S_c$ of the factors.

Let us take a concrete example and see how this method can be made use of. Suppose we have to factorise $x^3+6x^2+11x+6$ and that, by some method, we know $(x+1)$ to be a factor. We first use the *Ādyamādyena* formula and thus mechanically put down $x^2$ and 6 as the first and the last coefficients in the quotient, i.e. the product of the remaining two binomial factors. But we know already that the $S_c$ of the given expression is 24; and, as the $S_c$ of $(x+1)=2$ we therefore know that the $S_c$ of the quotient must be 12. And as the first and last digits thereof are already known to be 1 and 6, their total is 7. And therefore the middle term must be $12-7=5$. So, the quotient is $x^2+5x+6$.

This is a very simple and easy but absolutely certain and effective process.

The student will remember that the ordinary rule for divisibility of a dividend by a divisor as has been explained already in the section dealing with the "remainder theorem" is as follows:

If $E=DQ+R$, if $D=x-p$ and if $x=p$, then $E=R$.

### COROLLARIES

(*i*) So, if, in the dividend, we substitute 1 for x, the result will be that, as all the powers of 1 are unity itself, the dividend will now consist of the sum of all the coefficients.

Thus, if D is $x-1$, $R=a+b+c+d+$(where a, b, c d etc., are the successive coefficients); and then, if $a+b+c$ etc., $=0$, it will mean that as $R=0$, E is divisible by D. In other words, $x-1$ is a factor.

(*ii*) If, however, $D=x+1$ and if we substitute $-1$ for x in E, then, inasmuch as the odd powers of $-1$ will all be $-1$ and the even powers thereof will all be 1, therefore it will follow that, in this case, $R=a-b+c-d$ etc.

So, if $R = 0$, i.e. if $a-b+c-d$ etc., $=0$, i.e. if $a-b+c-d$ etc.,

$=0$, i.e. $a+c+\ldots=b+d+\ldots$

i.e. if the sum of the coefficients of the odd powers of x and the sum of the coefficients of the even powers be equal, then $x+1$ will be a factor.

The following few illustrations will elucidate the actual application of the principle mainly by what may be called the Argumentation method, based on the simple multiplication formula to the effect that—

$(x+a)(x+b)(x+c)=x^3+x^2(a+b+c)+x(ab+ac+bc)+abc$, as follows:

(1) Factorise $x^3+6x^2+11x+6$

(*i*) Here, $S_c = 24$; and $t_1$ (the last term) is 6 whose factors are 1,

2, 3 or 1, 1, 6. But their total should be 6 (the coefficient of $x^2$). So we must reject the 1, 1, 6 group and accept the 1, 2, 3 group. And, testing for the third coefficient, we find $ab + bc + ca = 11$

$\therefore$ $E = (x + 1)(x + 2)(x + 3)$

or (*ii*) $S_o$ (the sum of the coefficients of the odd powers) $= 1 + 11 = 12$; and $S_e$ (the sum of the coefficients of the even powers) $= 6 + 6 = 12$. And as $S_o = S_e$

$\therefore$ $x + 1$ is a factor.

$\therefore$ Dividing E by that factor, we first use the '*Ādyamādyena*' *Sūtra* and put down 1 and 6 as the first and the last coefficients.

$\therefore$ The middle coefficient is $12 - (1 + 6) = 5$

$\therefore$ The $Q = x^2 + 5x + 6$ which (by *Ādyamādyena*) $= (x + 2)(x + 3)$

Thus $E = (x + 1)(x + 2)(x + 3)$

(2) Factorise $x^3 - 6x^2 + 11x - 6$

(*i*) Here $S_e = 0$ $\therefore$ $x - 1$ is a factor. But as $\frac{0}{0}$ is an indefinite figure, we cannot use the *Guṇita-Samuccaya* method here for the middle term but must divide out by mental '*Parāvartya*' and get the quotient as $x^2 - 5x + 6$ which by the '*Ādyamādyena*' rule $= (x - 2)(x - 3)$ $\therefore$ $E = (x - 1)(x - 2)(x - 3)$

or (*ii*) argue about $-1$, $-2$ and $-3$ having $-6$ as the total and $-6$ as the product; and test out and verify the 11. And therefore say, $E = (x - 1)(x - 2)(x - 3)$

(3) Factorise $x^3 + 12x^2 + 44x + 48$

(*i*) Here $S_e = 105$ whose factors are 1, 3, 5, 7, 15, 21, 35 and 105. And $t_1$ is 48 whose factors are 1, 2, 3, 4, 6, 8, 12, 16, 24 and 48. $\therefore$ $x + 1$ and $x - 1$ are out of court. And the only possible factors are $x + 2$, $x + 4$ and $x + 6$ (verify).

or (*ii*) argue that $2 + 4 + 6 = 12$ and $2 \times 4 \times 6 = 48$; and test for and verify 44 $\therefore$ $E = (x + 2)(x + 4)(x + 6)$

(4) Factorise $x^3 - 2x^2 - 23x + 60$

(*i*) Here $S_e = 36$ (with factors 1, 2, 3, 4, 6, 9, 12, 18 and 36); $t_1 = 60$ (which is $1 \times 2 \times 2 \times 3 \times 5$)

$\therefore$ Possible factors are 1, 2, 3, 4, 5, 6, 10, 12, 15, 20, 30 and 60.

But the sum of the coefficients in each factor must be a factor of the total $S_c$ (i.e. 36). Therefore, all the italicized numbers go out, and so do $x-1$, $x+4$, $x+6$ and $x+10$. Now, the only possible numbers here which when added, total −2 are −3, −4 and 5. Now, test for and verify $x-3$

$\therefore E = (x-3)(x^2+x-20) = (x-3)(x-4)(x+5)$

or, (*ii*) take the possibilities x−10, x−5, x+5, x−4, x+3, x−3, x+2 and x−2.

If $x-2$

$$\begin{array}{ll} x^3 - 2x^2 - 23x + 60 & \\ \quad 2 + \qquad\qquad 0 - 46 & \\ \hline 1 + 0 - 23 \qquad\quad 14 & \therefore R = 14 \end{array}$$

$\therefore x-2$ is *not* a factor.

But if $x-3$, $R=0 \therefore x-3$ is a factor
Then, argue as in the first method.

(5) Factorise $x^3-2x^2-x+6$. Here $S_c=0$

(*i*) $\therefore x-1$ is a factor; and the other part by division is $x^2-x-6$ which $=(x+2)(x-3)$

$\therefore E=(x-1)(x+2)(x-3)$

(*ii*) $t_1=6$ (whose factors are 1, 2 and 3). And the only combination which gives us the total −2, is −1, 2 and −3. Test and verify for −5. And put down the answer.

(6) Factorise $x^3+3x^2-17x-38$

Now $S_c=-51$ (with factors $\pm 1, +3, +17$ and $\pm 51$)

And −38 has the factors $\pm 1, \pm 2, \pm 19$ and $\pm 38$

$\therefore \pm 1$, $\pm 19$ and $\pm 38$ are not possible.

And only $\pm 2$ is possible. And if $x=-2$, $R=0$

$\therefore x+2$ is a factor $\therefore E=(x+2)(x^2+x-19)$
which has no further factors.

(7) Factorise $x^3+8x^2+19x+12$

(*i*) Here $S_c=40$; and $L_t=12 \therefore 1+3+4$ are the proper numbers. Now test for and verify 19.

$\therefore E=(x+1)(x^2+7x+12)=(x+1)(x+3)(x+4)$

or (*ii*) $\because 1+19=8+12 \therefore x+1$ is a factor. Then the quotient is obtainable by the '*Ādyamādyena*' and '*Samuccaya*' *Sūtras*. And that again can be factorised with the aid of the former.

$\therefore E = (x+1)\ (x+3)\ (x+4)$

(8) Factorise $x^3 - 7x + 6$

(*i*) $\because S_e = 0, \therefore x - 1$ is a factor.

$\therefore$ By '*Parāvartya*' method of division (mental),

$E = (x-1)\ (x^2+x-6) = (x-1)\ (x-2)\ (x+3)$

or (*ii*) (by a different kind of application of '*Ādyamādyena*')

$x^3 - 7x + 6 = (x^3 - 1) - 7x + 7 = (x-1)\ (x^2 + x + 1 - 7)$

$= (x-1)\ (x-2)\ (x+3)$

*Note*:

(1) This method is always applicable when $x^2$ is absent; and this means that the three independent terms together total zero.

(2) Note the note on this and other allied points in the section relating to cubic equations in a later chapter.

(3) Note that this method of factorisation by Argumentation is equally applicable to biquadratics also.

(4) The relationship between the binomial factors of a polynomial and its differentials first, second and so on is an interesting and intriguing subject which will be dealt with in a later chapter.

(5) The use of differentials for finding out repeated factors will also be dealt with later.

# 10

# Highest Common Factor

In the current system of mathematics, we have two methods which are used for finding the H.C.F. of two or more given expressions.

The first is by means of factorisation which is not always easy; and the second is by a process of continuous division like the method used in the G.C.M. chapter of Arithmetic. The latter is a mechanical process and can therefore be applied in all cases. But it is rather *too* mechanical and, consequently, long and cumbrous.

The Vedic method provides a third method which is applicable to *all* cases and is, at the same time, free from this disadvantage.

It is, mainly, an application of the '*Lopana-Sthāpana*' *Sūtra*, the '*Saṅkalana-Vyavakalana*' process and the '*Ādyamādya*' rule. The procedure adopted is one of alternate destruction of the highest and the lowest powers by a suitable multiplication of the coefficients and the addition or subtraction of the multiples. A concrete example will elucidate the process:

(1) Suppose we have to find the H.C.F. of $(x^2+7x+6)$ and $(x^2-5x-6)$

(*i*) $x^2+7x+6=(x+1)(x+6)$; and $x^2-5x-6=(x+1)(x-6)$

$\therefore$ The H.C.F. is $(x+1)$.

(*ii*) The second method the G.C.M. one is well-known and need not be put down here.

(*iii*) The third process of '*Lopana-Sthāpana*', i.e. of Elimination and Retention, or Alternate destruction of the highest and the lowest powers is explained below:
Let $E_1$ and $E_2$ be the two expressions. Then, for destroying the highest power, we should subtract $E_2$ from $E_1$; and for destroying the lowest one, we should add the two. The chart is as follows:

$$\left.\begin{array}{r} x^2+7x+6 \\ x^2-5x-6 \end{array}\right\} \text{Subtraction} \qquad \left.\begin{array}{r} x^2-5x-6 \\ x^2+7x+6 \end{array}\right\} \text{Addition}$$

$$12)\ \underline{12x+12} \qquad\qquad 2x)\ \underline{2x^2+2x}$$

$$x+1 \qquad\qquad x+1$$

We then remove the common factor if any from each; and we find $x+1$ staring us in the face.

$\therefore$ $x+1$ is the H.C.F.

The *Algebraical principle or proof* hereof is as follows:

Let P and Q be the two expressions; H their H.C.F. and A and B the quotients after their division by the H.C.F.

$$\therefore \frac{P}{H} = A; \text{ and } \frac{Q}{H} = B \therefore P = HA \text{ and } Q = HB$$

$$\therefore P \pm Q = H\ (A \pm B); \text{ and } MP \pm NQ = H(MA \pm NB)$$

$\therefore$ The H.C.F. of P and Q is also the H.C.F. of $P \pm Q$,

$2P \pm Q$, $P \pm 2Q$ and $MP \pm NQ$

All that we have therefore to do is to select our M and N in such a way that the highest and the lowest powers are removed and the H.C.F. appears and *shows itself* before us.

A few more illustrative examples may be seen below:

(1) (*i*) $x^3 - 3x^2 - 4x + 12 = (x+2)(x-2)(x-3)$;
and $x^3 - 7x^2 + 16x - 12 = (x-2)^2 (x-3)$
$\therefore$ the H.C.F. is $(x-2)(x-3) = x^2 - 5x + 6$

But the factorisation of the two expressions will be required.

or (*ii*) The G.C.M. method.

or (*iii*) The '*Lopana-Sthāpana*' method:

$$\begin{array}{rr} x^3-3x^2-4x+12 & x^3-7x^2+16x-12 \\ -(x^3-7x^2+16x-12) & +(x^3-3x^2-4x+12) \\ \hline 4)\ 4x^2-20x+24 & 2x)\ 2x^3-10x^2+12x \\ \hline x^2-5x+6 & x^2-5x+6 \end{array}$$

$\therefore$ The H.C.F. is $(x^2 - 5x + 6)$

(2) (*i*) $4x^3 + 13x^2 + 19x + 4 = (4x+1)(x^2+3x+4)$;
and $2x^3 + 5x^2 + 5x - 4 = (2x+1)(x^2+3x+4)$
$\therefore$ The H.C.F. is $x^2 + 3x + 4$

But the factorisation of the two cubics will be cumbrous.

or (*ii*) The G.C.M. method.

or (*iii*) The Vedic method:

$$\begin{array}{c} 4x^3+13x^2+19x+4 \\ -(4x^3+10x^2+10x-8) \\ \hline 3)\ 3x^2+9x+12 \\ \hline x^2+3x+4 \end{array}$$

$$\begin{array}{c} 2x^3+5x^2+5x-4 \\ +(4x^3+13x^2+19x+4) \\ \hline 6x)\ 6x^3+18x^2+24x \\ \hline x^3+3x+4 \end{array}$$

$\therefore$ The H.C.F. is $(x^2+3x+4)$

(3) (*i*) $x^4+x^3-5x^2-3x+2=(x+1)(x-2)(x^2+2x-1)$;
and $x^4-3x^3+x^2+3x-2=(x+1)(x-2)(x-1)^2$
$\therefore$ The H.C.F. is $x^3-x-2$
But this factorisation of the two biquadratics is bound to be a comparatively laborious process.

(*ii*) The cumbrous G.C.M. method.

(*iii*) The Vedic method:

$$\begin{array}{c} x^4+x^3-5x^2-3x+2 \\ -(x^4-3x^3+x^2+3x-2) \\ \hline 2)\ 4x^3-6x^2-6x+4 \\ \hline 2x^3-3x^2-3x+2 \\ -(2x^3-2x^2-4x) \\ \hline -1(-x^2+x+2) \\ \hline x^2-x-2 \\ \hline \end{array}$$

$$\begin{array}{c} x^4+3x^3+x^2+3x-2 \\ x^4+x^3-5x^2-3x+2 \\ \hline 2x^2)\ 2x^4-2x^3-4x^2 \\ \hline x^2-x-2 \end{array}$$

(*N.B.* multiply this by $2x$ and take it over to the left for subtraction)

$\therefore$ The H.C.F. is $x^2-x-2$

(4) (*i*) The Vedic method:

$$\begin{array}{c} 6x^4-7x^3-5x^2+14x+7 \\ -(6x^4-10x^3\qquad\quad+14x) \\ \hline 3x^3-5x^2\qquad\quad+7 \\ \hline \end{array}$$

$3x^3-5x^2+7$
(*N.B.* multiply this by $2x$ and subtract from L.H.S.)

(*ii*) The factorisation of the big biquadratic will be "harder".

(*iii*) The G.C.M. method is, in this case, easy. But how should one know this beforehand and start monkeying or experimenting with it?

(5) (i) *The Vedic method:*

$$\begin{array}{c} 6x^4-11x^3+16x^2-22x+8 \\ -(6x^4-11x^3-\ 8x^2+22x-8) \\ \hline 4)\ 24x^2-44x+16 \\ \hline 6x^2-11x+4 \end{array}$$

$$\begin{array}{c} 6x^4-11x^3-\ 8x^2+22x-8 \\ +(6x^4-11x^3+16x^2-22x+8) \\ \hline 2x^2)\ 12x^4-22x^3+8x^2 \\ \hline 6x^2-11x+4 \end{array}$$

$\therefore$ The H.C.F. is $6x^2-11x+4$

(*ii*) $6x^4 - 11x^3 + 16x^2 - 22x + 8 = (2x-1)(3x-4)(x^2+2)$; and
$6x^4 - 11x^3 - 8x^2 + 22x - 8 = (2x-1)(3x-4)(x^2-2)$
$\therefore$ The H.C.F. is $(2x-1)(3x-4) = 6x^2 - 11x + 4$

(*iii*) The cumbersome G.C.M. method.

(6) (*i*) $2x^3 + x^2 - 9 = (2x-3)(x^2+2x+3)$;
and $x^4 + 2x^2 + 9 = (x^2+2x+3)(x^2-2x+3)$
$\therefore$ The H.C.F. is $x^2+2x+3$

But the factorisation-work especially of the former expression will be a tough job

(*ii*) The G.C.M. method will be cumbrous as usual.

(*iii*) The Vedic method.

$$\begin{array}{r} 2x^3 + x^2 - 9 \\ \underline{x^4 + 2x^2 + 9} \\ x^2)\ \underline{x^4 + 2x^3 + 3x^2} \\ x^2 + 2x + 3 \end{array}$$

$$\begin{array}{r} 2x^4 + 4x^2 + 18 \\ \underline{2x^4 + x^3 - 9x} \\ x^3 - 4x^2 - 9x - 18 \\ \underline{x^3 + 2x^2 + 3x} \\ -6)\ \underline{-6x^2 - 12x - 18} \\ x^2 + 2x + 3 \end{array}$$

*N.B.* As this has no further factors, it must be in the R.H.S. Multiply it by x and take it over to the right for subtraction.

$\therefore$ The H.C.F. is $x^2+2x+3$

(7) (*i*) $4x^4 + 11x^3 + 27x^2 + 17x + 5$ and $3x^4 + 7x^3 + 18x^2 + 7x + 5$

$$\begin{array}{r} \therefore 12x^4 + 33x^3 + 81x^2 + 51x + 15 \\ \underline{12x^4 + 28x^3 + 72x^2 + 28x + 20} \\ 5x^3 + 9x^2 + 23x - 5 \\ \underline{5x^3 + 20x^2 + 45x + 50} \\ -11)\ \underline{11x^2 - 22x - 55} \\ \underline{x^2 + 2x + 5} \end{array}$$

$$\begin{array}{r} 4x^4 + 11x^3 + 27x^2 + 17x + 5 \\ \underline{3x^4 + 7x^3 + 18x^2 + 7x + 5} \\ x)\ \underline{x^4 + 4x^3 + 9x^2 + 10x} \\ x^3 + 4x^2 + 9x + 10 \\ \underline{10x^3 + 18x^2 + 46x - 10} \\ 11x)\ \underline{11x^3 + 22x^2 + 55x} \\ \underline{x^2 + 2x + 5} \end{array}$$

(*ii*) The G.C.M. method will be cumbrous as usual.

(*iii*) $4x^4 + 11x^3 + 27x^2 + 17x + 5 = (x^2+2x+5)(4x^2+3x+1)$ and
$3x^4 + 7x^3 + 18x^2 + 7x + 5 = (x^2+2x+5)(3x^2+x+1)$

But the factorisation of the two big biquadratics into two further *factorless* quadratics each, will entail greater waste of time and energy.

So, the position may be analysed thus:

(*i*) The G.C.M. method is mechanical and reliable but too cumbrous;

(*ii*) The Factorisation method is more intellectual but harder to work out and therefore less dependable;

(*iii*) The Vedic method is free from all these defects and is not only intellectual but also simple, easy and reliable.

# 11

# Simple Equations (First Principles)

As regards the solution of equations of various types, the Vedic sub-*Sūtras* give us some First Principles which are theoretically not unknown to the western world but are not utilised in actual practice as basic and fundamental first principles of a practically Axiomatic character in mathematical computations.

In order to solve such equations, the students do not generally use these basic sub-*Sūtras* as such but almost invariably go through the whole tedious work of practically *proving* the formula in question instead of taking it for granted and *applying* it! Just as if on every occasion when the expression $a^3+b^3+c^3-3abc$ comes up; one should not take it for granted that its factors are $(a+b+c)$ and $(a^2+b^2+c^2-ab-bc-ca)$ but should go through the long process of multiplying these two, showing the product and then applying it to the case on hand, similarly for Pythagoras Theorem etc.!

The Vedic method gives us these sub formulae in a condensed form like *Parāvartya* etc., and enables us to perform the necessary operation by mere *application* thereof. The underlying principle behind all of them is परावर्त्य योजयेत् (*Parāvartya Yojayet*) which means: "Transpose and adjust". The applications, however, are numerous and splendidly useful. A few examples of this kind are cited hereunder, as illustrations:

(1) $2x+7=x+9 \therefore 2x-x=9-7 \therefore x=2$. The student has to perform hundreds of such transposition-operations in the course of his work; but he should by practice obtain such familiarity with and master over it as to assimilate and assume the general form as that if $ax+b=cx+d,\ x=\frac{d-b}{a-c}$ and apply it by mental arithmetic automatically to the particular case actually before him and say:

$2x + 7 = x + 9 \therefore x = \frac{9-7}{2-1} = \frac{2}{1} = 2$; and

the whole process should be a short and simple mental process.

### SECOND GENERAL TYPE

(2) The above is the commonest kind of transpositions. The second common type is one in which each side the L.H.S. and the R.H.S. contains two binomial factors.

In general terms, let $(x + a)(x + b) = (x + c)(x + d)$. The usual method is to work out the two multiplications and do the transpositions and say :

$$(x + a)(x + b) = (x + c)(x + d)$$

$$\therefore x^2 + ax + bx + ab = x^2 + cx + dx + cd$$

$$\therefore ax + bx - cx - dx = cd - ab$$

$$\therefore x(a + b - c - d) = cd - ab$$

$$\therefore x = \frac{cd - ab}{a + b - c - d}$$

It must be possible for the student, by practice, to assimilate and assume the whole of this operation and say immediately:

$$x = \frac{cd - ab}{a + b - c - d}$$

As examples, the following may be taken:

(1) $(x + 1)(x + 2) = (x - 3)(x - 4)$ $\quad \therefore x = \frac{12 - 2}{1 + 2 + 3 + 4} = \frac{10}{10} = 1$

(2) $(x - 6)(x + 7) = (x + 3)(x - 11)$ $\quad \therefore x = \frac{-33 + 42}{-6 + 7 - 3 + 11} = \frac{9}{9} = 1$

(3) $(x - 2)(x - 5) = (x - 1)(x - 4)$ $\quad \therefore x = \frac{4 - 10}{-2 - 5 + 1 + 4} = \frac{-6}{-2} = 3$

(4) $(x - 7)(x - 9) = (x - 3)(x - 22)$ $\quad \therefore x = \frac{66 - 63}{-7 - 9 + 3 + 22} = \frac{3}{9} = \frac{1}{3}$

(5) $(x + 7)(x + 9) = (x + 3)(x + 22)$ $\quad \therefore x = \frac{66 - 63}{7 + 9 - 3 - 22} = \frac{3}{-9} = -\frac{1}{3}$

(6) $(x + 7)(x + 9) = (x - 8)(x - 11)$ $\quad \therefore x = \frac{88 - 63}{7 + 9 + 8 + 11} = \frac{25}{35} = \frac{5}{7}$

(7) $(x + 7)(x + 9) = (x + 3)(x + 21)$ $\quad \therefore x = \frac{63 - 63}{7 + 9 - 3 - 21} = \frac{0}{-8} = 0$

This gives rise to a general corollary to the effect that, if $cd - ab = 0$, i.e. if $cd = ab$, i.e. if the product of the absolute terms be the same on both the sides, the numerator becomes zero; and $\therefore x = 0$.

### THIRD GENERAL TYPE

The third type is one which may be put into the general form: $\frac{ax+b}{cd+d}=\frac{p}{q}$; and, after doing all the cross-multiplications and transpositions etc., we get $x=\frac{pd-bq}{aq-cp}$. The student should (by practice) be able to assimilate and assume this also and do it all *mentally* as a single operation.

*Note*: The only rule to remember for facilitating this process is that all the terms involving x should be conserved on to the left side and that all the independent terms should be gathered together on the right side and that every transposition for this purpose must invariably produce a change of sign, i.e. from + to − and conversely; and from × into ÷ and conversely.

### FOURTH GENERAL TYPE

The fourth type is of the form $\frac{m}{x+a}+\frac{n}{x+b}=0$.

After all the L.C.M's, the cross-multiplications and the transpositions etc., are over, we get $x=\frac{-mb-na}{m+n}$. This is simple enough and easy enough for the student to assimilate; and it should be assimilated and readily applied mentally to any case before us.

In fact, the application of this process may, in due course, by means of practice, be extended so as to cover cases involving a larger number of terms. For instance.

$$\frac{m}{x+a}+\frac{n}{x+b}+\frac{p}{x+c}=0$$

$$\therefore \frac{m(x+b)(x+c)+n(x+c)(x+a)+p(x+a)(x+b)}{(x+a)(x+b)(x+c)}=0$$

$$\therefore m[x^3+x(b+c)+bc]+n[x^2+x(c+a)+ca]+p[x^2+x(a+b)+ab]=0$$

$$\therefore x^2(m+n+p)+x[m(b+c)+n(c+a)+p(a+b)+(mbc+nca+pab)=0$$

If $m+n+p=0$, then

$$x=\frac{-mbc-nca-pab}{m(b+c)+n(c+a)+p(a+b)}$$

But if $m+n+p\neq 0$, then it will be a quadratic equation and will have to be solved as such as explained in a later chapter.

And this method can be extended to any number of terms on the same lines as explained above.

## Linking Note
## Special Types of Equations

The above types may be described as general types. But there are, as in the case of multiplications, divisions etc., particular *types* which possess certain specific characteristics of a special character which can be more easily tackled than the ordinary ones with the aid of certain very short special processes practically what one may describe as *mental* one-line method.

As already explained in a previous context, all that the student has to do is to look for certain characteristics, spot them out, identify the particular type and apply the formula which is applicable thereto.

We will discuss these special types of equations, in the next few chapters.

# 12

# Simple Equations

By Sūtra Śūnyam etc.

We begin this section with an exposition of several special types of equations which can be solved practically at sight with the aid of a beautiful special *Sūtra* which reads: शून्यं साम्यसमुच्चये (*Śūnyaṃ Sāmyasamuccaye*) and which, in cryptic language which renders it applicable to a large number of different cases merely says: "when the *Samuccaya* is the same, that *Samuccaya* is zero", i.e. it should be equated to zero.

'*Samuccaya*' is a technical term which has several meanings in different contexts; and we shall explain them, one by one.

First Meaning and Application

'*Samuccaya*' first means a term which occurs as a common factor in all the terms concerned.

Thus $12x + 3x = 4x + 5x \quad \therefore 12x + 3x - 4x - 5x = 0$

$\therefore 6x = 0 \quad \therefore x = 0$

All these detailed steps are unnecessary; and, in fact, *no one* works it out in this way. The mere fact that x occurs as a common factor in all the terms on both sides [or on the L.H.S. (with zero on the R.H.S.)] is sufficient for the inference that x is zero; and no intermediate step is necessary for arriving at this conclusion. This is practically axiomatic.

And this is applicable not only to x or other such "unknown quantity" but to every such case. Thus, if $9(x+1) = 7(x+1)$ we need not say: $9(x+1) = 7(x+1)$

$\therefore 9x+9 = 7x+7 \quad \therefore 9x-7x = 7-9 \quad \therefore 2x = -2 \quad \therefore x = -1$

On the contrary, we can straightaway say: $9(x+1) = 7(x+1)$

$\therefore x+1 = 0 \quad \therefore x = -1.$

### Second Meaning And Application

The word '*Samuccaya*' has, as its second meaning, the product of the independent terms. Thus, $(x+7)(x+9)=(x+3)(x+21)$

$\therefore$ Here $7 \times 9 = 3 \times 21$. Therefore $x=0$

This is also practically axiomatic, as has been dealt with in a previous section of this very subject of equations and need not be gone into again.

### Third Meaning and Application

'*Samuccaya*' thirdly means the sum of the denominators of two fractions having the same numerical numerator. Thus,

$$\frac{1}{2x-1}+\frac{1}{3x-1}=0 \therefore 5x-2=0$$

This is axiomatic too and needs no elaboration.

### Fourth Meaning and Application

Fourthly '*Samuccaya*' means combination or total. In this sense, it is used in several different contexts; and they are explained below:

(*i*) If the sum of the numerators and the sum of the denominators be the same, then that sum = zero. Thus,

$$\frac{2x+9}{2x+7}=\frac{2x+7}{2x+9}$$
$$\therefore (2x+9)(2x+9)=(2x+7)(2x+7)$$
$$\therefore 4x^2+36x+81=4x^2+28x+49$$
$$\therefore 8x=-32$$
$$\therefore x=-4$$

This is the current meth·:ᐧ. But the "*Śūnyaṃ Sāmya-Samuccaye*" formula tells us that, inasmuch as $N_1+N_2=4x+16$ and $D_1+D_2$ is also $4x+16 \therefore 4x+16=0 \therefore x=-4$. In fact, as soon as this characteristic is noted and the type recognised, the student can at once mentally say $x=-4$.

*Note*: If in the algebraical total, there be a numerical factor, that should be removed. Thus:

$$\frac{3x+4}{6x+7}=\frac{x+1}{2x+3}$$

Here $N_1+N_2=4x+5$; and $D_1+D_2 = 8x+10$. Removing the numerical factor, we have $4x+5$ on both sides here too.

$\therefore 4x+5=0 \therefore = -5/4$

No laborious cross-multiplications of $N_1$ by $D_2$ and $N_2$ by $D_1$ and transpositions etc., are necessary in the Vedic method.

At sight, we can at once say $4x+5=0$ and be done with it.

## Fifth Meaning and Application for Quadratics

With the same meaning, i.e. total of the word समुच्चय '*Samuccaya*', there is a fifth kind of application possible of this *Sūtra*. And this has to do with quadratic equations. None need, however, go into a panic over this. It is as simple and as easy as the fourth application; and even little children can understand and readily apply this *Sūtra* in this context, as explained below.

In the two instances given above, it will be observed that the cross-multiplications of the coefficients of x give us the same coefficient for $x^2$. In the first case, we had $4x^2$ on both sides; and in the second example, it was $6x^2$ on both sides. The two cancelling out, we had simple equations to deal with.

But there are other cases where the coefficients of $x^2$ are not the same on the two sides; and this means that we have a quadratic equation before us.

But it does not matter. For, the same *Sūtra* applies although in a different direction here too and gives us also the second root of the quadratic equation. The only difference is that inasmuch as Algebraic '*Samuccaya*' includes subtraction too, we therefore now take into account, not only the sum of $N_1$ and $N_2$ and the sum of $D_1$ and $D_2$ but also the differences between the numerator and the denominator on each side; and, if they be equal, we at once equate that difference to Zero.

Let us take a concrete example and suppose we have to solve the equation $\frac{3x+4}{6x+7}=\frac{5x+6}{2x+3}$

(*i*) We note that $N_1+N_2=8x+10$ and $D_1+D_2$ is also $8x+10$; we therefore use the method described in the fourth application given above and equate $8x+10$ to zero and say $x=-5/4$.

(*ii*) But mental cross-multiplication reveals that the $x^2$ coefficients on the L.H.S. and the R.H.S. are 6 and 30 respectively and not the same. So, we decide that it is a quadratic equation; and we observe that $N_1 \sim D_1 = 3x+3$ and that $N_2 \sim D_2$ is also $3x+3$. And so, according to the present application of the same *Sūtra*, we at one say $3x+3=0 \quad \therefore \quad x=-1$.

Thus the two roots are –5/4 and –1; and we have solved a quadratic equation at mere sight (without the usual paraphernalia of cross-multiplication, transposition etc.). We shall revert to this at a later stage (when dealing with quadratic equations themselves, as such).

### Sixth Meaning and Application

With the same sense 'total' of the word '*Samuccaya*' but in a different application, we have the same *Sūtra* coming straight to our rescue, in the solution of what the various text-books everywhere describe as "Harder Equations", and deal with in a very late chapter thereof under that caption. In fact, the label "Harder" has stuck to this type of equations to such an extent that they devote a separate section thereto and the Matriculation examiners everywhere would almost seem to have made it an invariable rule of practice to include one question of this type in their examination papers.

Now, suppose the equation before us is:

$$\frac{1}{x-7}+\frac{1}{x-9}=\frac{1}{x-6}+\frac{1}{x-10}$$

In all the text-books, we are told to transpose two of the terms so that each side may have a plus term and a minus term, take the L.C.M. of the denominators, cross-multiply, equate the denominators, expand them, transpose and so on and so forth. And, after 10 or more steps of working, they tell you that 8 is the answer.

The Vedic *Sūtra*, however, tells us that, if other elements being equal, the sum-total of the denominators on the L.H.S. and the total on the R.H.S. be the same, then that total is zero!

In this instance, as $D_1+D_2$ and $D_3+D_4$ both total $2x - 16$, $\therefore 2x-16=0 \therefore x=8$! And that is all there is to it! A few more instances may be noted:

(1) $\frac{1}{x+7}+\frac{1}{x+9}=\frac{1}{x+6}+\frac{1}{x+10} \quad \therefore x=-8$

(2) $\frac{1}{x-7}+\frac{1}{x+9}=\frac{1}{x+11}+\frac{1}{x-9} \quad \therefore x=-1$

(3) $\frac{1}{x-8}+\frac{1}{x-9}=\frac{1}{x-5}+\frac{1}{x-12} \quad \therefore x=8\frac{1}{2}$

### Disguised Specimens

The above were plain, simple cases which could be readily recognised as belonging to the type under consideration. There are however several cases which really belong to this type but come under various kinds of disguises thin, thick or ultra-thick! But, however thick the disguise may be, there are simple devices by which we can penetrate and see through the disguises and apply the '*Śūnya Samuccaye*' formula:

### THIN DISGUISES

(1) $$\frac{1}{x-8}-\frac{1}{x-5}=\frac{1}{x-12}-\frac{1}{x-9}$$

Here, we should transpose the *minuses*, so that all the 4 terms are *plus* ones:

$$\frac{1}{x-8}+\frac{1}{x-9}=\frac{1}{x-12}+\frac{1}{x-5} \quad \therefore x=8\frac{1}{2}$$

The transposition-process here is very easy and can be done *mentally* in less than the proverbial trice.

(2) $$\frac{1}{x+1}-\frac{1}{x+3}=\frac{1}{x+2}-\frac{1}{x+4} \quad \therefore x=-2\frac{1}{2}$$

(3) $$\frac{1}{x+1}-\frac{1}{x-3}=\frac{1}{x-4}-\frac{1}{x-8} \quad \therefore x=3\frac{1}{2}$$

(4) $$\frac{1}{x-b}-\frac{1}{x-b-d}=\frac{1}{x-c+d}-\frac{1}{x-c} \quad \therefore x=\frac{1}{2}(b+c)$$

(5) $$\frac{1}{x+b}-\frac{1}{x+b+d}=\frac{1}{x+c-d}-\frac{1}{x+c} \quad \therefore x=-\frac{1}{2}(b+c)$$

*Note*: If the last two examples with so many literal coefficients involved were to be done according to the current system, the labour entailed over the L.C.M.'s, the multiplications etc., would have been terrific; and the time taken would have been proportionate too! But, by this Vedic method, the equation is solved *at sight*!

### MEDIUM DISGUISES

The above were cases of thin disguises, where mere transposition was sufficient to enable us to penetrate them. We now turn to cases of disguises of medium thickness:

(1) $$\frac{x-2}{x-3}+\frac{x-3}{x-4}=\frac{x-1}{x-2}+\frac{x-4}{x-5}$$

By dividing the Numerators out by the Denominators, we have:

$$1+\frac{1}{x-3}+1+\frac{1}{x-4}=1+\frac{1}{x-2}+1+\frac{1}{x-5}$$

Cancelling out the two ones from both sides, we have the equation before us in its undisguised shape and can at once say, $\therefore x = 3\frac{1}{2}$

Now, this process of division can be mentally performed very easily, thus:

(*i*) $\frac{x}{x}+\frac{x}{x}=\frac{x}{x}+\frac{x}{x} \because (1+1=1+1)$

(*ii*) Applying the *Parāvartya* method mentally and transferring the independent term of the denominator with its sign changed to the Numerator, we get 1 as the result in each of the 4 cases.

With the help of these two tests, we know that "the other elements are the same", and, as $D_1 + D_2 = D_3 + D_4$, we therefore identify the case before us as coming completely within the jurisdiction of the "*Śūnyam Samuccaye*" formula:

$$\because 2x - 7 = 0 \therefore x = 3\frac{1}{2}$$

(2) $\frac{x}{x-2}+\frac{x-9}{x-7}=\frac{x+1}{x-1}+\frac{x-8}{x-6}$

Here, $\frac{1}{1}+\frac{1}{1}=\frac{1}{1}+\frac{1}{1}$;

Secondly, by *Parāvartya*,

$$\frac{2}{x-2}-\frac{2}{x-7}=\frac{2}{x-1}+\frac{2}{x-6}$$

We transpose the *minus* terms and find that all the tests have been satisfactorily passed. All this argumentation can of course be done mentally.

So, we say: $2x - 8 = 0 \therefore x = 4$

(3) $\frac{2x-3}{x-2}+\frac{3x-20}{x-7}=\frac{x-3}{x-4}+\frac{4x-19}{x-5}$

Here, $\frac{2}{1}+\frac{3}{1}=\frac{1}{1}+\frac{4}{1}$; the numerators all become 1; and $D_1 + D_2 = D_3 + D_4 = 2x - 9 = 0 \therefore x = 4\frac{1}{2}$

(4) $\frac{3x-8}{x-3}+\frac{4x-35}{x-9}=\frac{2x-9}{x-5}+\frac{5x-34}{x-7}$

Here, $\frac{3}{1}+\frac{4}{1}=\frac{2}{1}+\frac{5}{1}$; and the other 2 tests are all right too.
$\therefore x = 6$

(5) $\frac{3x-13}{x-4}+\frac{4x-41}{x-10}=\frac{2x-13}{x-6}+\frac{5x-41}{x-8}$

All the tests are found satisfactorily passed.

$\therefore\ 2x - 14 = 0 \therefore x = 7$

(6) $\dfrac{4x+21}{x+5} + \dfrac{5x-69}{x-14} = \dfrac{3x-5}{x-2} + \dfrac{6x-41}{x-7}$

All the tests are all right $\therefore\ 2x - 9 = 0$

(7) $\dfrac{x^2+3x+3}{x+2} + \dfrac{x^2-15}{x-4} = \dfrac{x^2+7x+11}{x+5} + \dfrac{x^2-4x-20}{x-7}$

Either by simple division or by simple factorisation both of them, mental, we note:

(*i*) $(x+1)+(x+4) = (x+2)+(x+3)$

(*ii*) the numerators are all unity; and

(*iii*) $D_1 + D_2 = D_3 + D_4 = 2x - 2 = 0 \ \therefore\ x = 1$

### Thicker Disguises

(1) $\dfrac{2}{2x+3} + \dfrac{3}{3x+2} = \dfrac{1}{x+1} + \dfrac{6}{6x+7}$

(*i*) At first sight, this does *not* seem to be of the type which we have been dealing with in this section. But we note that the coefficient of x in the four denominators is not the same. So, by suitable multiplication of the numerator and the denominator in each term, we get 6 (the L.C.M. of the four coefficients) uniformly as the coefficient of x in all of them. Thus, we have:

$$\frac{6}{6x+9} + \frac{6}{6x+4} = \frac{6}{6x+6} + \frac{6}{6x+7}$$

Now, we can readily recognise the type and say:

$$12x + 13 = 0 \therefore x = \frac{-13}{12}$$

But we cannot gamble on the possible *chance* of its being of this type and go through all the laborious work of L.C.M., the necessary multiplications etc., and perhaps find at the end of it all, that we have drawn a blank ! There must therefore be some valid and convincing test whereby we can satisfy ourselves beforehand on this point and, if convinced, *then and only then* should we go through all the toil involved.

And that test is quite simple and easy:

$\frac{2}{2} + \frac{3}{3} = \frac{1}{1} + \frac{6}{6}$. But even then, only the *possibility* or the *probability* and not the certainty of it follows therefrom.

(*ii*) A second kind of test—with guarantee of certainty—is available too. And this is by cross-multiplication of $N_1$ by $D_2$ and of $N_2$ by $D_1$ on the one hand and of $N_3$ by $D_4$ and of $N_4$ by $D_3$ on the other. And this too can be done mentally.

Thus, in the case dealt with, we get from each side—the same $12x+13$ as the total $\therefore 12x+13=0 \therefore x=\frac{-13}{12}$

(2) $\frac{3}{3x+1}-\frac{6}{6x+1}=\frac{3}{3x+2}-\frac{2}{2x+1}$

(*i*) We transpose mentally and note :

$\frac{3}{3}+\frac{2}{2}=\frac{3}{3}+\frac{6}{6}$. So, we may try the L.C.M. method.

$$\frac{6}{6x+2}+\frac{6}{6x+3}=\frac{6}{6x+1}+\frac{6}{6x+4}$$

$$\therefore 12x+5=0 \therefore x=\frac{-5}{12}$$

(*ii*) Even here, after the preliminary testing of $\frac{3}{3}+\frac{2}{2}$ being equal to $\frac{3}{3}+\frac{6}{6}$, we may straightaway cross-multiply and say:

$\therefore 12x+5=0 \therefore x=-5/12$

(3) $\frac{3}{3x+1}+\frac{2}{2x-1}=\frac{3}{3x-2}+\frac{2}{2x+1}$

$\therefore$ By either of the two methods, we get $12x-1=0$

$\therefore x=\frac{1}{12}$

(4) $\frac{1}{x+3}+\frac{3}{3x-1}=\frac{1}{x+5}+\frac{3}{3x-7}$

$\therefore$ By either method, $6x+8=0 \therefore x=\frac{-4}{3}$

(5) $\frac{2x+11}{x+5}-\frac{9x-9}{3x-4}=\frac{4x+13}{x+3}-\frac{15x-47}{3x-10}$

Here $\frac{2}{1}-\frac{9}{3}=\frac{4}{1}-\frac{15}{3} \therefore$ Yes.

By simple division, we put this into proper shape, as follows:

$$\frac{1}{x+5}+\frac{3}{3x-10}=\frac{1}{x+3}+\frac{3}{3x-4}$$

Here $\frac{1}{1}+\frac{3}{3}=\frac{1}{1}+\frac{3}{3}$ $\therefore$ Yes.

$\therefore$ By either method, $6x+5=0$ $\therefore$ $x=\frac{-5}{6}$

(6) $\frac{5-6x}{3x-1}+\frac{2x+7}{x+3}=\frac{31-12x}{3x-7}+\frac{4x+21}{x+5}$

$\therefore \frac{1}{x+3}+\frac{3}{3x-1}=\frac{1}{x+5}+\frac{3}{3x-7}$

$\therefore$ By either method, $6x+8=0$ $\therefore$ $x=\frac{-4}{3}$

## Further Applications of the Formula

(1) *In the case of a special type of seeming "cubics"*

There is a certain type of equations which look like cubic equations but, which after wasting a lot of our time and energy turn out to be simple equations of the first degree and which come within the range of the "*Śūnyam Samuccaye*" formula. Thus, for instance

$$(x-3)^3+(x-9)^3=2(x-6)^3$$

The current system works this out at enormous length by expanding all the three cubes, multiplying, transposing etc., and finally gives us the answer $x = 6$

The Vedic *Sūtra* now under discussion is, however, applicable to this kind of case too and says:

$(x-3)+(x-9)=2x-12$. Taking away the numerical factor, we have $x-6$. And $x-6$ is the factor under the cube on R.H.S.

$\therefore x-6=0 \therefore x=6$

*The Algebraical proof* of it is as follows:

$(x-2a)^3+(x-2b)^3=2(x-a-b)^3$

$\therefore x^3-6x^2a+12xa^2-8a^3+x^3-6x^2b+12xb^2-8b^3$

$=2(x^3-3x^2a-3x^2b+3xa^2+3xb^2+6xab-a^3-3a^2b-3ab^2-b^3)$

$=2x^3-6x^2a-6x^2b+6xa^2+6xb^2+12xab-2a^3-6a^2b-6ab^2-2b^3$

Cancelling out the common terms from both sides, we have:

$12xa^2+12xb^2-8a^3-8b^3=6xa^2+6xb^2+12xab-2a^3-6a^2b-6ab^2-2b^3$

$6xa^2+6xb^2-12xab=6a^3-6a^2b-6ab^2+6b^3$

$\therefore$ $6x(a-b)^2=6(a+b)(a-b)^2$

$\therefore x=a+b$

Obviously this particular combination was not thought of and worked out by the mathematicians working under the current system. At any rate, it is not found listed in their books under any known formula or as a conditional identity and so on. The Vedic mathematicians, however, seem to have worked it all out and have given us the benefit thereof by the application of this formula to examples of this type.

We need hardly point out that the expansions, multiplications, additions, transpositions, factorisations in each particular case of this type must necessarily involve the expenditure of tremendous time and energy, while the Vedic formula gives us the answer *at sight*!

Three more illustrations may be taken:

(*i*) $(x-149)^3+(x-51)^3=2(x-100)^3$

The very prospect of the squaring, cubing etc., of these numbers must appeal the student. But, by the present *Sūtra* we can at once say: $2x-200=0 \therefore x=100$

(*ii*) $(x-249)^3+(x+247)^3=2(x-1)^3$

This is still more terrific. But, with the aid of this *Sūtra*, we can at once say: $x = 1$; and

(*iii*) $(x+a+b-c)^3+(x+b+c-a)^3=2(x+b)^3$

The literal coefficients make this still worse. But the Vedic one-line mental answer is: $x = -b$.

(2) *In the case of a special type of seeming "biquadratics"*

There is also similarly, a special type of *seemingly* "biquadratic" equations which are really of the first degree and which the same *Sūtra* solves for us, at sight. Thus, for example:

$$\frac{(x+3)^3}{(x+5)^3}=\frac{x+1}{x+7}$$

According to the current method, we cross-multiply and say: $(x+7)(x+3)^3=(x+1)(x+5)^3$

Expanding the two sides with the aid of the usual formula $[(x+a)(x+b)(x+c)(x+d)=x^4+x^3(a+b+c+d)$

$+x^2(ab+ac+ad+bc+bd+cd)$

$+x(abc+abd+acd+bcd)+abcd)]$ twice over, we will next say:

$$x^4+16x^3+90x^2+216x+189=x^4+16x^3+90x^2+200x+125$$

Cancelling out the common terms and transposing, we then say:

$\therefore 16x=-64 \therefore x=-4$

According to the Vedic formula, however, we do not crossmultiply the binomial factors and so on but simply *observe* that $N_1+D_1$* and $N_2+D_2$ are both $2x+8$ and $\therefore\ 2x+8=0$

$\therefore\ x=-4$

The Algebraic proof hereof is as follows:

$$\frac{(x+a+d)^3}{(x+a+2d)^3}=\frac{x+a}{x+a+3d}$$

$\therefore$ By the usual process of cross-multiplications,

$$(x+a+3d)(x+a+d)^3=(x+a)(x+a+2d)^3$$

$\therefore$ By expansion of both sides

$$x^4+x^3(4a+6d)+x^2(6a^2+18ad+12d^2)+x(4a^3+18a^2d+24ad^2+10d^3)+(a^4+6a^3d+12a^2d^2+12ad^3+3d^4)=x^4+x^3(4a+6d)+x^2(6a^2+18ad+12d^2)+x(4a^3+18a^2d+24ad^2+8d^3)+\text{etc.}$$

$\therefore$ Cancelling common terms out, we have :

$$x(10d^3)+10ad^3+3d^4=x(8d^3)+8ad^3$$

$\therefore\ 2d^3x+2ad^3+3d^4=0$

$\therefore$ Cutting $d^3$ out, we have $2x+2a+3d=0\ \therefore\ x=-\frac{1}{2}(2a+3d)$

At this point, the student will note that $N_1+D_1$ under the cubes and $N_2+D_2$ are both $(2x+2a+3d)$. And this gives us the required clue to the particular characteristic which characterises this type of equations, i.e. that $N_1+D_1$ under the cube and $N_2+D_2$ must be the same; and, obviously, therefore, the '*Śūnyaṃ Samuccaye*' *Sūtra* applies to this type. And, while the current system has evidently not tried, experienced and listed it, the Vedic seers had doubtless experimented on, observed and listed this particular combination also and listed it under the present *Sūtra*.

*Note*: (1) The condition noted above about the 4 binomials is interesting. The sum of the first+the second must be the same as the sum of the third and fourth.

(2) The most obvious and readily understandable condition fulfilling this requirement is that the absolute terms in $N_2$, $N_1$, $D_1$ and $D_2$ binomials should be in Arithmetical Progression.

(3) This may also be postulated in this way, i.e. that the difference between the two binomials on the R.H.S. must be equal to three times the difference between those on the L.H.S. This, however, is only a corollary—result arising from the A.P. relationship amidst the four binomials namely, that if $N_2$, $N_1$, $D_1$ and $D_2$ are in A.P. it is obvious that

*(within the cubes)

$D_2 - N_2 = 3(D_1 - N_1)$.

(4) In any case, the formula in this special type may be enunciated—in general terms—thus: if N+D on both sides be the same, N+D should be equal to zero.

Two more examples of this type may be taken:

(1) $\frac{(x-5)^3}{(x-7)^3} = \frac{(x-3)}{(x-9)} \therefore 2x - 12 = 0 \therefore x = 6$

(2) $\frac{(x-a)^3}{(x+b)^3} = \frac{x-2a-b}{x+a+2b}$

Working this out with all the literal coefficients and with cross-multiplications, expansions, cancellations, transpositions etc., galore would be a horrid task even for the most laborious student. The Vedic formula, however, tells us that $(x-a)+(x+b)$ and $(x-2a-b)$ +$(x+a+2b)$ both total up to $2x-a+b$

$\therefore x = \frac{1}{2}(a-b)$

*Note*: In all the above examples, it will be observed that the 4 binomials are not merely in Arithmetical Progression but are also so related that their cross totals are also the same.

Thus, in the first example worked out above, by cross-multiplication, we have $(x+7)(x+3)^3 = (x+1)(x+5)^3$; and the cross-addition of these factors gives us $4x + 16$ as the total on both sides; and this tallies with the value $x = -4$ obtained above.

In the second example:

$$(x+a+3d)(x+a+d)^3 = (x+a)(x+a+2d)^3$$

And here too, cross-addition gives us$4x+4a+6d$ as the total on both sides. And this too gives us the same answer as before.

In the third example, we have:

$$(x-9)(x-5)^3 = (x-3)(x-7)^3$$

And here too the cross-addition process gives us $4x - 24$ as the total on both sides. And we get the same answer as before.

In the fourth case, we have:

$$(x+a+2b)(x-a)^3 = (x-2a-b)(x+b)^3$$

And cross-addition again gives us the total $4x - 2a + 2b$ on both sides and, therefore, the same value of x as before.

The student should not, however, fall into the error of imagining that this is an *additional* test or sufficient condition for the application of the formula. This really comes in as a corollary-consequence of the A.P. relationship between the Binomial factors. But it is not a *sufficient* condition by itself for the applicability of the present formula. The rule about $N_1+D_1$ and $N_2+D_2$ being the same, is the only condition sufficient for this purpose.

An instance in point is given below:

$$\frac{(x+3)^3}{(x+5)^3}=\frac{x+2}{x+8}$$

Here, cross-addition gives 4x+17 as the total on both sides, and the condition $D_2-N_2=3(D_1-N_1)$ is also satisfied as $6=2\times 3$. But $3+5\neq 2+8$; and, as this essential condition is lacking, this particular equation does *not* come within the purview of this *Sūtra.*

On actual cross-multiplication and expansion etc., we find:

$$x^4+17x^3+105x^2+275x+250=x^4+17x^3+99x^2+243x+216$$

$\therefore 6x^2+32x+34=0$, which is a quadratic equation with the two Irrational Roots $\dfrac{-16\pm+\sqrt{52}}{6}$ and not simple equation at all, of the type we are here dealing with.

And this is in conformity with the lack of the basic condition in question, i.e. that $N_1+D_1$ and $N_2+D_2$ should be the same.

(3) *In the case of another special type of "Biquadratics"*

There is also another special type of "Biquadratics" which are really simple equations of the first degree, and to which the "*Śūnyaṃ Samuccaye*" *Sūtra* is applicable which we now go on to. This section may, however, be held over for a later reading.

(1) $\dfrac{(x+1)(x+9)}{(x+2)(x+4)}=\dfrac{(x+6)(x+10)}{(x+5)(x+7)}$

or $(x+1)(x+9)(x+5)(x+7)=(x+2)(x+4)(x+6)(x+10)$

We first note that cross-addition gives us the same total (4x+22) on both sides. This gives us the assurance that, on cross-multiplication, expansion etc., the $x^4$ and the $x^3$ coefficients will cancel out. But what about the $x^2$ coefficients?

For them too to vanish, it is necessary that the sum of the products of the independent terms taken two at a time should be the same on both sides. And this is the case when if (x+a) (x+b) (x+c) (x+d) = (x+e) (x+f) (x+g) (x+h), we have not merely a+b+c+d= e+f+g+h but also two other conditions fulfilled:

(*i*) that the sum of *any* 2 binomials on the one side is equal to the sum of *some* two binomials on the other; and

(*ii*) ab+cd on the left = ef+gh on the right.

In the example actually before us, we find all these conditions fulfilled:

(*i*) $(x+1)+(x+9)=(x+4)+(x+6)$; $(x+1)+(x+5)=(x+2)+(x+4)$; $(x+1)+(x+7)=(x+2)+(x+6)$; $(x+9)+(x+5)=(x+4)+(x+10)$; $(x+9)+(x+7)=(x+6)+(x+10)$; and $(x+5)+(x+7)=(x+2)+(x+10)$; and

(*ii*) $(5+63)$ and $(8+60)$ are both equal to 68.

So, by this test, at sight, we know the equation comes under the range of this *Sūtra* $\therefore 4x+22=0 \therefore x=-5\frac{1}{2}$

Similar is the case with regard to the equation:

(2) $\dfrac{(x+2)(x+4)}{(x+1)(x+3)}=\dfrac{(x-1)(x+7)}{(x-2)(x+6)}$

$\therefore (x-2)(x+2)(x+4)(x+6)=(x-1)(x+1)(x+3)(x+7)$; and

(*i*) By cross-addition, the total on both sides is $4x+10$;

(*ii*) The sum of *each* pair of binomials on the one side is equal to the sum of *some* pair thereof on the other; and

(*iii*) $ab+cd=ef+gh$, i.e. $-4+24=-1+21\ (=20)$

$\therefore$ The *Sūtra* applies; and $4x+10=0 \therefore x=-2\frac{1}{2}$

This however is *not* the case with the equation:

(3) $(x-1)(x-6)(x+6)(x+5)=(x-4)(x-2)(x+3)(x+7)$

Here we observe:

(*i*) The total on both sides is $4x+4$; *but*

(*ii*) The totals of pairs of binomials on the two sides do *not* tally; *and*

(*iii*) $ab+cd \neq ef+gh$

This equation is therefore a quadratic and not within the scope of the present *Sūtra*.

*The Algebraical Explanation* for this type of equations is:

$(x+a)(x+b)(x+c)(x+d)=(x+e)(x+f)(x+g)(x+h)$

The data are:

(*i*) $a+b+c+d=e+f+g+h$;

(*ii*) The sum of *any* pair of binomials on the one side must be the same as the sum of *some* pair of binomials on the other. Suppose $a+b=e+f$; and $c+d=g+h$; and

(*iii*) $ab+cd=ef+gh$

$\therefore x^4+x^3(a+b+c+d)+x^2(ab+ac+ad+bc+bd+cd)$
$+x(abc)+abd+acd+bcd)+abcd$

$= x^4 + x^3(e + f + g + h) + x^2(ef + eg + eh + fg + fh + gh)$
$+ x(efg + efh + egh + fgh) + efgh$

∴ The $x^4$ and $x^3$ cancel out; and, owing to the data in the case, the $x^2$ coefficients are the same on both sides; and therefore they too cancel out. And there is no quadratic equation left for us to solve herein.

*Proof* : The $x^2$ coefficients are:

L.H.S. $ab + ac + ad + bc + bd + cd$
R.H.S. $ef + eg + eh + fg + fh + gh$
i.e. $(ab + cd) + a(c + d) + b(c + d) = ab + cd + (a + b)(c + d)$
and $(ef + gh) + e(g + h) + f(g + h) = ef + gh + (e + f)(g + h)$
But $(ab + cd) = (ef + gh)$; and $a + b = e + f$; and $c + d = g + h$
∴ the L.H.S. = the R.H.S.; and $x^2$ vanishes !

FURTHER EXTENSION OF THE SŪTRA

In the beginning of this chapter, it was noted that if a function containing the unknown x, y etc., occurs as a common factor in all the terms on both sides or on the L.H.S. with zero on the R.H.S. that function can be removed therefrom and equated to Zero. We now proceed to deal with certain types of cases which do not *seem* to be of this kind but are really so. All that we have to do is to re-arrange the terms in such a manner as to unmask the masked terms, so to say and make the position transparently clear on the surface. For example,

(1) $\frac{x+a}{b+c} + \frac{x+b}{c+a} + \frac{x+c}{a+b} = -3$

Taking −3 over from the R.H.S. to the L.H.S. distributing it amongst the 3 terms there, we have:

$\frac{x+a}{b+c} + 1 + \frac{x+b}{c+a} + 1 + \frac{x+c}{a+b} + 1 = 0$

i.e. $\frac{x+a+b+c}{b+c} + \frac{x+b+c+a}{c+a} + \frac{x+c+a+b}{a+b} = 0$

∴ By virtue of the *Samuccaya* rule,

$x + a + b + c = 0$ ∴ $x = -(a + b + c)$

This whole working can be done, at sight, e.g. *mentally*.

(2) $\frac{x+a}{b+c} + \frac{x+b}{c+a} + \frac{x+c}{a+b} = \frac{x+2a}{b+c-a} + \frac{x+2b}{c+a-b} + \frac{x+2c}{a+b-c}$

Add unity to each of the 6 terms; and observe:

∴ $\frac{x+a+b+c}{D_1} + \frac{x+a+b+c}{D_2} + \frac{x+a+b+c}{D_3}$

$$= \frac{x+a+b+c}{D_4} + \frac{x+a+b+c}{D_5} + \frac{x+a+b+c}{D_6}$$

$\therefore x+a+b+c=0 \;\therefore x=-(a+b+c)$

(3) $\frac{x-a}{b+c} + \frac{x-b}{c+a} + \frac{x-c}{a+b} = \frac{x+a}{2a+b+c} + \frac{x+b}{2b+c+a} + \frac{x+c}{2c+a+b}$

Subtract unity from each of the 6 terms; and we have:

$x-a-b-c=0 \;\therefore x=(a+b+c)$

(4) $\frac{x+a^2}{(a+b)(a+c)} + \frac{x+b^2}{(b+c)(b+a)} + \frac{x+c^2}{(c+a)(c+b)}$

$$= \frac{x-bc}{a(b+c)} + \frac{x-ca}{b(c+a)} + \frac{x-ab}{c(a+b)}$$

Subtracting 1 from each of the 6 terms, we have:

$x-ab-ac-bc=0 \;\therefore x=(ab+bc+ca)$

(5) $\frac{x-bc}{b+c} + \frac{x-ca}{c+a} + \frac{x-ab}{a+b}$

$$= \frac{x+2a^2-bc}{2a+b+c} + \frac{x+2b^2-ca}{2b+c+a} + \frac{x+2c^2-ab}{2c+a+b}$$

Subtracting a from the first terms, b from the second terms and c from the third terms on both sides, we have :

$x-ab-bc-ca=0 \;\therefore x=ab+bc+ca$

(6) $\frac{x+a^2+2c^2}{b+c} + \frac{x+b^2+2a^2}{c+a} + \frac{x+c^2+2b^2}{a+b} = 0$

As $(b-c)+(c-a)+(a-b)=0$, we add $b-c$, $c-a$ and $a-b$ to the first, second and third terms respectively; and we have:

$x+a^2+b^2+c^2=0 \;\therefore x=-(a^2+b^2+c^2)$

(7) $\frac{ax+a(a^2+2bc)}{b-c} + \frac{bx+b(b^2+2ca)}{c-a} + \frac{cx+c(c^2+2ab)}{a-b} = 0$

As $a(b-c)+b(c-a)+c(a-b)=0$

$\therefore$ We add a(b–c) to the first term, b(c–a) to the second and c(a–b) to the last; and we have:

$$t_1 = \frac{ax+a(a^2+2bc)}{b-c} + a(b-c)$$

$$= \frac{ax+a(a^2+2bc)+a(b-c)^2}{b-c}$$

$$= \frac{ax+a(a^2+b^2+c^2)}{b-c} = \frac{a}{b-c}\{x+(a^2+b^2+c^2)\}$$

$$\text{Similarly, } t_2 = \frac{b}{c-a}\{x + (a^2 + b^2 + c^2)\}$$

$c$

$$\text{and } t_3 = \frac{c}{a-b}\{x + (a^2 + b^2 + c^2)\} = 0$$

$$\therefore x + a^2 + b^2 + c^2 = 0 \ \therefore x = -(a^2 + b^2 + c^2)$$

(8) $\dfrac{x + a^3 + 2b^3}{b-c} + \dfrac{x + b^3 + 2c^3}{c-a} + \dfrac{x + c^3 + 2a^3}{a-b}$

$$= 2a^2 + 2b^2 + 2c^2 + ab + ac + bc$$

Splitting the R.H.S. into $(b^2 + bc + c^2) + (c^2 + ca + a^2) + (a^2 + ab + b^2)$, transposing the three parts to the left and combining the first with the first, the second with the second and the third with the third by way of application of the '*Ādyamādyena*' formula, we have:

$$t_1 = \frac{x + a^3 + 2b^3}{b-c} - (b^2 + bc + c^2)$$

$$= \frac{x + a^3 + 2b^3 - b^3 + c^3}{b-c} = \frac{x + a^3 + b^3 + c^3}{b-c}$$

$$\text{Similarly, } t_2 = \frac{\text{the same N}}{c-a}$$

$$\text{and } t_3 = \frac{\text{the same N}}{a-b}$$

$$\therefore x = -(a^3 + b^3 + c^3)$$

# 13

# Merger Type of Easy Simple Equations

BY THE PARĀVARTYA METHOD

Having dealt with various sub-divisions under a few special types of simple equations which the *Śūnyaṃ Sāmyasamuccaye* formula helps us to solve easily, we now take up another special type of simple equations which the *Parāvartya Sūtra* (dealt with already in connection with Division etc.) can tackle for us.

This is what may be described as the merger type; and this too includes several sub-headings under that heading.

### THE FIRST TYPE

The first variety is one in which a number of terms on the left hand side is equated to a single term on the right side, in such manner that $N_1 + N_2 + N_3$ etc., the sum of the numerators on the left and the single numerator on the right are the same.

For instance,

$\frac{3}{x+1} + \frac{4}{x+2} = \frac{7}{x+3}$. Here $N_1 + N_2$, i.e. $(3 + 4) = N$ (i.e. 7). So the *Sūtra* applies.

The procedure is one of merging of the R.H.S. fraction into the left, so that only two terms remain. The process is as follows:

As we mean to merge the R.H.S. into the L.H.S., we subtract the independent term of the to-be-merged binomial from the absolute terms in the binomials on the left and multiply those remainders by the numerators of the terms on the left. And the process is complete.

(*i*) We first put down the two to-be-retained denominators down thus:

$\overline{x+1}$ $\qquad\qquad$ $\overline{x+2}$

(*ii*) Then, as 3 from the R.H.S. is to be merged, we subtract that 3 from the 1 in the first term, obtain –2 as the remainder, multiply it by the numerator, i.e. 3, get –6 as the product and put that down as the new numerator for our first term.

(*iii*) And we do the same thing with the second term, obtain –4 as the product and set it down as our numerator for the second term of the new, i.e. the derived equation.

(*iv*) As the work of merging has been completed, we put zero on the right hand side. So the resultant new equation after the merger now reads:

$$\frac{-6}{x+1} - \frac{4}{x+2} = 0$$

Then, by simple cross-multiplication, we say

$$\therefore\ 4x + 4 = -6x - 12\ \therefore\ 10x = -16\ \therefore\ x = -8/5$$

or, by the general formula (– mb – na)/(m+n) explained already in the chapter on simple equations and first principles, we say at once:

$$x = \frac{12+4}{-6-4} = -8/5$$

*The Algebraical Proof* hereof is:

$$\frac{3}{x+1} + \frac{4}{x+2} = \frac{7}{x+3} = \frac{3}{x+3} + \frac{4}{x+3}$$

$$\therefore \frac{3}{x+1} - \frac{3}{x+3} = \frac{4}{x+3} - \frac{4}{x+2}$$

$$\therefore \frac{3(x+3-x-1)}{(x+1)(x+3)} = \frac{4(x+2-x-3)}{(x+3)(x+2)}$$

$$\therefore \frac{6}{x+1} = \frac{-4}{x+2} \therefore 6x + 12 = 4x - 4 \therefore 10x = -16$$

$$\therefore x = -8/5$$

*The General Algebraical Proof* hereof is:

$$\frac{p}{x+a} + \frac{q}{x+b} = \frac{p+q}{x+c}$$

$$\therefore \frac{p}{x+a} + \frac{q}{x+b} = \frac{p}{x+c} + \frac{q}{x+c}$$

$$\therefore \frac{p}{x+a} - \frac{p}{x+c} = \frac{q}{x+c} - \frac{q}{x+b}$$

$$\therefore \frac{p(x+c-x-a)}{(x+a)\,(x+c)} = \frac{q(x+b-x-c)}{(x+c)\,(x+b)}$$

$$\therefore \frac{p(c-a)}{(x+a)} = \frac{q(b-c)}{x+b}$$

$$\therefore x\{p(c-a)+q(c-b)\} = bp(a-c)+aq(b-c)$$

$$\therefore x = \frac{bp(a-c)+aq(b-c)}{p(c-a)+q(c-b)}$$

The Algebraical explanation may look frightfully long. But the application of the '*Parāvartya*' *Sūtra* as hereinabove explained and illustrated is simple enough and easy enough and should be welcomed by the student with delight.

A few more examples of this sort may be noted:

(1) $\frac{3}{x-2} + \frac{5}{x-6} = \frac{8}{x+3}$

Here $3+5=8$ ∴ The *Sūtra* applies.

$$\therefore \frac{(3)(-5)}{x-2} + \frac{(-9)(5)}{(x-6)} = \frac{-15}{x-2} + \frac{-45}{x-6} = 0 \;\therefore x = \frac{-90-90}{-15-45} = 3$$

(2) $\frac{2}{x+2} + \frac{3}{x+3} = \frac{5}{x+5}$

Here $2+3=5$ ∴ The formula applies.

$$\therefore \frac{-6}{x+2} + \frac{-6}{x+3} = 0 \;\therefore x = -2\tfrac{1}{2}$$

*Note*: At this stage, when both the numerators are found to be –6 and can therefore be removed, the formula "*Śūnyaṃ Samuccaye*" may be readily applied; and we may say:

$(x+2)+(x+3)=0 \;\therefore x = -2\frac{1}{2}$

But if $\frac{2}{x+2} + \frac{3}{x+3} = \frac{5}{x+7}$, as $2+3=5$,

the merger *Sūtra* applies; but after the merger, the numerators are different (i.e. –10 and –12) and therefore the '*Śūnyam*' *Sūtra* will not apply.

### Disguises

Here too, we have often to deal with disguises, by seeing through and penetrating them, in the same way as in the previous chapter with regard to the '*Śūnyaṃ Samuccaye*' formula.

A few illustrations will make this clear:

(1) $\frac{5}{x-2} + \frac{2}{3-x} = \frac{3}{x-4}$

Here, mere transposition will do the trick. Thus:

$\frac{2}{x-3}+\frac{3}{x-4}=\frac{5}{x-2}$ Now, $2+3=5$ $\therefore$ The *Sūtra* applies.

$\therefore \frac{-2}{x-3}+\frac{-6}{x-4}=0 \therefore x=\frac{-26}{-8}=\frac{13}{4}$

(2) $\frac{4}{2x+1}+\frac{9}{3x+2}=\frac{15}{3x+1}$

$4+9 \neq 15$ $\therefore$ Doubt arises; but the coefficients of x being different in the three denominators, we try the L.C.M. method and get:

$\frac{12}{6x+3}+\frac{18}{6x+4}=\frac{30}{6x+2}$

And here, on noting $N_1(12)+N_2(18)=N_3(30)$, we say: "Yes; the *Sūtra* applies" and proceed to apply it:

$\therefore 12/(6x+3)+36/(6x+4)=0 \therefore x=-13/24$

But how should we know *before-hand* that the *Sūtra does* apply ? The test is very simple and merely consists in the division of each numerator by the x-coefficient in the denominator as in the '*Śūnyaṃ*' case.

Thus $\frac{4}{2}+\frac{9}{3}=2+3=5$; and $\frac{15}{3}$: is also 5.

Say, "yes" and go ahead, with the merging.

(3) $\frac{4}{2x-1}+\frac{9}{3x-1}=\frac{25}{5x-1}$

Here $\left(\frac{4}{2}+\frac{9}{3}\right)$ and $\frac{25}{5}$ are the same (i.e. 5) $\therefore$ Yes.

$\therefore \frac{60}{30x-15}+\frac{90}{30x-10}=\frac{150}{30x-6}$ Note, $60+90=150$

$\therefore \frac{2}{30x-15}+\frac{3}{30x-10}=\frac{5}{30x-6}$ Note $2+3=5$ $\therefore$ Yes

Proceed therefore and say:

$\frac{-18}{30x-15}-\frac{12}{30x-10}=0 \therefore x=\frac{2}{5}$

(4) $\frac{7}{7x+1}+\frac{6}{4x+1}=\frac{15}{6x+5}$

Here $\frac{7}{7}+\frac{6}{4}=1+1\frac{1}{2}=2\frac{1}{2}$; and $\frac{15}{6}$ is also $2\frac{1}{2}$ $\therefore$ Yes

Do the merging therefore and say:

$\therefore \frac{84}{84x+12}+\frac{126}{84x+21}=\frac{210}{84x+70}$

$\therefore \frac{2}{84x+12}+\frac{3}{84x+21}=\frac{5}{84x+70} \therefore$ Yes

$\therefore \frac{-116}{84x+12}+\frac{-147}{84x+21}=0 \therefore x=\frac{-50}{263}$

(5) $\frac{4}{5x+1}+\frac{7}{10x+1}=\frac{3}{2x+1}$

Here $\frac{4}{5}+\frac{7}{10}=\frac{15}{10}=3/2 \therefore$ Yes

$\therefore \frac{8}{10x+2}+\frac{7}{10x+1}=\frac{15}{10x+5} \therefore$ Yes

$\therefore \frac{-24}{10x+2}+\frac{-28}{10x+1}=0$

$\therefore -520x-80=0$

$\therefore x=-2/13$

(6) $\frac{7}{7x+1}+\frac{6}{4x+1}=\frac{15}{6x+1}$

Here $\frac{7}{7}+\frac{6}{4}=2\frac{1}{2}$; and $\frac{15}{6}$ is also $2\frac{1}{2} \therefore$ Yes

$\therefore \frac{84}{84x+12}+\frac{126}{84x+21}=\frac{210}{84x+14} \therefore$ Yes

$\therefore \frac{2}{84x+12}+\frac{3}{84x+21}=\frac{5}{84x+14} \therefore$ Yes

$\therefore \frac{-4}{84x+12}+\frac{21}{84x+21}=0$

$\therefore 1428x+168=0$

$\therefore x=\frac{-168}{1428}=\frac{-2}{17}$

## Extension of Merger Method
(*Multiple Merger*)

We now take up and deal with equations wherein $N_1+N_2+N_3$ (of the L.H.S.) =N of the R.H.S. and wherein the same '*Parāvartya*' (Merger) formula can be applied in exactly the same way as before. Thus:

(1) $\frac{1}{x+2}+\frac{3}{x+3}+\frac{5}{x+5}=\frac{9}{x+4}$

Test : $1+3+5=9 \therefore$ Yes.

$\therefore \frac{-2}{x+2}+\frac{-3}{x+3}+\frac{5}{x+5}=0 \therefore$ Yes, again.

i.e. $\frac{2}{x+2}+\frac{3}{x+3}=\frac{5}{x+5}$

$$\therefore \frac{-6}{x+2}+\frac{-6}{x+3}=0$$

$\therefore$ (*i*) By the Basic Formula $\left(x=\frac{-mb-na}{m+n}\right)$

$$x=\frac{18+12}{-6-6}=\frac{30}{-12}=-2\tfrac{1}{2}$$

or (*ii*) By the '*Śūnyaṃ Samuccaye*' formula:

$$(x+2)+(x+3)=0 \therefore x=-2\tfrac{1}{2}$$

*Note*: These two steps of successive merging can be combined into one by multiplying $N_1$ first by $2-4$ and then by $2-5$, i.e. by 6 and similarly $N_2$ first by $3-4$ and then by $3-5$, i.e. by 2 and proceeding as before

$$\therefore \frac{6}{x+2}+\frac{6}{x+3}=0$$

$\therefore$ By either method Basic or *Sūnyaṃ*, $x=-2\frac{1}{2}$

*The Algebraic Proof* hereof is this:

$$\frac{m}{x+a}+\frac{n}{x+b}+\frac{p}{x+c}=\frac{m+n+p}{x+d}$$

$$\therefore \frac{m(a-d)}{x+a}+\frac{n(b-d)}{x+b}+\frac{p(c-d)}{x+c}=0$$

$$\therefore \frac{m(a-d)(a-c)}{x+a}+\frac{n(b-d)(b-c)}{x+b}=0$$

which is the exact shape of the formula required for the single-step merger (vide *supra*).

Similarly, the merger-formula can be extended to any number of terms as follows:

$$\frac{m}{x+a}+\frac{n}{x+b}+\frac{p}{x+c}+\frac{q}{x+d}+\frac{r}{x+e}+\ldots$$
$$=\frac{m+n+p+q+r+\ldots}{x+w}$$

$$\therefore \frac{m(a-w)(---)(a-e)(a-d)(a-c)}{x+a}$$
$$+\frac{n(b-w)(---)(b-e)(b-d)(b-c)}{x+b}=0$$

which is the general formula for the purpose. Thus, in the above example

$$x=\frac{(-3)(-2)(-3)+(-3)(-2)(-2)}{(-3)(-2)(+1)+3(-1)(-2)}=\frac{-18-12}{6+6}=\frac{-30}{12}=-2\tfrac{1}{2}$$

A few more illustrations of this type are given below:

(1) $\frac{3}{3x+1}+\frac{4}{4x+1}+\frac{48}{8x+1}=\frac{48}{6x+1}$

Here $\frac{3}{3}+\frac{4}{4}+\frac{48}{8}=8$; and $\frac{48}{6}=8 \therefore$ Yes

$\therefore \frac{24}{24x+8}+\frac{24}{24x+6}+\frac{144}{24x+3}=\frac{192}{24x+4} \therefore$ Yes

$\therefore \frac{1}{24x+8}+\frac{1}{24x+6}+\frac{6}{24x+3}=\frac{8}{24x+4} \therefore$ Yes

$\therefore \frac{4}{24x+8}+\frac{2}{24x+6}=\frac{6}{24x+3} \quad \therefore$ Yes

$\therefore \frac{20}{24x+8}+\frac{6}{24x+6}=0$

$\therefore 624x+168=0 \ \therefore x=\frac{-7}{26}$

(2) $\frac{2}{2x+1}+\frac{18}{3x+1}+\frac{75}{5x+1}=\frac{88}{4x+1}$

Here $\frac{2}{2}+\frac{18}{3}+\frac{75}{5}=22$; and $\frac{88}{4}$ is also 22

$\therefore \frac{60}{60x+30}+\frac{360}{60x+20}+\frac{900}{60x+12}=\frac{1320}{60x+15} \therefore$ Yes

$\therefore \frac{1}{60x+30}+\frac{6}{60x+20}+\frac{15}{60x+12}=\frac{22}{60x+15} \therefore$ Yes

$\therefore \frac{15}{60x+30}+\frac{30}{60x+20}=\frac{45}{60x+12} \therefore$ Yes

$\therefore \frac{1}{60x+30}+\frac{2}{60x+20}=\frac{3}{60x+12} \therefore$ Yes

$\therefore \frac{18}{60x+30}+\frac{16}{60x+20}=0$

$\therefore 2040x+840=0 \ \therefore x=\frac{-7}{17}$

*Note*: Any change of sequence of the terms on the L.H.S. will cause no change in the working or the result.

(3) $\frac{4}{2x-1}+\frac{27}{3x-1}+\frac{125}{5x-1}=\frac{144}{4x-1}$

Here $\frac{4}{2}+\frac{27}{3}+\frac{125}{5}=2+9+25=36$;

and $\frac{144}{4}$ is also 36 $\therefore$ Yes

$\therefore \frac{120}{60x-30}+\frac{540}{60x-20}+\frac{1500}{60x-12}=\frac{2160}{60x-15} \therefore$ Yes

$$\therefore \frac{2}{60x-30}+\frac{9}{60x-20}+\frac{25}{60x-12}=\frac{36}{60x-15} \quad \therefore \text{ Yes}$$

$$\therefore \text{(By merger)} \frac{-30}{60x-30}+\frac{-45}{60x-20}=\frac{-75}{60x-12} \quad \therefore \text{ Yes}$$

$$\therefore \frac{2}{60x-30}+\frac{3}{60x-20}=\frac{5}{60x-12} \quad \therefore \text{ Yes}$$

$$\therefore \frac{-36}{60x-30}+\frac{-24}{60x-20}=0$$

$$\therefore \frac{1}{20x-10}+\frac{1}{30x-10}=0$$

$\therefore$ By basic rule or by cross-multiplication or by *Śūnyaṃ* formula, $50x-20=0 \therefore x=\frac{2}{5}$

*or* by Multiple simultaneous merger

$$60x=\frac{(-40)(-18)(-15)+(-270)(-8)(-5)}{(-2)(-18)(-5)+(-9)(-8)(-5)}=24$$

$$\therefore x=\frac{2}{5}$$

*Note*: Again any change of sequence of the terms on the L.H.S. will cause no change in the working or the result.

# 14

# Complex Mergers

There is still another type—a special and complex type of equations which are usually dubbed 'harder' but which can be readily tackled with the aid of the *Parāvartya Sūtra*. For instance:

$$\frac{10}{2x+1}+\frac{3}{3x-2}=\frac{2}{2x-3}+\frac{15}{3x+2}$$

Note the tests: (1) $\frac{10}{2}+\frac{3}{3}=\frac{2}{2}+\frac{15}{3}$; and

(2) $10 \times 3 = 2 \times 15$

i.e. 10 : 15 : : 2 : 3 (or 10 : 2 : : 15 : 3)

Transposing, $\frac{10}{2x+1}-\frac{15}{3x+2}=\frac{2}{2x-3}-\frac{3}{3x-2}$; and taking the L.C.M.

$$\therefore \frac{30}{6x+3}-\frac{30}{6x+4}=\frac{6}{6x-9}-\frac{6}{6x-4}$$

Simple cross-multiplication leads us to the main test:

$$\therefore \frac{30}{(6x+3)(6x+4)}=\frac{30}{(6x-9)(6x-4)}$$

Here comes the third test, i.e. that the numerator of the final derived equation is the same on both sides—

$$\therefore (6x+3)(6x+4)=(6x-9)(6x-4)$$

$$\therefore 6x=\frac{36-12}{3+4+9+4}=\frac{24}{20}=\frac{6}{5} \quad \therefore x = 1/5$$

*Clue*—This gives us the necessary clue, namely, that, after putting up the L.C.M. coefficient for x in all the denominators, $(D_1)(D_2) = (D_3)(D_4)$. As the transposition, the L.C.M. etc., can be done *mentally*, this clue amounts to a solution of the equation *at sight*.

In these examples, we should transpose the 4 fractions in such a manner that, after the cross-multiplication etc., are over, all the four denominators of the final derived equation have the same L.C.M. coefficient for x and the numerator is the same on the L.H.S. and the R.H.S. of the same equation.

A few more illustrations will be found helpful:

(1) $\frac{2}{6x+1}+\frac{2}{2x-1}=\frac{9}{9x-5}+\frac{1}{3x+1}$

(*i*) By transposing etc., we have

$$\frac{6}{18x+3}-\frac{6}{18x+6}=\frac{18}{18x-10}-\frac{18}{18x-9}$$

Here the N on both sides of the final derived equation is 18.

$\therefore$ The *Sūtra* applies.

$\therefore$ $(18x+3)(18x+6)=(18x-10)(18x-9)$

$\therefore$ $18x=\frac{90-18}{3+6+10+9}=\frac{72}{28}$ $\therefore$ $x=\frac{4}{28}=\frac{1}{7}$

*Note*: In some cases details of which we need not now enter into but which will be dealt with later, the original fractions themselves after the transposition fulfil the conditions of the test. In such cases, we need not bother about the L.C.M. etc., but may straightaway transpose the terms and apply the '*Parāvartya*' formula. In fact, the case just now dealt with is of this type, as will be evident from the following:

(*ii*) $\frac{2}{6x+1}-\frac{1}{3x+1}=\frac{9}{9x-5}-\frac{2}{2x-1}$

Here $\frac{2}{6}=\frac{1}{3}$; $\frac{9}{9}=\frac{2}{2}$; and the numerator on both sides of the final derived equation is 1.

$\therefore$ The *Sūtra* applies and can be applied immediately without bothering about the L.C.M. etc.

$\therefore$ $(6x+1)(3x+1)=(9x-5)(2x-1)$

$\therefore$ $18x^2+9x+1=18x^2-19x+5$ $\therefore$ $28x=4$ $\therefore$ $x=\frac{1}{7}$

(2) $\frac{2}{2x+3}+\frac{3}{3x+2}=\frac{1}{x+1}+\frac{6}{6x+7}$

$\therefore$ $\frac{2}{2x+3}-\frac{1}{x+1}=\frac{6}{6x+7}-\frac{3}{3x+2}$

(*i*) By L.C.M. method, $(6x+9)(6x+6)=(6x+7)(6x+4)$

$\therefore$ $6x=\frac{-26}{4}$ $\therefore$ $x=\frac{-13}{12}$

(*ii*) In this case, there is another peculiarity, i.e. that the transposition may be done in the other way too and yet the conditions are satisfied. So, we have:

$(6x+9)(6x+7)=(6x+6)(6x+4)$

$\therefore\ 6x=\frac{-39}{6}\quad \therefore\ x=\frac{-13}{12}$

(*iii*) And even, by cross-multiplication at the very outset, we get $12x+13=0$ by *Sāmya Samuccaye*. $\therefore x=-\frac{13}{12}$. In such cases, sequence in transposition does not matter! (This will be explained later.)

(3) $\frac{51}{3x+5}-\frac{68}{4x+11}=\frac{52}{4x-15}-\frac{39}{3x-7}$

Tests: $\frac{51}{3}$ and $\frac{68}{4}$ are both 17; and $\frac{52}{4}$ and $\frac{39}{3}$ are both 13.

This equation can be solved in several ways all of them very simple and easy.

(*i*) By the L.C.M. process:

$$\frac{204}{12x+20}-\frac{204}{12x+33}=\frac{156}{12x-45}-\frac{156}{12x-28}$$

$\therefore$ In the derived equation in its final form,

$N_1=204\times 13=12\times 13\times 17;$

and $N_2=156\times 17=12\times 13\times 17$

$\therefore$ The *Sūtra* applies.

$\therefore\ (12x+20)(12x+33)=(12x-45)(12x-28)$

$\therefore\ 12x=\frac{28\times 45-20\times 33}{20+33+45+28}=\frac{600}{126}\quad \therefore\ x=\frac{25}{63}$

(*ii*) or, removing the common factor 12:

$$\therefore \frac{17}{12x+20}-\frac{17}{12x+33}=\frac{13}{12x-45}-\frac{13}{12x-28}$$

$\therefore$ In the (final) derived equation,

$N_1=17\times 13;$ and $N_2=13\times 17\ \therefore$ The *Sūtra* applies—

$\therefore D_1\times D_2=D_3\times D_4\ \therefore 12x=\frac{600}{126}\ \therefore x=\frac{25}{63}$

(*iii*) or, at the very outset, i.e. without L.C.M. etc.:

$$\frac{51}{3x+5}-\frac{68}{4x+11}=\frac{52}{4x-15}-\frac{39}{3x-7}$$

$\therefore$ L.H.S. $N=561-340=221$; and

R.H.S. $N=-364+585=221$

$\therefore$ The *Sūtra* applies straightaway.

$\therefore (3x+5)(4x+11) = (4x-15)(3x-7)$

$\therefore 12x^2 + 53x + 55 = 12x^2 - 73x + 105$

$\therefore 126x = 50 \therefore x = \frac{25}{63}$

*Note:* In the second method, note that

$N_1 = N_2 = D_4 - D_3$ and $N_3 = N_4 = D_2 - D_1$ tests.

The general formula applicable in such cases is:

$$\frac{m-n}{x+p} + \frac{p-q}{x+n} = \frac{m-n}{x+q} + \frac{p-q}{x+m}$$

$$\therefore (m-n)\left(\frac{1}{x+p} - \frac{1}{x+q}\right) = (p-q)\left(\frac{1}{x+m} - \frac{1}{x+n}\right)$$

$$\therefore \frac{(m-n)(q-p)}{(x+p)(x+q)} = \frac{(p-q)(n-m)}{(x+m)(x+n)}$$

As the numerators are the same.

$\therefore$ The *Sūtra* applies

$\therefore (x+p)(x+q) = (x+m)(x+n)$

$$\therefore x = \frac{mn - pq}{p+q-m-n}$$

(4) $\frac{1}{2x-1} + \frac{8}{4x-1} = \frac{6}{3x-1} + \frac{3}{6x-1}$

(*i*) $\therefore \frac{6}{12x-6} - \frac{6}{12x-2} = \frac{24}{12x-4} - \frac{24}{12x-3}$

$\therefore$ In the final derived equation,

L.H.S. $N = 24$; and R.H.S. N is also 24

$\therefore$ The *Sūtra* applies.

$\therefore 12x = \frac{12-12}{D} = 0 \therefore x = 0$

(*ii*) '*Vilokanam*', i.e. mere observation too will suffice in this case.

(5) $\frac{3}{3x+1} + \frac{2}{2x-1} = \frac{3}{3x-2} + \frac{2}{2x+1}$

(*i*) Here the resultant N is the same 1 on both sides.

$\therefore$ Yes

$\therefore 6x^2 + 5x + 1 = 6x^2 - 7x + 2 \therefore 12x = 1 \therefore x = \frac{1}{12}$

(*ii*) or, by cross-multiplication at the very outset and *Śūnyaṃ*

*Samuccaye*, $12x - 1 = 0 \quad \therefore x = \frac{1}{12}$

(6) $\frac{5}{5x+2} + \frac{3}{3x+1} = \frac{5}{5x+3} + \frac{15}{15x+2}$

(*i*) $\therefore \frac{15}{15x+6} - \frac{15}{15x+9} = \frac{15}{15x+2} - \frac{15}{15x+5}$

$\therefore$ The resultant numerator on both sides is 45.

$\therefore$ The *Sūtra* applies.

$\therefore 15x = \frac{10-54}{8} = \frac{-44}{8} \quad \therefore x = -\frac{11}{30}$

(*ii*) or, by cross-multiplication at the very outset and *Śūnyaṃ* etc., formula, we get $30x + 11$ and $150x + 55$ on the L.H.S. and the R.H.S. respectively; and the numerical factor 5 being removed, both give us $30x + 11 + 0 \therefore x = -\frac{11}{30}$

(7) $\frac{2x+11}{x+5} + \frac{6x+11}{2x+3} = \frac{4x+4}{2x+1} + \frac{3x+19}{x+6}$

(*i*) $\therefore$ By *Parāvartya* division:

$\frac{1}{x+5} + \frac{2}{2x+3} = \frac{2}{2x+1} + \frac{1}{x+6}$

$\therefore \frac{2}{2x+10} - \frac{2}{2x+12} = \frac{2}{2x+1} - \frac{2}{2x+3}$

$\therefore$ 4 is the N on both sides of the derived equation.

$\therefore$ The *Sūtra* applies.

$\therefore (2x+10)(2x+12) = (2x+1)(2x+3)$

$\therefore 2x = \frac{3-120}{18} = \frac{-117}{18} \quad \therefore x = -\frac{13}{4}$

(*ii*) or by cross-multiplication at the very outset and *Śūnyaṃ Sūtra*, we have

$4x + 13 = 0 \therefore x = -\frac{13}{4}$

(8) $\frac{2x+11}{x+5} + \frac{15x-47}{3x-10} = \frac{9x-9}{3x-4} + \frac{4x+13}{x+3}$

(*i*) $\therefore \frac{3}{3x+15} + \frac{3}{3x-10} = \frac{3}{3x-4} + \frac{3}{3x+9}$

$\therefore \frac{3}{3x+15} - \frac{3}{3x+9} = \frac{3}{3x-4} - \frac{3}{3x-10}$

In the resultant equation,

$\therefore -18$ is the numerator on both sides.

$\therefore$ The *Sūtra* applies.

$\therefore$ $(3x + 15)(3x + 9) = (3x - 4)(3x - 10)$

$\therefore$ $3x = \frac{40 - 135}{38} = \frac{-5}{2}$ $\therefore$ $x = -\frac{5}{6}$

(*ii*) or by cross-multiplication at the very outset and *Śūnyaṃ* formula,

$\therefore$ $18x + 15 = 0$ $\therefore$ $x = -\frac{5}{6}$

(9) $\frac{12x^2 + 19x + 7}{3x + 4} + \frac{12x^2 + x + 3}{4x - 1} = \frac{24x^2 + 14x + 13}{12x + 1} + \frac{5x^2 + 6x + 2}{x + 1}$

$\therefore$ By *Parāvartya* division twice over.

$$\frac{3}{3x + 4} + \frac{4}{4x - 1} = \frac{12}{12x + 1} + \frac{1}{x + 1}$$

$$\frac{12}{12x + 16} + \frac{12}{12x - 3} = \frac{12}{12x + 1} + \frac{12}{12x + 12}$$

$\therefore$ By *Śūnyaṃ Sūtra*, we immediately obtain:

$24x + 13 = 0$ $\therefore$ $x = -\frac{13}{24}$

*Note*: The cross-multiplication and '*Śūnyaṃ*' method is so simple, easy and straight here that there is no need to try any other process at all. The student may, however, for the sake of practice try the other methods also and get further *verification* therefrom for the correctness of the answer just hereinabove arrived at.

# 15

# Simultaneous Simple Equations

Here too, we have the general formula applicable to all cases under the '*Parāvartya*' *Sūtra* and also the special *Sūtras* applicable only to special types of cases.

### The General Formula

The current system has a fairly satisfactory method—known as the cross-multiplication method—for solving of simultaneous simple equations, which is somewhat akin to the Vedic '*Parāvartya*' method and comes very near thereto.

But even here, the unfortunate drawback still remains that, in spite of all the arrow-directions etc., intended to facilitate its use, the students and sometimes even the teachers of Mathematics often get confused as regards the *plus* and the *minus* (+ and –) and how exactly they should be used; and consequently, we find most of them preferring—in actual daily practice—the substitution method or the elimination method by which they frame new equations involving only x or only y. And this, of course, does not permit a one-line mental-method answer; and it entails the expenditure of more time and toil.

The Vedic method by the *Parāvartya* rule enables us to give the answer immediately by mere mental Arithmetic. Thus,

$$\left.\begin{array}{l} 2x + 3y = 8 \\ 4x + 5y = 14 \end{array}\right\}$$

The rule followed is the "Cyclic" one.

(*i*) For the value of x, we start with the y-coefficients and the independent terms and cross-multiply forward, i.e. rightward, i.e. we start from the upper row and multiply across by the lower one, and conversely; and the connecting link between the two cross-products is always a *minus*. And this gives us our numerator;

(*ii*) For finding the denominator, we go from the upper row across to the lower one, i.e. the x-coefficient but backward, i.e. leftward. Thus,

$\left.\begin{array}{l}2x + 3y = 8\\ 4x + 5y = 14\end{array}\right\}$ $\therefore$ for the value of x, the numerator is $3 \times 14 - 5 \times 8 = 2$; and the denominator is $3 \times 4 - 2 \times 5 = 2$

In other words $x = \frac{2}{2} = 1$.

And, as for the value of y, we follow the cyclic system, i.e. start with the independent term on the upper row towards the x-coefficient on the lower row. So, our numerator is:

$$8 \times 4 - 14 \times 2 = 32 - 28 = 4$$

And note that the denominator is invariably the same as before (for x) and thus we avoid the confusion caused in the current system by another set of multiplications, a change of sign etc. In other words, $y = \frac{4}{2} = 2$

(2) $\left.\begin{array}{l}x - y = 7\\ 5x + 2y = 42\end{array}\right\}$ $\therefore x = \frac{-42-14}{-5-2} = \frac{-56}{-7} = 8;$ and $y = \frac{35-42}{-7} = \frac{-7}{-7} = 1$

(3) $\left.\begin{array}{l}2x + y = 5\\ 3x - 4y = 2\end{array}\right\}$ $\therefore x = \frac{2+20}{3+8} = \frac{22}{11} = 2;$ and $y = \frac{15-4}{11} = \frac{11}{11} = 1$

(4) $\left.\begin{array}{l}5x - 3y = 11\\ 6x - 5y = 9\end{array}\right\}$ $\therefore x = \frac{-27+55}{-18+25} = \frac{28}{7} = 4;$ and $y = \frac{66-45}{7} = \frac{21}{7} = 3;$

(5) $\left.\begin{array}{l}11x + 6y = 28\\ 7x - 4y = 10\end{array}\right\}$ $\therefore x = \frac{60+112}{42+44} = \frac{172}{86} = 2;$ and $y = \frac{196-110}{86} = \frac{86}{86} = 1$

## A Special Type

There is a special type of simultaneous simple equations which may involve big numbers and may therefore seem "hard" but which, owing to a certain ratio between the coefficients, can be readily, i.e. mentally solved with the aid of the *Sūtra* आनुरूप्ये शून्यं अन्यत् (*Śūnyaṃ Anyat*) which cryptically says : If one is in ratio, the other one is zero.

An example will make the meaning and the application clear:

$$\left.\begin{array}{r} 6x + 7y = 8 \\ 19x + 14y = 16 \end{array}\right\}$$

Here we note that the y-coefficients are in the same ratio to each other as the independent terms are to each other. And the *Sūtra* says that, in such a case, the other one, namely, x=0. This gives us two simple equations in y, which gives us the same value $\frac{8}{7}$ for y. Thus $x = 0; y = \frac{8}{7}$

*N.B.*: Look for the ratio of the coefficients of *one* of the unknown quantities being the same as that of the independent terms on the R.H.S.; and if the four are in proportion, put the other unknown quantity down as zero; and equate the first unknown quantity to the absolute term on the right.

*The Algebraical Proof* is this:

$$\left.\begin{array}{l} ax + by = bm \\ cx + dy = dm \end{array}\right\}$$

$$\therefore \left.\begin{array}{r} adx + bdy = bdm \\ bcx + bdy = bdm \end{array}\right\}$$

$$\therefore x(ad - bc) = 0 \therefore \left.\begin{array}{r} x = 0 \\ \text{and } y = m \end{array}\right\}$$

A few more illustrations may be taken:

(1) $\left.\begin{array}{r} 12x + 8y = 7 \\ 16x + 16y = 14 \end{array}\right\}$ Here, $\because 8:16::7:14$ mentally $\therefore \left.\begin{array}{r} x = 0 \\ \text{and } y = \frac{7}{8} \end{array}\right\}$

(2) $\left.\begin{array}{r} 12x + 78y = 12 \\ 16x + 16y = 16 \end{array}\right\}$ Here $\because 12:16::12:16$ mentally $\therefore \left.\begin{array}{r} x = 1 \\ \text{and } y = 0 \end{array}\right\}$

(3) $\left.\begin{array}{r} 499x + 172y = 212 \\ 9779x + 387y = 477 \end{array}\right\}$

Here $\left.\begin{array}{r} 172 = 4 \times 43 \text{ and } 387 = 9 \times 43 \\ \text{and } 212 = 4 \times 53 \text{ and } 477 = 9 \times 53 \end{array}\right\} \therefore$ The ratio is the same.

$$\left.\begin{array}{r} \therefore x = 0 \\ \text{and } y = \frac{53}{43} \end{array}\right\}$$

*Note*: The big coefficients of x need not frighten us !

*N.B.*: This rule is also capable of infinite extension and may be extended to any number of unknown quantities.

Thus,

(1) $\left.\begin{aligned} ax + by + cz &= a \\ bx + cy + az &= b \\ cx + ay + bz &= c \end{aligned}\right\}$ $\quad \therefore \left.\begin{aligned} x &= 1 \\ y &= 0 \\ \text{and } z &= 0 \end{aligned}\right\}$

(2) $\left.\begin{aligned} ax + by + cz &= cm \\ ax + ay + fz &= fm \\ mx + py + qz &= qm \end{aligned}\right\}$ $\quad \therefore \left.\begin{aligned} x &= 0 \\ y &= 0 \\ \text{and } z &= m \end{aligned}\right\}$

(3) $\left.\begin{aligned} 97x + ay + 43z &= am \\ 49979x + by + (p + q)z &= bm \\ 49x(a - d)^3 + cy + (m - n)^8 z &= cm \end{aligned}\right\}$ $\quad \therefore \left.\begin{aligned} x &= 0 \\ y &= m \\ \text{and } z &= 0 \end{aligned}\right\}$

*N.B.*: The coefficients have been deliberately made big and complex but need not frighten us.

## A Second Special Type

There is another special type of simultaneous linear equations where the x-coefficients and the y-coefficients are found interchanged. No elaborate multiplications etc., are needed here. The axiomatic *Upasūtra* संकलन-व्यवकलनाभ्यां ('*Saṅkalana- Vyavakalanābhyām*') which means "by addition and by subtraction" gives us immediately two equations giving the values of $(x + y)$ and $(x - y)$. And a repetition of the same process gives us the values of x and y! And the whole work can be done mentally. Thus,

$$\left.\begin{aligned} 45x - 23y &= 113 \\ 23x - 45y &= 91 \end{aligned}\right\}$$

$\therefore$ By addition, $68x - 68y = 68(x - y) = 204 \therefore x - y = 3$
And by subtraction, $22x + 22y = 22(x + y) = 22 \therefore x + y = 1$

$$\left.\begin{aligned} \therefore x &= 2 \\ \text{and } y &= -1 \end{aligned}\right\}$$

*Note*: However big and complex the coefficients may be, there is no multiplication involved but only simple addition and simple subtraction.

The other special types of simultaneously linear equations will be discussed at a later stage.

# 16

# Miscellaneous (Simple) Equations

There are other types of miscellaneous linear equations which can be treated by the Vedic *Sūtras*. A few of them are shown below.

### First Type

Fractions of a particular cyclical kind are involved here. And, by the *Parāvartya Sūtra*; we write down the numerator of the sum-total of all the fractions in question and equate it to zero. Thus,

(1) $\frac{1}{(x-1)(x-2)}+\frac{2}{(x-2)(x-3)}+\frac{3}{(x-3)(x-1)}=0$

Here, each numerator is to be multiplied by the factor absent from its denominator. This is usually and actually done everywhere but not as a rule of *mental* practice. This, however, should be regularly practised; and the resultant numerator equated to zero.

In the present instance,

$(x-3)+(2x-2)+(3x-6)=6x-11=0 \therefore x=\frac{11}{6}$

*The Algebraical proof* is well-known and is as follows:

$$\frac{p}{(x+a)(x+b)}+\frac{q}{(x+b)(x+c)}+\frac{r}{(x+c)(x+a)}$$

$$=\frac{p(x+c)+q(x+a)+r(x+b)}{(x+a)(x+b)(x+c)}$$

$$=\frac{x(p+q+r)+(pc+qa+rb)}{(x+a)(x+b)(x+c)}=0$$

$$\therefore x=\frac{-(pc+qa+rb)}{p+q+r}$$

In other words,

$$x = \frac{\text{Each N multiplied by the absent number with sign reversed}}{N_1 + N_2 + N_3}$$

As this is simple and easy to remember and to apply, the work can be done *mentally*. And we can say, $x = \frac{11}{6}$.

A few more examples are noted:

(1) $$\frac{1}{(x-1)(x-3)} + \frac{3}{(x-3)(x-5)} + \frac{5}{(x-5)(x-1)} = 0$$

$$\therefore x = \frac{5+3+15}{1+3+5} = \frac{23}{9}$$

(2) $$\frac{2}{(x-1)(x+2)} + \frac{3}{(x+2)(x-4)} + \frac{4}{(x-4)(x-1)} = 0$$

$$\therefore x = \frac{8+3-8}{2+3+4} = \frac{3}{9} = \frac{1}{3}$$

(3) $$\frac{1}{(x-3)(x-4)} + \frac{3}{(x-4)(x-9)} + \frac{5}{(x-9)(x-3)} = 0$$

$$\therefore x = \frac{9+9+20}{1+3+5} = \frac{38}{9}$$

A few disguised samples may also be taken:

(1) $$\frac{1}{x^2+3x+2} + \frac{5}{x^2+5x+6} + \frac{3}{x^2+4x+3} = 0$$

$\therefore$ mentally

$$\frac{1}{(x+1)(x+2)} + \frac{5}{(x+2)(x+3)} + \frac{3}{(x+3)(x+1)} = 0$$

$$\therefore x = \frac{-3-5-6}{1+5+3} = \frac{-14}{9}$$

(2) $$\frac{1}{6x^2+5x+1} + \frac{1}{12x^2+7x+1} + \frac{1}{8x^2+6x+1} = 0$$

$\therefore$ mentally

$$\frac{1}{(2x+1)(3x+1)} + \frac{1}{(3x+1)(4x+1)} + \frac{1}{(4x+1)(2x+1)} = 0$$

$$\therefore 9x + 3 = 0 \therefore x = -\frac{1}{3}$$

(3) $$\frac{3}{(x+3)^2 - 2^2} + \frac{2}{(x+4)^2 - 1^2} + \frac{1}{(x+2)^2 - 1^2} = 0$$

$\therefore$ mentally $\dfrac{3}{(x+1)(x+5)}+\dfrac{2}{(x+5)(x+3)}+\dfrac{1}{(x+3)(x+1)}=0$

$\therefore x=\dfrac{-9-2-5}{3+2+1}=\dfrac{-16}{6}=\dfrac{-8}{3}$

(4) $\dfrac{x+4}{(x+1)(x+3)}+\dfrac{x+8}{(x+3)(x+5)}+\dfrac{x+6}{(x+5)(x+1)}=\dfrac{3}{x}$

$\therefore$ mentally

$\dfrac{x^2+4x}{(x+1)(x+3)}-1+\dfrac{x^2+8x}{(x+3)(x+5)}-1+\dfrac{x^2+6x}{(x+5)(x+1)}-1=0$

$\therefore$ mentally $\dfrac{-3}{(x+1)(x+3)}+\dfrac{-15}{(x+3)(x+5)}+\dfrac{-5}{(x+5)(x+1)}=0$

$\therefore x=\dfrac{15+15+15}{-3-15-5}=\dfrac{-45}{23}$

(5) $\dfrac{x-3}{(x-1)(x-2)}+\dfrac{x-5}{(x-2)(x-3)}+\dfrac{x-4}{(x-3)(x-1)}=\dfrac{3}{x}$

$\therefore$ mentally $\dfrac{-2}{(x-1)(x-2)}+\dfrac{-6}{(x-2)(x-3)}+\dfrac{-3}{(x-3)(x-1)}=0$

$\therefore x=\dfrac{-6-6-6}{-2-6-3}=\dfrac{18}{11}$

(6) $\dfrac{x-4}{(x-1)(x-3)}+\dfrac{x-9}{(x-3)(x-6)}+\dfrac{x-7}{(x-6)(x-1)}=\dfrac{3}{x}$

$\therefore$ mentally $\dfrac{-3}{(x-1)(x-3)}+\dfrac{-18}{(x-3)(x-6)}+\dfrac{-6}{(x-6)(x-1)}=0$

$\therefore x=\dfrac{-18-18-18}{-3-18-6}=\dfrac{54}{27}=2$

(7) $\dfrac{x-6}{(x-2)(x-3)}+\dfrac{x-8}{(x-3)(x-4)}+\dfrac{x-7}{(x-4)(x-2)}=\dfrac{3}{x+1}$

$\therefore$ mentally $\dfrac{-12}{(x-2)(x-3)}-\dfrac{-20}{(x-3)(x-4)}-\dfrac{-15}{(x-4)(x-2)}=0$

$\therefore x=\dfrac{-48-40-45}{-12-20-15}=\dfrac{133}{47}$

(8) $\dfrac{35x+23}{(5x-1)(7x-1)}+\dfrac{63x+47}{(7x-1)(9x-1)}+\dfrac{45x+31}{(9x-1)(5x-1)}=\dfrac{3}{x-1}$

$$\therefore \left\{\frac{35x^2 - 12x - 23}{(5x-1)(7x-1)} - 1\right\} + \left\{\frac{63x^2 - 16x - 47}{(7x-1)(9x-1)} - 1\right\} + \left\{\frac{45x^2 - 14x - 31}{(9x-1)(5x-1)} - 1\right\} = 0$$

$$\therefore \frac{3}{(5x-1)(7x-1)} + \frac{6}{(7x-1)(9x-1)} + \frac{4}{(9x-1)(5x-1)} = 0$$

$\therefore 85x - 13 = 0 \; \therefore x = \frac{13}{85}$

### Second Type

A second type of such special simple equations is one where we have $\frac{1}{AB} + \frac{1}{AC} = \frac{1}{AD} + \frac{1}{BC}$ and the factors (A, B, C and D) of the denominators are in Arithmetical Progression. The *Sūtra* सोपान्त्यद्वयमन्त्यं (*Sopāntyadvayamantyaṁ*) which means "the ultimate and twice the penultimate" gives us the answer immediately, for instance:

$$\frac{1}{(x+2)(x+3)} + \frac{1}{(x+2)(x+4)} = \frac{1}{(x+2)(x+5)} + \frac{1}{(x+3)(x+4)}$$

Here, according to the *Sūtra*, L+2p (the last+twice the penultimate) $= (x+5) + 2(x+4) = 3x + 13 = 0 \; \therefore x = -4\frac{1}{3}$

The proof of this is as follows:

$$\frac{1}{(x+2)(x+3)} + \frac{1}{(x+2)(x+4)} = \frac{1}{(x+2)(x+5)} + \frac{1}{(x+3)(x+4)}$$

$$\therefore \frac{1}{(x+2)(x+3)} - \frac{1}{(x+2)(x+5)} = \frac{1}{(x+3)(x+4)} - \frac{1}{(x+2)(x+4)}$$

$$\therefore \frac{1}{x+2}\left\{\frac{2}{(x+3)(x+5)}\right\} = \frac{1}{x+4}\left\{\frac{-1}{(x+2)(x+3)}\right\}$$

Removing the factors (x+2) and (x+3)

$\frac{2}{x+5} = \frac{-1}{x+4}$, i.e. $\frac{2}{L} = \frac{-1}{P} \therefore L + 2P = 0$

The General Algebraical Proof is as follows:

$\frac{1}{AB} + \frac{1}{AC} = \frac{1}{AD} + \frac{1}{BC}$ (where A, B, C and D are in A.P.)

Let d be the common difference.

$$\therefore \frac{1}{A(A+d)} + \frac{1}{A(A+2d)} = \frac{1}{A(A+3d)} + \frac{1}{(A+d)(A+2d)}$$

$$\therefore \frac{1}{A(A+d)} - \frac{1}{A(A+3d)} = \frac{1}{(A+d)(A+2d)} - \frac{1}{A(A+2d)}$$

$$\therefore \frac{1}{A}\left\{\frac{2d}{(A+d)(A+3d)}\right\} = \frac{1}{A+2d}\left\{\frac{-d}{A(A+d)}\right\}$$

Cancelling the factors A(A+d) of the denominators and d of the numerators:

$$\therefore \frac{2}{A+3d} = \frac{1}{A+2d}$$

In other words $\frac{2}{L} = \frac{-1}{P}$ $\therefore L + 2P = 0$

Another Algebraical proof :

$$\frac{1}{AB} + \frac{1}{AC} = \frac{1}{AD} + \frac{1}{BC}$$

$$\therefore \frac{1}{AB} - \frac{1}{AD} = \frac{1}{BC} - \frac{1}{AC}$$

$$\therefore \frac{1}{A}\left\{\frac{D-B}{BD}\right\} = \frac{1}{C}\left\{\frac{A-B}{AB}\right\}$$

But $\because$ A, B, C and D are in AP

$$\therefore D - B = -2(A - B)$$

$$\therefore \frac{2}{D} = \frac{-1}{C} \therefore 2C + D = 0, \text{ i.e. } 2P + L = 0$$

A few more samples may be tried:

(1) $$\frac{1}{x^2+7x+12} + \frac{1}{x^2+8x+15} = \frac{1}{x^2+9x+18} + \frac{1}{x^2+9x+20}$$

$\therefore$ mentally

$$\frac{1}{(x+3)(x+4)} + \frac{1}{(x+3)(x+5)} = \frac{1}{(x+3)(x+6)} + \frac{1}{(x+4)(x+5)}$$

$$\therefore 2P + L = (2x+10) + (x+6) = 0 \therefore x = -5\tfrac{1}{3}$$

(2) $$\frac{1}{(2x+1)(3x+2)} + \frac{1}{(2x+1)(4x+3)} = \frac{1}{(2x+1)(5x+4)} + \frac{1}{(3x+2)(4x+3)}$$

$$\therefore 2P + L = (8x+6) - (5x+4) = 13x + 10 = 0 \therefore x = -\tfrac{10}{13}$$

## Third Type

A third type of equations are those where numerator and denominator on the L.H.S. barring the independent terms stand in the same ratio to each other as the entire numerator and the entire denominator of the R.H.S. stand to each other and these can be readily solved with the aid of the *Upasūtra* अन्त्ययोरेव (*Antyayoreva*) which means, "only the last terms", i.e. the absolute terms. Thus,

$$\frac{x^2+x+1}{x^2+3x+3}=\frac{x+1}{x+3}$$

Here, $(x^2+x)=x(x+1)$ and $(x^2+3x)=x(x+3)$

∴ The rule applies, and we say:

$$\frac{x+1}{x+3}=\frac{1}{3}\therefore x=0$$

The Algebraical proof is as follows:

$$\frac{AC+D}{BC+E}=\frac{A}{B}=\frac{AC}{BC}=\frac{D}{E}\text{(by Dividend)}$$

Another Algebraic proof is this:

$$\frac{AC+D}{BC+E}=\frac{A}{B}\therefore ABC+AE=ABC+BD$$

$$\therefore AE=BD\therefore\frac{A}{B}=\frac{D}{E}$$

A few more examples may be taken:

(1) $\frac{3x^2+5x+8}{5x^2+6x+12}=\frac{3x+5}{5x+6}=\frac{8}{12}\therefore 4x=12\therefore x=3$

(2) $\frac{2-2x-3x^2}{2-5x-6x^2}=\frac{3x+2}{6x+5}=\frac{2}{2}\therefore x=-1$

(3) $\frac{81x^2+108x+2}{54x^2+27x+5}=\frac{3x+4}{2x+1}=\frac{2}{5}\therefore x=\frac{-18}{11}$

(4) $\frac{58x^2+87x+7}{87x^2+145x+11}=\frac{2x+3}{3x+5}=\frac{7}{11}\therefore x=2$

(5) $\frac{158x^2+237x+4}{395x^2+474x+4}=\frac{2x+3}{5x+6}=\frac{4}{4}\therefore x=-1$

(6) $\frac{1-px}{1-qx}=\frac{2+pqx-p^2qx^2}{2+pqx-pq^2x^2}=\frac{2}{2}\therefore x=0$

(7) $\frac{(2x+3)^2}{(2x+5)^2}=\frac{x+3}{x+5}\therefore\frac{4x^2+12x+9}{4x^2+20x+25}=\frac{x+3}{x+5}=\frac{9}{25}\therefore x-\frac{-15}{8}$

(8) $\frac{(x+1)(x+6)}{(x+3)(x+5)}=\frac{x+7}{x+8}$

$$\therefore \frac{x^2+7x+6}{x^2+8x+15} = \frac{x+7}{x+8} = \frac{6}{15} \therefore x = -6\tfrac{1}{3}$$

*Note*: By cross-multiplication,

$(x+1)(x+6)(x+8) = (x+3)(x+5)(x+7)$

Here, the total of the binomials is 3x+15 on each side. But the *Śūnyaṃ Samuccaye' Sūtra* does *not* apply because the number of factors in the original shape is 2 on the L.H.S. and only one on the R.H.S. '*Antyayoreva*' is the *Sūtra* to be applied.

(9) $(x+1)(x+2)(x+9) = (x+3)(x+4)(x+5)$

The total on each side is the same, i.e. 3x+12. But the '*Śūnyaṃ Samuccaye*' *Sūtra* does *not* apply. The '*Antyayoreva*' formula is the one to be applied.

$$\frac{(x+1)(x+2)}{(x+4)(x+5)} = \frac{x+3}{x+9} = \frac{2}{20} \therefore x = \frac{-7}{3}$$

(10) $(x+2)(x+3)(x+11) = (x+4)(x+5)(x+7)$

The case is exactly like the one above.

$$\therefore \frac{(x+2)(x+3)}{(x+4)(x+7)} = \frac{x+5}{x+11} = \frac{6}{28} \therefore x = \frac{-74}{22} = -\frac{37}{11}$$

## Fourth Type

Another type of special fraction-additions in connection with simple equations is often met with, wherein the factors of the denominators are in Arithmetical Progression or related to one another in a special manner as in summation of series. These we can readily solve with the aid of the same "*Antyayoreva*" *Sūtra* but in a different context, and in a different sense. We therefore deal with this special type here.

(1) The first sub-section of this type is one in which the factors are in A.P. Thus,

$$\frac{1}{(x+1)(x+2)} + \frac{1}{(x+2)(x+3)} + \frac{1}{(x+3)(x+4)}$$

The *Sūtra* tells us that the sum of this series is a fraction whose numerator is the sum of the numerators in the series and whose denominator is the product of the two ends, i.e. the first and the last binomials.

So, in this case, $S_3 = \dfrac{3}{(x+1)(x+4)}$ and so on.

The Algebraical proof of this is as follows:

$$t_1 + t_2 = \frac{1}{(x+1)(x+2)} + \frac{1}{(x+2)(x+3)} = \frac{x+3+x+1}{(x+1)(x+2)(x+3)}$$

$$= \frac{2(x+2)}{(x+1)(x+2)(x+3)} = \frac{2}{(x+1)(x+3)}$$

wherein the numerator is the sum of the original numerators and the denominator is the product of the first and the last binomial factors.

Adding $t_3$ to the above, we have $\frac{2}{(x+1)(x+3)} + \frac{1}{(x+3)(x+4)}$

$$= \frac{2x+8+x+1}{(x+1)(x+3)(x+4)} = \frac{3(x+3)}{(x+1)(x+3)(x+4)} = \frac{3}{(x+1)(x+4)}$$

Continuing this process to any number of terms, we find that the numerator continuously increases by one and the denominator invariably drops the middle binomial and retains only the first and the last, thus proving the correctness of the rule in question.

$$t_1 = \frac{1}{(x+1)(x+2)} = \frac{1}{(x+1)} - \frac{1}{x+2}$$

$$t_2 = \frac{1}{(x+2)(x+3)} = \frac{1}{x+2} - \frac{1}{x+3}$$

and so on to any number of terms.

*Note*: The second term of each step on the R.H.S. and the first term on the next step of the L.H.S. cancel each other and that, consequently, whatever may be the number of terms we take, all the terms on the R.H.S. except the very first and the last cancel out and the numerator being the difference between the first and the last binomial, i.e. the only binomials surviving is the sum of the original numerators on the L.H.S. And this proves the proposition in question.

A few more illustrations are taken.

(1) $\frac{1}{(x+3)(x+4)} + \frac{1}{(x+4)(x+5)} + \ldots$

$$\therefore S_4 = \frac{4}{(x+3)(x+7)}$$

(2) $\frac{1}{x^2-3x+2} + \frac{1}{x^2-5x+6} + \ldots$

$$= \frac{1}{(x-1)(x-2)} + \frac{1}{(x-2)(x-3)} + \ldots$$

$$\therefore S_5 = \frac{5}{(x-1)(x-6)}$$

(3) $\frac{1}{x^2+5x+4}+\frac{1}{x^2+11x+28}+\ldots$

$$=\frac{1}{(x+1)(x+4)}+\frac{1}{(x+4)(x+7)}+\ldots$$

$$\therefore S_4=\frac{4}{x^2+14x+13}$$

(4) $\frac{1.}{(x+a)(x+2a)}+\frac{1}{(x+2a)(x+3a)}+\ldots$

$$\therefore S_4=\frac{4}{(x+a)(x+5a)}$$

(5) $\frac{1}{(x+1)(3x+1)}+\frac{1}{(3x+1)(5x+1)}+\frac{1}{(5x+1)(7x+1)}+\ldots$

Here, there is a slight difference in the *structure* of the denominator, i.e. the A.P. is not in respect of the independent term in the binomials as in the previous examples but in the x-coefficient itself. But this makes no difference as regards the applicability of the *Sūtra.*

$$\therefore S_3=\frac{3}{(x+1)(7x+1)}$$

The first Algebraical proof of this is exactly as before:

$$t_1+t_2=\frac{2}{(x+1)(5x+1)};t_1+t_2+t_3=\frac{3}{(x+1)(7x+1)}\text{ and so on.}$$

The addition of each new term automatically establishes the proposition.

The second Algebraical proof is slightly different but follows the same lines and leads to the same result:

$$t_1=\frac{1}{(x+1)(3x+1)}=\frac{1}{2x}\left(\frac{1}{x+1}-\frac{1}{3x+1}\right)$$

$$t_2=\frac{1}{(3x+1)(5x+1)}=\frac{1}{2x}\left(\frac{1}{3x+1}-\frac{1}{5x+1}\right)\text{ and so on.}$$

*Note*: The cancellations take place exactly as before, with the consequence that the sum-total of the fractions

$$=\frac{1}{2x}\frac{(2cx)}{D_1\times D_c}=\frac{c}{D_1\times D_c}$$

which proves the proposition.

(6) $\frac{1}{(x+a)(2x+3a)}+\frac{1}{(2x+3a)(3x+5a)}+\ldots$

Here, the progression is with regard to both the items in the binomials, i.e. the x-coefficients and the absolute terms. But this too makes no difference to the applicability of the formula under discussion,

$$\therefore S_3 = \frac{3}{(x+a)(4x+7a)}$$

(7) $\frac{1}{(3x+a)(5x+a)} + \frac{1}{(5x+a)(7x+a)} + \ldots$

$$\therefore S_3 = \frac{3}{(3x+a)(9x+a)}$$

(8) $\frac{1}{(x^2+x+1)(x^2+2x+2)} + \frac{1}{(x^2+2x+2)(x^2+3x+3)} + \ldots$

Seemingly, there is a still greater difference in the *structure* of the denominators. But even this makes no difference to the applicability of the aphorism. So we say:

$$S_4 = \frac{4}{(x^2+x+1)(x^2+5x+5)}$$

Both the Algebraical explanations apply to this case also. And we may extend the rule indefinitely to as many terms and to as many varieties as we may find necessary.

We may conclude this sub-section with a few examples of its application to Arithmetical numbers:

(1) $\frac{1}{7\times 8} + \frac{1}{8\times 9} + \frac{1}{9\times 10} + \frac{1}{10\times 11} + \ldots$

In a sum like this, the finding of the L.C.M. and the multiplications, divisions, additions, cancellations etc., will be tiresome and disgusting. But our recognition of this series as coming under its right particular classification enables us to say at once:

$S_4 = \frac{4}{7\times 11} = \frac{4}{77}$ and so on.

*Note*: The principle explained above is in constant requisition in connection with the "Summation of Series" in Higher Algebra etc., and therefore is of utmost importance to the mathematician and the statistician, in general.

## Fifth Type

There is also a fifth type of fraction-additions dealing with simple equations which we often come across, which are connected with the "Summation of Series" as in the previous type and which we may readily tackle with the aid of the same *Antyayoreva* formula.

The characteristic peculiarity here is that each numerator is the difference between the two binomial factors of its denominator. Thus,

(1) $\frac{a-b}{(x+a)(x+b)} + \frac{b-c}{(x+b)(x+c)} + \frac{c-d}{(x+c)(x+d)} + \ldots$

$$\therefore S_3 = \frac{a-d}{(x+a)(x+d)}$$

Both the Algebraical explanations hereof are exactly as before and need not be repeated here.

(2) $$\frac{x-y}{(a+x)(a+y)} + \frac{y-z}{(a+y)(a+z)} + \frac{z-w}{(a+z)(a+w)} + \ldots$$

$$\therefore S_3 = \frac{x-w}{(a+x)(a+w)}$$

(3) $$\frac{1}{(x+7)(x+8)} + \frac{2}{(x+8)(x+10)} + \frac{14}{(x+10)(x+24)}$$

$$\therefore S_3 = \frac{17}{(x+7)(x+24)}$$

(4) $$\frac{3}{(x+7)(x+10)} + \frac{9}{(x+10)(x+19)} + \frac{27}{(x+19)(x+46)} + \frac{99}{(x+46)(x+145)} + \ldots$$

$$\therefore S_4 = \frac{138}{(x+7)(x+145)}$$

(5) $$\frac{a-b}{(px+a)(px+b)} + \frac{b-c}{(px+b)(px+c)} + \frac{c-d}{(px+c)(px+d)}$$

$$\therefore S_3 = \frac{a-d}{(px+a)(px+d)}$$

*Note:* (*i*) If, instead of d, there be an a in the last term in this case, the numerator in the answer becomes zero; and consequently the L.H.S., i.e. the sum of the various fractions is zero.

(*ii*) The difference between the binomial factors of the denominator in the L.H.S. is the numerator of each fraction; and this characteristic will be found to characterise the R.H.S. also.

(*iii*) The note at the end of the previous sub-section regarding the summation of series holds good here too.

# 17

# Quadratic Equations

In the Vedic mathematic *Sūtras*, calculus comes in at a very early stage. As it so happens that differential calculus is made use of in the Vedic *Sūtras* for breaking a quadratic equation down at sight into two simple equations of the first degree and as we now go on to our study of the Vedic *Sūtras* bearing on quadratic equations, we shall begin this chapter with a brief exposition of the calculus.

Being based on basic and fundamental first principles relating to limiting values, they justifiably come into the picture at a very early stage. But these have been expounded and explained with enormous wealth of details covering not merely the *Sūtras* themselves but also the sub-*sūtras*, axioms, corollaries, implications etc. We do not propose to go into the arguments by which the calculus has been established but shall contend ourselves with an exposition of the rules enjoined therein and the actual *modus operandi.* The principal rules are briefly given below:

(*i*) In every quadratic expression put in its standard form, i.e. with 1 as the coefficient of $x^2$, the sum of its two binomial factors is its first differential.

Thus, as regards the quadratic expression $x^2-5x+6$, we know its binomial factors are $(x-2)$ and $(x-3)$. And therefore, we can at once say that $(2x-5)$ which is the sum of these two factors is its $D_1$, i.e. first differential.

(*ii*) This first differential of each term can also be obtained by multiplying its ध्वज•(*Dhvaja*) घात (*Ghāta*), i.e. the power by the अङ्क (*Aṅka,* i.e. its coefficient) and reducing it by one.

Thus, as regards $x^2-5x+6$

$x^2$ gives $2x$; $-5x$ gives $-5$; and 6 gives zero.

$\therefore D_1 = 2x - 5.$

(*iii*) Defining the discriminant as the square of the coefficient of the middle term *minus* the product of double the first coefficient and double the independent term, the text then lays down the very important proposition that the first differential is equal to the square root of the discriminant.

In the above case $x^2-5x+6=0$

$\therefore 2x-5 = \pm\sqrt{25-24} = \pm 1$

Thus the given quadratic equation is broken down at sight into the above two simple equations, i.e. $2x-5=1$ and $2x-5=-1$

$\therefore x = 2$ or $3$

The current modern method dealing with its standard quadratic equation $ax^2+bx+c=0$ tells us that,

$x = \dfrac{-b \pm \sqrt{b^2-4ac}}{2a}$. This is no doubt all right, so far as it goes; but it is still a very crude and clumsy way of stating that the first differential is the square root of the discriminant.

Another Indian method of medieval times well-known as Shree Shreedharacharya's method is a bit better than the current modern methods; but that too comes nowhere near the Vedic method which gives us (1) the relationship of the differential with the original quadratic as the sum of its factors and (2) its relationship with the discriminant as its square root ! and thirdly, breaks the original quadratic equation, at sight, into two simple equations which immediately give us the two values of x!

A few more illustrations are shown hereunder:

(1) $4x^2-4x+1=(2x-1)(2x-1)=0 \therefore 8x-4=0$

(2) $7x^2-5x-2=(x-1)(7x+2)=0 \therefore 14x-5=\pm\sqrt{81}=\pm 9$

(3) $x^2-11x+10=(x-10)(x-1)=0 \therefore 2x-11=\pm\sqrt{81}=\pm 9$

(4) $6x^2+5x-3=0 \therefore 12x+5=\pm\sqrt{97}$

(5) $7x^2-9x-1=0 \therefore 14x-9=\pm\sqrt{109}$

(6) $5x^2-7x-5=0 \therefore 10x-7=\pm\sqrt{149}$

(7) $9x^2-13x-2=0 \therefore 18x-13=\pm\sqrt{241}$

(8) $11x^2+7x+7=0 \therefore 22x+7=\pm\sqrt{-259}$

(9) $ax^2+bx+c=0 \therefore 2ax+b=\pm\sqrt{b^2-4ac}$

This portion of the Vedic *Sūtras* deals also with the binomial theorem, factorisation, factorials, repeated factors, continued fractions, differentiations, integrations, successive differentiations, integrations by means of continued fractions etc. But just now we are concerned only with the just hereinabove explained use of the differential calculus in the solution of quadratic equations in general because of the relationship $D_1 = \pm\sqrt{\text{the discriminant}}$. The other applications just referred to will be dealt with at later stages.

This calculus method is perfectly general, i.e. it applies to *all* cases of quadratic equations. There are, however, certain special types of quadratic equations which can be still more easily and still more rapidly solved with the help of the special *Sūtras* applicable to them. Some of these formulas are old friends but in a new garb and a new set-up, a new context and so on. And they are so efficient in the facilitating of mathematics work and in reducing the burden of the toil therein. We therefore go on to some of the most important amongst these special types.

### First Special Type
### (*Reciprocals*)

This deals with reciprocals. The equations have, under the current system, to be worked upon laboriously, before they can be solved. For example:

(1) $x+\frac{1}{x}=\frac{17}{4}$

According to the current system, we say:

$$\therefore \frac{x^2+1}{x}=\frac{17}{4}$$

$$\therefore 4x^2+4=17x$$

$$\therefore 4x^2-17x+4=0$$

$$\therefore (x-4)(4x-1)=0 \therefore x=4 \text{ or } 1/4$$

$$\text{or } \therefore x=\frac{17\pm\sqrt{289-64}}{8}=\frac{17\pm 15}{8}=4 \text{ or } 1/4$$

But, according to the *Vilokanam* sub-*Sūtra* of Vedic mathematics, we observe that the L.H.S. is the sum of two reciprocals, split the $\frac{17}{4}$ of the R.H.S. into $4+\frac{1}{4}$ and at once say,

$x+\frac{1}{x}=4+\frac{1}{4} \therefore x=4 \text{ or } 1/4.$ It is a matter of simple observation and no more.

(2) $x+\frac{1}{x}=\frac{26}{5}=5\frac{1}{5} \therefore x=5 \text{ or } \frac{1}{5}$

(3) $\frac{x}{x+1}+\frac{x+1}{x}=82/9=9\frac{1}{9} \therefore \frac{x}{x+1}=9 \text{ or } 1/9 \therefore x=\frac{1}{8} \text{ or } -\frac{9}{8}$

(4) $\frac{x+1}{x+2}+\frac{x+2}{x+1}=\frac{37}{6}=6\frac{1}{6} \therefore \frac{x+1}{x+2}=6 \text{ or } \frac{1}{6} \therefore x=-\frac{11}{5} \text{ or } -\frac{4}{5}$

(5) $\frac{x+4}{x-4}+\frac{x-4}{x+4}=\frac{10}{3}=3\frac{1}{3} \therefore \frac{x+4}{x-4}=3 \text{ or } \frac{1}{3} \therefore x=8 \text{ or } -8$

(6) $x+\frac{1}{x}=\frac{13}{6}$

Here the R.H.S. does not readily seem to be of the same sort as the previous examples. But a little observation will suffice to show that $\frac{13}{6}$ can be split up into $\frac{3}{2}+\frac{2}{3}$

$$\therefore x+\frac{1}{x}=\frac{3}{2}+\frac{2}{3}\therefore x=\frac{3}{2} \text{ or } \frac{2}{3}$$

(7) $x+\frac{1}{x}=\frac{25}{12}=\frac{4}{3}+\frac{3}{4}\therefore x=\frac{4}{3} \text{ or } \frac{3}{4}$

(8) $\frac{x+5}{x+6}+\frac{x+6}{x+5}=\frac{29}{10}=\frac{5}{2}+\frac{2}{5}\therefore \frac{x+5}{x+6}=\frac{5}{2} \text{ or } \frac{2}{5}\therefore x=-\frac{13}{3} \text{ or } -\frac{20}{3}$

(9) $\frac{x}{x+1}+\frac{x+1}{x}=\frac{169}{60}=\frac{12}{5}+\frac{5}{12}\therefore \frac{x}{x+1}=\frac{12}{5} \text{ or } \frac{5}{12}\therefore x=\frac{5}{7} \text{ or } -\frac{12}{7}$

(10) $\frac{2x+11}{2x-11}+\frac{2x-11}{2x+11}=\frac{193}{84}=\frac{12}{7}+\frac{7}{12}\therefore \frac{2x+11}{2x-11}=\frac{12}{7} \text{ or } \frac{7}{12}$

$$\therefore x=\frac{\pm 20}{10}$$

(11) $x-\frac{1}{x}=\frac{5}{6}$

Here the connecting symbol is a *minus*. Accordingly, we say:

$$x-\frac{1}{x}=\frac{3}{2}-\frac{2}{3}\therefore x=\frac{3}{2} \text{ or } -\frac{2}{3}$$

*N.B.* : Note the *minus* of the second root very carefully. For, the value $x=\frac{2}{3}$ will give us not $\frac{5}{6}$ but $-\frac{5}{6}$ on the R.H.S. and will therefore be wrong!

(12) $x-\frac{1}{x}=\frac{45}{14}=\frac{7}{2}-\frac{2}{7}\therefore x=\frac{7}{2} \text{ or } -\frac{2}{7}$

(13) $\frac{x}{x+3}-\frac{x+3}{x}=\frac{15}{56}=\frac{8}{7}-\frac{7}{8}\therefore \frac{x}{x+3}=\frac{8}{7} \text{ or } -\frac{7}{8}\therefore x=-24 \text{ or } -\frac{7}{5}$

(14) $\frac{x+7}{x+9}-\frac{x+9}{x+7}=\frac{32}{63}=\frac{9}{7}-\frac{7}{9}\therefore \frac{x+7}{x+9}=\frac{9}{7} \text{ or } -\frac{7}{9}\therefore x=-16 \text{ or } -\frac{63}{8}$

(15) $\frac{5x+9}{5x-9}-\frac{5x-9}{5x+9}=\frac{56}{45}=\frac{9}{5}-\frac{5}{9}\therefore \frac{5x+9}{5x-9}=\frac{9}{5} \text{ or } -\frac{5}{9}\therefore x=\frac{63}{10} \text{ or } -\frac{18}{35}$

*Note*: In the above examples, the L.H.S. was of the form $\frac{a}{b}\pm\frac{b}{a}$; and, consequently, we had to split the R.H.S. into the same form $\left(\frac{a}{b}\pm\frac{b}{a}\right)$

And this, when simplified, $= \frac{a^2 \pm b^2}{ab}$

In other words, the denominator on the R.H.S. had to be factorised into two factors, the sum of whose squares or their difference, as the case may be is the numerator.

As this factorisation and the addition or subtraction of the squares will not always be easy and readily possible, we shall, at a later stage, expound certain rules which will facilitate this work of expressing a given number as the sum of two squares or as the difference of two squares.

## Second Special Type

(*Under the Śūnyaṃ Samuccaye Formula*)

We now take up a second special type of quadratic equations which the *Śūnyaṃ Samuccaye Sūtra* can help us to solve, at sight, a sort of problem which the mathematicians all regard as "Hard"!

We may first remind the student of that portion of an earlier chapter wherein, referring to various applications of the *Samuccaye Sūtra*, we dealt with the easy method by which the oneness of the sum of the numerator on the one hand and the denominator on the other gave us one root and the oneness of the difference between the numerator and the denominator on both sides gave us another root, of the same quadratic equation. We need not repeat all of it but only refer back to that portion of this volume and remind the student of the kind of illustrative examples with which we illustrated our theme:

(1) $\frac{3x+4}{6x+7} = \frac{5x+6}{2x+3} \therefore 8x+10=0$; or $3x+3=0$

(2) $\frac{7x+5}{9x-5} = \frac{9x+7}{7x+17} \therefore 16x+12=0$; or $2x-10=0$

(3) $\frac{7x-9}{2x-9} = \frac{9x-7}{14x-7} \therefore 16x-16=0$; or $5x=0$

(4) $\frac{16x-3}{7x+7} = \frac{2x-15}{11x-25} \therefore 18x-18=0$; or $9x-10=0$

## Third Special Type

There is a third special type of quadratic equations which is also generally considered "very hard" but whereof one root is readily yielded by the same old friend the "*Sāmya Samuccaye*" *Sūtra* and the other is given by another friend—not so old, i.e. the "*Śūnyaṃ Anyat*" *Sūtra* which was used for a special type of simultaneous equations.

Let us take a concrete instance of this type. Suppose we have to solve the equation:

Let us take a concrete instance of this type. Suppose we have to solve the equation:

$$\frac{2}{x+2}+\frac{3}{x+3}=\frac{4}{x+4}+\frac{1}{x+1}$$

The nature of the characteristics of this special type will be recognisable with the help of the usual old test and an additional new test.

The tests are: $\frac{2}{1}+\frac{3}{1}=\frac{4}{1}+\frac{1}{1}$; and $\frac{2}{2}+\frac{3}{3}=\frac{4}{4}+\frac{1}{1}$

In all such cases, "*Śūnyaṃ Anyat*" formula declares that one root is zero; and the "*Śūnyaṃ Samuccaye*" *Sūtra* says:

$D_1+D_2=0 \therefore 2x+5=0 \therefore x=-2\frac{1}{2}$

The Algebraical proof hereof is as follows:

$\frac{2}{x+2}$(by simple division) $=1-\frac{x}{x+2}$; and so on.

So, $\left\{1-\frac{x}{x+2}\right\}+\left\{1-\frac{x}{x+3}\right\}=\left\{1-\frac{x}{x+4}\right\}+\left\{1-\frac{x}{x-1}\right\}$

$\therefore$ (Removing 1–1 and 1–1 from both sides) x, the common factor of all terms=0;

and on its removal, $\frac{1}{x+2}+\frac{1}{x+3}=\frac{1}{x+4}+\frac{1}{x+1}$

$\therefore$ By the *Samuccaye* formula, 2x+5=0;

*Note*: In all these cases, *Vilokanam,* i.e. mere observation gives us both the roots.

A few more illustrations of this special type are given

(1) $\frac{3}{x+3}+\frac{4}{x+4}=\frac{2}{x+2}+\frac{5}{x+5} \therefore x=0$ or $-3\frac{1}{2}$

(2) $\frac{1}{2x+1}+\frac{1}{3x+1}=\frac{2}{3x+2}+\frac{1}{6x+1}$

Now, $\because \frac{1}{1+2x}=1-\frac{2x}{1+2x}$ and so on.

$$\therefore \frac{2x}{2x+1}+\frac{3x}{3x+1}=\frac{3x}{3x+2}+\frac{6x}{6x+1}$$

$\therefore x=0$

or by cross-multiplication 12x+5 or (36x+15)=0

$\therefore x=-5/12$

(3) $\frac{a}{x+a}+\frac{b}{x+b}=\frac{a-c}{x+a-c}+\frac{b+c}{x+b+c}$

$\therefore x = 0 \text{ or } -\frac{1}{2}(a+b)$

(4) $\frac{a-b}{x+a-b} + \frac{b-c}{x+b-c} = \frac{a+b}{x+a+b} - \frac{b+c}{x-b-c}$

$\therefore x = 0 \text{ or } \frac{1}{2}(c-a)$

(5) $\frac{a+b}{x+a+b} + \frac{b+c}{x+b+c} = \frac{2b}{x+2b} + \frac{a+c}{x+a+c}$

$\therefore x = 0 \text{ or } -\frac{1}{2}(a+2b+c)$

### Fourth Special Type

And again, there is still another special type of quadratics which are "harder" but our old friends "*Śūnyaṃ Anyat*" and "*Parāvartya*" (Merger) can help us to solve easily.

*Note*: Apropos of the subject-matter of the immediately preceding sub-section the 3rd special type, let us now consider the equation $\frac{2}{x+2} + \frac{3}{x+3} = \frac{5}{x+5}$. This may *look*, at the outset, alike, but really is *not*, a quadratic equation of the type dealt with in the previous sub-section under *Śūnyaṃ Anyat* and *Śūnyaṃ Sāmya Samuccaye* but only a simple merger because, not only is the number of terms on the R.H.S. one short of the number required but also $\frac{2}{2} + \frac{3}{3} \neq \frac{5}{5}$. It is really a case under *Śūnyaṃ Anyat* and *Parāvartya* (merger).

Here, the test is the usual one for the merger process, i.e. $N_1 + N_2$ (on the L.H.S.) = $N_3$ (on the R.H.S.). Thus,

$$\frac{2}{x+2} + \frac{3}{x+3} = \frac{5}{x+5}$$

$\therefore$ By merger method $\frac{-6}{x+2} + \frac{-6}{x+3} = 0$ $\quad\quad \therefore 2x + 5 = 0$

$\therefore x = -2\frac{1}{2}$

A few true illustrations are given below:

(1) $\frac{4}{x+2} + \frac{9}{x+3} = \frac{25}{x+5}$

Here $\because \frac{4}{2} + \frac{9}{3} = \frac{25}{5} \therefore$ Yes

$\therefore$ By Division $2 - \frac{2x}{x+2} + 3 - \frac{3x}{x+3} = 5 - \frac{5x}{x+5}$

$\therefore x = 0$. This can be verified by mere observation.

or $\frac{2}{x+2}+\frac{3}{x+3}=\frac{5}{x+5}$ $\therefore$ by merger, $x=-2\frac{1}{2}$

This result can be readily put down, by putting up *each numerator* over the absolute term of the denominator as the numerator of each term of the resultant equation and retaining the denominator as before. Or by taking the square root of each numerator in the present case.

Thus $\frac{4}{2}=2; \frac{9}{3}=3;$ and $\frac{25}{5}=5$. And these will be our new numerators. Thus, we have the newly derived equation:

$$\frac{2}{x+2}+\frac{3}{x+3}=\frac{5}{x+5}$$

By merger $x=-2\frac{1}{2}$

(2) $\frac{2}{x+1}+\frac{9}{x+3}=\frac{25}{x+5}$

Here $\because \frac{2}{1}+\frac{9}{3}=\frac{25}{5}$ $\therefore$ Yes

The derived equation is:

$$\frac{2}{x+1}+\frac{3}{x+3}=\frac{5}{x+5}$$

which is the same as in all the three preceding cases.

$\therefore x=0$ or $-2\frac{1}{2}$

*Note*: In the last two cases, the first term alone is different and yet, since the quotients $\frac{4}{2}$ and $\frac{2}{1}$ are the same, therefore it makes no difference to the result; and we get the same two roots in all the three cases!

(3) $\frac{6}{2x+3}+\frac{4}{3x+2}=\frac{4}{4x+1}$

Here $\therefore \frac{6}{3}+\frac{4}{2}=\frac{4}{1}$ $\therefore$ Yes $\therefore x=0$

$\therefore$ or by division, $\frac{+4}{2x+3}+\frac{+6}{3x+2}=\frac{+16}{4x+1}$

Note that $\frac{2\times 6}{3}=4, \frac{3\times 4}{2}=6$ and $\frac{4\times 4}{1}=16$

and that these are the new numerators for the derived equation.

$\therefore$ By L.C.M. $\frac{24}{12x+18}+\frac{24}{12x+8}=\frac{48}{12x+3}$ $\therefore$ Yes

or $\frac{1}{12x+18}+\frac{1}{12x+8}=\frac{2}{12x+3}$ $\therefore$ Yes

$\therefore$ By merger $\frac{15}{12x+18}+\frac{5}{12x+8}=0 \therefore x=-\frac{7}{8}$

*N.B.*: The remaining examples in this chapter may be heldover for a later reading.

(4) $\frac{a}{x+a}+\frac{b}{x+b}=\frac{2c}{x+c}$

Here $\because \frac{a}{a}+\frac{b}{b}=\frac{2c}{c} \therefore$ Yes $\therefore x=0$

$\therefore \frac{1}{x+a}+\frac{1}{x+b}=\frac{2}{x+c} \therefore$ Yes

$\therefore$ By merger $\frac{a-c}{x+a}+\frac{b-c}{x+b}=0$

$\therefore x=\frac{bc+ca-2ab}{a+b-2c}$

(5) $\frac{a^2-b^2}{x+a+b}+\frac{b^2-c^2}{x+b+c}=\frac{a^2-c^2}{x+a+c}$

Here $\because \frac{a^2-b^2}{a+b}+\frac{b^2-c^2}{b+c}=\frac{a^2-c^2}{a+c} \therefore$ Yes $\therefore x=0$

or $\frac{a-b}{x+a+b}+\frac{b-c}{x+b+c}=\frac{a-c}{x+a+c} \therefore$ Yes

$\therefore$ By merger $\frac{(a-b)(b-c)}{x+a+b}+\frac{(b-c)(b-a)}{x+b+c}=0$ and so on.

(6) $\frac{1}{ax+d}+\frac{1}{bx+d}=\frac{2}{cx+d}$

Here by division, we have : $d+ax)\ 1\ (\frac{1}{d}.$

$$\begin{array}{r} 1+\frac{ax}{d} \\ \hline \frac{-ax}{d} \end{array}$$

$\therefore \frac{a}{ax+d}+\frac{b}{bx+d}=\frac{2c}{cx+d}$

$\therefore \frac{abc}{abcx+bcd}+\frac{abc}{abcx+acd}=\frac{2abc}{abcx+abd} \therefore$ Yes

$\therefore \frac{1}{D_1}+\frac{1}{D_2}=\frac{2}{D_3}$

$\therefore \frac{bcd-abd}{abcx+bcd}+\frac{acd-abd}{abcx+acd}=0$

$$\therefore \frac{bd(c-a)}{bc(ax+d)} = \frac{ad(b-c)}{ac(bx+d)}$$

$$\therefore \frac{c-a}{ax+d} = \frac{b-c}{bx+d}$$

$$\therefore x = \frac{ad+bd-2cd}{ac+bc-2ab}$$

(7) $\frac{ax+2d}{ax+d} + \frac{bx+3d}{bx+d} = \frac{2cx+5d}{cx+d}$

Here $\therefore \frac{2d}{d} + \frac{3d}{d} = \frac{5d}{d} \therefore$ Yes $\therefore x = 0$

Or by division

$$\frac{1}{abcx+bcd} + \frac{2}{abcx+acd} = \frac{3}{abcx+abd}$$

$\therefore$ By merger, $\frac{bcd-abd}{D_1} + \left(\frac{acd-abd}{D_2}\right) = 0$

$$\therefore \frac{bd(a-c)}{bc(ax+d)} + \frac{2ad(b-c)}{ac(bx+d)} = 0$$

$$\therefore \frac{a-c}{ax+d} + \frac{2(b-c)}{(bx+d)} = 0$$

$$\therefore x = \frac{ad+2bd-3cd}{bc+2ac-3ab}$$

or by mere division *Parāvartya* at the very first step.

$$\therefore \frac{d}{ax+d} + \frac{2d}{bx+d} = \frac{3d}{cx+d}$$

which is the same as No. 6 supra.

$$\therefore x = \frac{ad+2bd-3cd}{bc+2ac-3ab}$$

## Concluding Linking Note
### (*On Quadratic Equations*)

In addition to the above, there are several other special types of quadratic equations, for which the Vedic *Sūtras* have made adequate provision and also suggested several interesting devices and so forth. But these we shall go into and deal with, at a later stage.

Just at present, we address ourselves to our next appropriate subject for this introductory and illustrative volume, namely the solution of cubic and biquadratic equations etc.

# 18

# Cubic Equations

We solve cubic equations in various ways:

(*i*) with the aid of the *Parāvartya Sūtra*, the *Lopana-Sthāpana Sūtra*, the पूरणापूरणाभ्यां formula (*Pūraṇa-Apūrṇābhyām*) which means "by the completion or non-completion" of the square, the cube, the fourth power etc.

(*ii*) by the method of Argumentation and Factorisation as explained in a previous chapter.

### The Pūraṇa Method

The *Pūraṇa* method is well-known to the current system. In fact, the usually-in-vogue general formula $x = \frac{-b \pm \sqrt{b^2 - 4ac}}{2a}$ for the standard quadratic $(ax^2 + bx + c = 0)$ has been worked out by this very method. Thus,

$$ax^2 + bx + c = 0$$

$\therefore$ Dividing by a, $x^2 + \frac{bx}{a} + \frac{c}{a} = 0$

$$\therefore x^2 + \frac{bx}{a} = \frac{-c}{a}$$

$\therefore$ completing the square on the L.H.S.

$$\therefore x^2 + \frac{bx}{a} = \frac{b^2}{4a^2} = -\frac{c}{a} + \frac{b^2}{4a^2} = \frac{b^2 - 4ac}{4a^2}$$

$$\therefore \left(x + \frac{b}{2a}\right)^2 = \frac{b^2 - 4ac}{4a^2}$$

$$\therefore x + \frac{b}{2a} = \frac{\pm\sqrt{b^2 - 4ac}}{2a}$$

$$\therefore x = \frac{-b \pm \sqrt{b^2 - 4ac}}{2a}$$

This method of "completing the square" is thus quite well-known to the present-day mathematicians, in connection with the solving of quadratic equations. But this is only a fragmentary and fractional application of the general formula which in conjunction with the *Parāvartya*, and the *Lopana-Sthāpana Sūtras* is equally applicable to cubic, biquadratic and other higher-degree equations as well.

## COMPLETING THE CUBIC

With regard to cubic equations, we combine the *Parāvartya Sūtra* as explained in the 'Division by *Parāvartya*' chapter and the *Pūraṇa* sub-formula. Thus,

(1) $x^3 - 6x^2 + 11x - 6 = 0$

$\therefore x^3 - 6x^2 = -11x + 6$

But $(x-2)^3 = x^3 - 6x^2 + 12x - 8$

$\therefore$ Substituting the value of $x^3 - 6x^2$ from above, we have:

$(x-2)^3 = -11x + 6 + 12x - 8 = x - 2$

Let $x - 2 = y$ (and $\therefore$ let $x = y + 2$)

$\therefore y^3 = y \therefore y = 0$ cr $\pm 1 \therefore x = 3$ or 1 or 2

*N.B.*: It need hardly be pointed out that, by argumentation re : the coefficients of $x^3$, $x^2$ etc., we can arrive at the same answer as explained in the previous chapter dealing with factorisation by Argumentation and that this holds good in all the cases dealt with in the present chapter.

(2) $x^3 + 6x^2 + 11x + 6 = 0$

$\therefore x^3 + 6x^2 = -11x - 6$

But $(x+2)^3 = x^3 + 6x^2 + 12x + 8 = -11x - 6 + 12x + 8 = x + 2$

$\therefore y^3 = y$ (where y stands for $x + 2$)

$\therefore y = 0$ or $\pm 1 \therefore x = -2, -3$ or $-1$

(3) $x^3 + 6x^2 - 37x + 30 = 0$

$\therefore x^3 + 6x^2 = 37x - 30$

$\therefore (x+2)^3 = x^3 + 6x^2 + 12x + 8 = 49x - 22 = 49(x+2) - 120$

*N.B.*: The object is to bring (x+2) on the R.H.S. and thus help to formulate an equation in y, obtain the three roots and then, by substitution of the value of x in terms of y, obtain the three values of x.

$\therefore y^3 - 49y + 120 = 0 \therefore (y-3)(y^2 + 3y - 40) = 0$

$\therefore (y-3)(y-5)(y+8) = 0 \therefore y = 3$ or 5 or $-8$

$\therefore x = 1$ or 3 or $-10$.

(4) $x^3 + 9x^2 + 23x + 15 = 0$

$\therefore x^3 + 9x^2 = -23x - 15$

$\therefore (x + 3)^3 = (x^3 + 9x^2 + 27x - 27) = 4x + 12 = 4(x + 3)$

$\therefore y^3 = 4y \therefore y = 0$ or $\pm 2 \therefore x = -3$ or $-1$ or $-5$

(5) $x^3 + 9x^2 + 24x + 16 = 0$

$\therefore x^3 + 9x^2 = -24x - 16$

$\therefore (x + 3)^3 = (x^3 + 9x^2 + 27x + 27) = 3x + 11 = 3(x + 3) + 2$

$\therefore y^3 = 3y + 2 \therefore y^3 - 3y - 2 = 0$

$\therefore (y + 1)^2 (y - 2) = 0 \therefore y = -1$ or $2$; $x = -4$ or $-1$

(6) $x^3 + 7x^2 + 14x + 8 = 0$

$\therefore x^3 + 7x^2 = -14x - 8$

$\therefore (x + 3)^3 = (x^3 + 9x^2 + 27x + 27) = 2x^2 + 13x + 19 = (x + 3)(2x + 7) - 2$

$\therefore y^3 = y(2y + 1) - 2 \therefore y^3 - y(2y + 1) + 2 = 0 = (y - 1)(y + 1)(y - 2)$

$\therefore y = 1$ or $-1$ or $2 \therefore x = -2$ or $+4$ or $\pm 1$

(7) $x^3 + 8x^2 + 17x + 10 = 0 \therefore x^3 + 8x^2 = -17x - 10$

$\therefore (x + 3)^3 = (x^3 + 9x^2 + 27x + 27) = x^2 + 10x + 17 = (x + 3)(x + 7) - 4$

$\therefore y^3 = y(y + 4) - 4 \therefore y^3 - y^2 - 4y - 4 = 0 \therefore y = 1$ or $\pm 2$

$\therefore x = -2$ or $-1$ or $-5$

(8) $x^3 + 10x^2 + 27x + 18 = 0$

Now $\therefore (x + 4)^3 = (x^3 + 12x^2 + 48x + 64)$

Hence the L. H. S. $= (x + y)^3 - (2x^2 + 21x + 46) = (x + y)^3 \{(x + 4)(2x + 13) - 6\}$

$\therefore y^3 = y(2y + 5) - 6 \therefore (y - 1)(y + 2)(y - 3) = 0$

$\therefore y = 1$ or $-2$ or $3$

$\therefore x = -3$ or $-6$ or $-1$

*Note*: Expressions of the form $x^3 - 7x + 6$ can be split into $x^3 - 1 - 7x + 7$ etc., and readily factorise. This is always applicable to all such cases (where $x^2$ is absent) and should be fully utilised.

The *Pūraṇa* method explained in this chapter for the solution of cubic equations will be found of great help in factorisation; and vice versa.

"Harder" cubic equations will be taken up later.

# 19

# Biquadratic Equations

The procedures *Pūraṇa* etc., expounded in the previous chapter for the solution of cubic equations can be equally well applied in the case of biquadratics etc., too. Thus,

(1) $x^4 + 4x^3 - 25x^2 - 16x + 84 = 0$

$$\therefore x^4 + 4x^3 = 25x^2 + 16x - 84$$

$$\therefore (x+1)^4 = x^4 + 4x^3 + 6x^2 + 4x + 1$$

$$= (25x^2 + 16x - 84) + (6x^2 + 4x + 1) = 31x^2 + 20x - 83$$

$$= (x+1)(31x - 11) - 72$$

$$\therefore y^4 = y(31y - 42) - 72 \therefore y^4 - 31y^2 + 42y + 72 = 0$$

$$\therefore y = -1, 3, 4 \text{ or } -6$$

$$\therefore x = -2, 2, 3 \text{ or } -7$$

(2) $x^4 + 8x^3 + 14x^2 - 8x - 15 = 0$

$$\therefore x^4 + 8x^3 = -14x^2 + 8x + 15$$

$$\therefore (x+2)^4 = x^4 + 8x^3 + 24x^2 + 32x + 16 = 10x^2 + 40x + 31$$

$$= (x+2)(10x + 20) - 9 = 10(x+2)^2 - 9$$

$$\therefore y^4 = 10y^2 - 9 \therefore y^2 + 1 \text{ or } 9 \therefore y = \pm 1 \text{ or } \pm 3$$

$$\therefore x = -1 \text{ or } -3 \text{ or } 1 \text{ or } -5$$

(3) $x^4 - 12x^3 + 49x^2 - 78x + 40 = 0$

$$\therefore x^4 - 12x^3 = -49x^2 + 78x - 40$$

$$\therefore (x-3)^4 = x^4 - 12x^3 + 54x^2 - 108x + 81$$

$$= 5x^2 - 30x + 41 = (x-3)(5x - 15) - 4 = 5(x-3)^2 - 4$$

$$\therefore y^4 - 5y^2 + 4 = 0 \therefore y^2 = 1 \text{ or } 4 \therefore y = \pm 1 \text{ or } \pm 2$$

$$\therefore x = 4 \text{ or } 5 \text{ or } 2 \text{ or } 1$$

(4) $x^4 + 16x^3 + 86x^2 + 176x + 105 = 0$

$\therefore x^4 + 16x^3 = -86x^2 - 176x - 105 = 0$

$\therefore (x+4)^4 = x^4 + 16x^3 + 96x^2 + 256x + 256$

$= 10x^2 + 80x + 151 = (x+4)(10x+40) - 9$

$= 10(x+4)^2 - 9$

$\therefore y^4 - 10y^2 + 9 = 0 \therefore y^2 = 1 \text{ or } 9 \therefore y = \pm 1 \text{ or } \pm 3$

$\therefore x = -3 \text{ or } -5 \text{ or } -1 \text{ or } -7$

(5) $x^4 - 16x^3 + 91x^2 - 216x + 180 = 0$

$\therefore x^4 - 16x^3 = -91x^2 + 216x - 180$

$\therefore (x-4)^4 = x^4 - 16x^3 + 96x^2 - 256x + 256$

$= 5x^2 - 40x + 76 = (x-4)(5x-20) - 4 = 5(x-4)^2 - 4$

$\therefore y^4 - 5y^2 + 4 = 0 \therefore y^2 = 1 \text{ or } 4 \therefore y = \pm 1 \text{ or } \pm 2$

$\therefore x = 3 \text{ or } 5 \text{ or } 6 \text{ or } 2$

(6) $x^4 - 20x^3 + 137x^2 - 382x + 360 = 0$

$\therefore x^4 - 20x^3 = -137x^2 + 382x - 360$

$\therefore (x-5)^4 = x^4 - 20x^3 + 150x^2 - 500x + 625$

$= 13x^2 - 118x + 265 = (x-5)(13x-53)$

$\therefore y^4 = y(13y+12) \therefore y = 0 \text{ or } y^3 - 13y - 12 = 0$

$\therefore y = 0 \text{ or } (y+1)(y+3)(y-4) = 0$

$\therefore y = 0 \text{ or } -1 \text{ or } -3 \text{ or } 4$

$\therefore x = 5, 4, 2 \text{ or } 9.$

*Note*: The student need hardly be reminded that all these examples which have all been solved by the *Pūraṇa* method hereinabove can also be solved by the Argumentation-cum-factorisation method.

## A Special Type

There are several special types of biquadratic equations dealt with in the Vedic *Sūtras*. But we shall here deal with only one such special type and hold the others over to a later stage.

This type is one wherein the L.H.S. consists of the sum of the fourth powers of two binomials and the R.H.S. gives us the equivalent thereof in the shape of an arithmetical number. The formula applicable to such cases is the व्यष्टि-समष्टि (*Vyaṣṭi Samaṣṭi*) *Sūtra* or the *Lopana-Sthāpana* one which teaches us how to use the average or the exact middle binomial for breaking the biquadratic down into a simple quadratic by the easy device of mutual cancellation of the odd powers, i.e. the $x^3$ and the x.

A single concrete illustration will suffice for explaining this process:

$(x+7)^4 + (x+5)^4 = 706$

Let x+6 the average of the two binomials=a

$\therefore (a+1)^4 + (a-1)^4 = 706$

$\therefore$ owing to the cancellation of the odd powers $x^3$ and x,

$$2a^4 + 12a^2 + 2 = 706 \therefore a^4 + 6a^2 - 352 = 0$$

$$\therefore a^2 = 16 \text{ or } -22 \therefore a = \pm 4 \text{ or } \pm\sqrt{-22}$$

$$\therefore x = -2 \text{ or } -10 \text{ or } \pm\sqrt{-22} - 6$$

*N.B.*: In simple examples like this, the integral roots are small ones and can be spotted out by mere inspection and the splitting up of 706 into 625 and 81 and for this purpose, the *Vilokanam* method will suffice. But, in cases involving more complex numbers, fractions, surds, imaginary quantities etc., and literal coefficients and so on: *Vilokanam* will not completely solve the equation. But here too, the *Vyaṣṭi-Samaṣṭi* formula will quite serve the purpose. Thus,

The General Formula will be as follows:

Given $(\overline{x+m}+n)^4 + (\overline{x+m}-n)^4 = p$

$$\therefore a^4 + 6a^2 + \left(n^4 - \frac{p}{2}\right) = 0$$

$$\therefore a^2 = \frac{-6 \pm \sqrt{36 - 4n^4 + 2p}}{2}$$

$$\therefore a = \pm\sqrt{\frac{-6 \pm \sqrt{36 - 4n^4 + 2p}}{2}}$$

$$\therefore x = -m \pm \sqrt{\frac{-6 \pm \sqrt{36 - 4n^4 + 2p}}{2}}$$

Applying this to the above example, we have

$$x = -6 \pm \sqrt{\frac{-6 \pm \sqrt{1444}}{2}} = -6 \pm \sqrt{\frac{-6 \pm 38}{2}}$$

$$= -6(\pm\sqrt{16} \text{ or } \sqrt{-22}) = -6(\pm 4 \text{ or } \pm\sqrt{-22})$$

which tallies with the above.

*N.B.*: "Harder" Biquadratics, Pentics etc., will be taken up later.

# 20

# Multiple Simultaneous Equations

We now go on to the solution of simultaneous equations involving three or more unknowns. The *Lopana-Sthāpana Sūtra*, the *Ānurūpya Sūtra* and the *Parāvartya Sūtra* are the ones that we make use of for this purpose.

### First Type

In the first type we have a significant figure on the R.H.S. in only one equation and zeroes in the other two. From the homogeneous zero equations, we derive new equations defining two of the unknowns in terms of the third; we then substitute these values in the third equation; and thus we obtain the values of all the three unknowns.

A second method is the judicious addition and subtraction of proportionate multiples for bringing about the elimination of one unknown and the retention of the other two.

In both these methods, we can make our own choice of the unknown to be eliminated, the multiples to be taken etc. Thus,

(1) $x + y - z = 0$ (A)
$4x - 5y + 2z = 0$ (B)
$3x + 2y + z = 10$ (C)

(*i*) $\left.\begin{array}{l} A + C \text{ gives us} : 4x + 3y = 10; \\ \& \; 2A + B \text{ gives us} : 6x - 3y = 0 \end{array}\right\}$ $\therefore 10x = 10$ $\therefore \left.\begin{array}{r} x = 1 \\ y = 2 \\ \text{and } z = 3 \end{array}\right\}$

(*ii*) $\left.\begin{array}{l} \text{from A, we have } x + y = z \\ \text{and from B, we have } 4x - 5y = -2z \end{array}\right\}$

$\therefore$ By *Parāvartya*, $x = \frac{1}{3}z$; and $y = \frac{2}{3}z$

$\therefore$ by substitution in C $z + 1\frac{1}{3}z + z = 10 \therefore z = 3, x = 1$ and $y = 2$

(2) $7y - 11z - 2x = 0$ ... A
$8y - 7z - 6x = 0$ ... B
$3x + 4y + 5z = 35$ ... C

(*i*) Adding B and 2C, we have $16y + 3z = 70$
Subtracting B from 3A, $13y - 26z = 0$

$\therefore y = \frac{1820}{455} = 4$; and $z = \frac{910}{455} = 2$

$\therefore$ Substituting these values in C $\therefore x = 3$

(*ii*) $\therefore 7y - 11z = 2x$, $8y - 7z = 6x$ $\therefore y = \frac{-52x}{-39} = 1\frac{1}{3}x$; and $z = \frac{-26x}{-39} = \frac{2}{3}x$

$\therefore 3x + 5\frac{1}{3}x + 3\frac{1}{3}x = 35$ $\therefore x = 3$, $y = 4$ and $z = 2$

(3) $2x - 3y + 4z = 0$ ... (A)
$7x + 2y - 6z = 0$ ... (B)
$4x + 3y + z = 37$ ... (C)

(*i*) A + C gives us : $6x + 5z = 37$
2A + 3B gives us : $25x - 10z = 0$

$\therefore x = \frac{370}{185} = 2$; and $z = \frac{925}{185} = 5$; and $y = 8$

From (A) and (B) we have

(*ii*) $\therefore -3y + 4z = -2x$, $2y - 6z = -7x$

$\therefore y = \frac{-40x}{-10} = 4x$; and $z = \frac{-25x}{-10} = 2\frac{1}{2}x$

$\therefore 4x + 12x + 2\frac{1}{2}x = 37 \therefore x = 2$; $y = 8$; and $z = 5$

### Second Type

This is one wherein the R.H.S. contains significant figures in all the three equations. This can be solved by *Parāvartya* cross-multiplication so as to produce two derived equations whose R.H.S. consists of zero only, or by the first or the second of the methods utilised in the previous sub-section. Thus,

$$\left.\begin{array}{l} 2x - 4y + 9z = 28 \\ 7x + 3y - 5z = 3 \\ 9x + 10y - 11z = 4 \end{array}\right\} \begin{array}{l} \ldots \; A \\ \ldots \; B \\ \ldots \; C \end{array}$$

(1) (*i*) $\left.\begin{array}{l} \therefore 196x + 84y - 140z = 84 \\ \text{and } 6x - 12y + 27z = 84 \end{array}\right\} \therefore 190x + 96y - 167z = 0$

$\left.\begin{array}{l} \therefore 28x + 12y - 20z = 12 \\ \text{and } 27x + 30y - 33z = 12 \end{array}\right\} \therefore x - 18y + 13z = 0$

Having thus derived two equations of this kind, i.e. of the first special type, we can now follow the first method under that type; and, after a lot of big multiplications, subtractions, additions and divisions, we can obtain the answer: $x = 2$, $y = 3$ and $z = 4$.

(*ii*) or, adopting the first method adopted in the last sub-section we have:

$$\left.\begin{array}{l} \therefore 3y - 5z = 3 - 7x \\ \text{and } 10y - 11z = 4 - 9x \end{array}\right\}$$

$\therefore$ By cross-multiplication,

$$\left.\begin{array}{l} y = \dfrac{-20 + 45x + 33 - 77x}{-17} = \dfrac{32x - 13}{17}; \\ \text{and } z = \dfrac{30 - 70x - 12 + 27x}{-17} = \dfrac{43x - 18}{17} \end{array}\right\}$$

$$\therefore 2x - \frac{128x - 52}{17} + \frac{387x - 162}{17} = 28$$

$\therefore 34x - 128x + 52 + 387x - 162 = 476 \; \therefore 293x = 586$

$\therefore x = 2;\ y = 3;$ and $z = 4$

This method too involves a lot of clumsy labour.

(*iii*) or, adopting the *Lopana-Sthāpana* method, we say:

$$\left.\begin{array}{l} C - A - B \text{ gives us } 11y - 15z = -27 \\ \text{and } 9B - 7C \text{ gives us } -43y + 32z = -1 \end{array}\right\}$$

$\therefore y = 3,\ z = 4$ and $x = 2$

(2) $\left.\begin{array}{l} x + 2y + 3z = 12 \quad \ldots \; A \\ 2x + 3y + 4z = 18 \quad \ldots \; B \\ 4x + 3y + 5z = 24 \quad \ldots \; C \end{array}\right\}$

(*i*) $\left.\begin{array}{l} \therefore 24x + 36y + 48z = 216 \\ 18x + 36y + 54z = 216 \end{array}\right\} \therefore 6x - 6z = 0 \; \therefore x - z = 0$

$$\text{Similarly } \left.\begin{aligned} 48x + 36y + 60z &= 288 \\ 24x + 48y + 72z &= 288 \end{aligned}\right\} \begin{aligned} &\therefore 24x - 12y - 12z = 0 \\ &\therefore 2x - y - z = 0 \end{aligned}$$

$$\therefore x = y = z = 2$$

(*ii*) $\therefore 2y + 3z = 12 - x$; and $3y + 4z = 18 - 2x \therefore y = 6 - 2x$; and $z = x \therefore x = y = z = 2$.

(*iii*) $$\left.\begin{aligned} &2A - B \text{ gives us: } y + 2z = 6 \\ &2B - C \text{ gives us: } 3y + 3z = 12 \end{aligned}\right\} \qquad \left.\begin{aligned} \therefore y &= 2 \\ z &= 2 \\ \text{and } x &= 2 \end{aligned}\right\}$$

or (*iv*) by mere observation.

(3) $$\left.\begin{aligned} x + 2y + 3z &= 14 \quad \ldots A \\ 2x + 3y + 4z &= 20 \quad \ldots B \\ 3x + y + 6z &= 23 \quad \ldots C \end{aligned}\right\}$$

(*i*) $$\left.\begin{aligned} \therefore 28x + 42y + 56z &= 280 \\ 20x + 40y + 60z &= 280 \\ \text{and } 42x + 14y + 84z &= 322 \\ 23x + 46y + 69z &= 322 \end{aligned}\right\} \qquad \left.\begin{aligned} &\therefore 8x + 2y - 4z = 0 \\ &\therefore 19x - 32y + 15z = 0 \end{aligned}\right\}$$

and so on as before.

(*ii*) $$\therefore \left.\begin{aligned} 2y + 3z &= 14 - x \\ 3y + 4z &= 20 - 2x \end{aligned}\right\} \quad \begin{aligned} &\therefore y = 60 - 6x - 56 + 4x = 4 - 2x \\ &z = 42 - 3x - 40 + 4x = x + 2 \end{aligned}$$

$$\therefore 3x + 4 - 2x + 6x + 12 = 23 \therefore x = 1,\ y = 2 \text{ and } z = 3$$

(*iii*) $$\left.\begin{aligned} 2A - B \text{ gives us: } y + 2z &= 8 \\ \text{and } 3A - C \text{ gives us: } 5y + 3z &= 19 \end{aligned}\right\} \quad \left.\begin{aligned} \therefore y &= 2 \\ z &= 3 \\ x &= 1 \end{aligned}\right\}$$

(4) $$\left.\begin{aligned} x + 2y + 3z &= 11 \quad \ldots A \\ 2x + 3y + 4z &= 16 \quad \ldots B \\ 3x + 5y + 6z &= 25 \quad \ldots C \end{aligned}\right\}$$

(*i*) $$\left.\begin{aligned} &(16x + 32y + 48z) - (22x + 33y + 44z) = -6x - y + 4z = 0 \\ \text{and } &(33x + 55y + 66z) - (25x + 50y + 75z) = 8x + 5y - 9z = 0 \end{aligned}\right\}$$

and so on.

(*ii*) $$\therefore \left.\begin{aligned} 2y + 3z &- 11 - x \\ 3y + 4z &= 16 - 2x \end{aligned}\right\} \text{ and so on, as before.}$$

(*iii*) $$\left.\begin{aligned} x + y + z &= 5 \\ \text{and } x + 2y + 2z &= 9 \end{aligned}\right\} \quad \begin{aligned} &\therefore y + z = 4; \text{ and } x + z = 3 \\ &\therefore x = 1, y = 2 \text{ and } z = 2 \end{aligned}$$

In all these processes, there is an element, more or less, of clumsiness and cumbrousness which renders them unfit to come under and fit satisfactorily into the Vedic category. Methods expounded in the Vedic *Sūtras* and free from the said drawback and also capable of universal application will be explained at a later stage.

# 21

# Simultaneous Quadratic Equations

The *Sūtras* needed for the solution of simultaneous quadratic equations have practically all been explained already. Only the actual applicational procedure, devices and *modus operandi* thereof have to be explained. Thus,

(1) $\left.\begin{array}{l} x+y=5 \\ \& \ xy=6 \end{array}\right\}$ $\therefore \left.\begin{array}{l} x^2+2xy+y^2=25 \\ \qquad 4xy \qquad = 24 \end{array}\right\}$ $\begin{array}{l} \therefore (x-y)^2=1 \\ \therefore x-y=\pm 1 \end{array}$

$$\therefore \left.\begin{array}{l} x=2 \\ y=3 \end{array}\right\} \text{or} \left.\begin{array}{l} x=3 \\ y=2 \end{array}\right\}$$

This is readily obtainable by *Vilokanam* (mere observation) and also because symmetrical values can always be reversed.

(2) $\left.\begin{array}{l} x-y=1 \\ \text{and } xy=6 \end{array}\right\}$ $\therefore \left.\begin{array}{l} x=3 \\ y=2 \end{array}\right\}$ or $\left.\begin{array}{l} -2 \\ -3 \end{array}\right\}$ Note the minus

(3) $\left.\begin{array}{l} 5x-y=17 \\ \text{and } xy=12 \end{array}\right\}$ $\therefore \left.\begin{array}{l} 25x^2-10xy+y^2=289 \\ \text{and} \quad 20xy \quad =240 \end{array}\right\}$ $\begin{array}{l} \therefore (5x+y)^2=529 \\ \therefore 5x+y=\pm 23 \end{array}$

$$\therefore 10x=40 \text{ or } -6 \ \therefore \left.\begin{array}{l} x=4 \\ y=3 \end{array}\right\} \text{or} \left.\begin{array}{l} -\frac{3}{5} \\ -20 \end{array}\right\}$$

*N.B.*: (1) When the value of x or y has been found, xy at once gives us the value of the other. Thus, if, here, $x=4$, $y=3$, no other substitution etc., is necessary.

(2) One set of values can be found out by *Vilokanam* alone.

(3) The internal relationship between the two sets of values will be explained later.

(4) $\left.\begin{array}{l} 4x-3y=7 \\ \text{and } xy=12 \end{array}\right\}$ $\therefore x=4$ and $y=3$ by mere *Vilokanam* observation.

(*ii*) $(4x-3y)^2 = 49 \therefore 4x+3y = \pm 25 \therefore 8x = 32$ or $-18$

$$\therefore \left.\begin{array}{l} x = 4 \\ y = 3 \end{array}\right\} \text{ or } \left.\begin{array}{l} -2\frac{1}{4} \\ -5\frac{1}{3} \end{array}\right\}$$

(5) $\left.\begin{array}{l} x^3 - y^3 = 19 \\ x - y = 1 \end{array}\right\} \therefore \left.\begin{array}{l} x^2+xy+y^2 = 19 \\ \text{and } x^2-2xy+y^2 = 1 \end{array}\right\} \therefore 3xy = 18 \therefore xy = 6$

$$\therefore \left.\begin{array}{l} x = 3 \\ y = 2 \end{array}\right\} \text{ or } \left.\begin{array}{l} -2 \\ -3 \end{array}\right\}$$

(6) $\left.\begin{array}{l} x^3 + y^3 = 61 \\ x + y = 1 \end{array}\right\} \therefore \left.\begin{array}{l} x^2-xy+y^2 = 61 \\ x^2+2xy+y^2 = 1 \end{array}\right\} \therefore 3xy = -60 \therefore xy = -20$

$$\therefore \left.\begin{array}{l} x = 5 \\ y = -4 \end{array}\right\} \text{ or } \left.\begin{array}{l} -4 \\ 5 \end{array}\right\}$$

*N.B.*: There is *plus* sign all through. Therefore it can all be simply reversed, i.e. one by *Vilokanam* and the other by reversal.

(7) $\left.\begin{array}{l} x + y = 4 \\ \text{and } x^2 + xy + 4x = 24 \end{array}\right\} \therefore$ (*i*) By *Vilokanam*, $x = 3$ and $y = 1$

Secondly $x(x+y) + 4x = 8x = 24 \quad \therefore \left.\begin{array}{l} x = 3 \\ y = 1 \end{array}\right\}$

(8) $\left.\begin{array}{l} x + 2y = 5 \\ \text{and } x^2 + 3xy + 2y^2 + 4x - y = 10 \end{array}\right\}$

$\therefore (x+y)(x+2y) + 4x - y = 5x + 5y + 4x - y = 10$

$\therefore \left.\begin{array}{l} 9x + 4y = 10 \\ \text{But } x + 2y = 5 \end{array}\right\}$

$\therefore$ By *Parāvartya* or by *Śūnyaṃ Anyat* $x = 0$ & $y = 2\frac{1}{2}$

(9) $x + 2y = 5$

and $x^2 + 3xy - 2y^2 + 4x + 3y = 0$

$\therefore (x+2y)(x+y) - 4y^2 + 4x + 3y = 0$

$\therefore 5x + 5y - 4y^2 + 4x + 3y = 9x + 8y - 4y^2 = 0$

$\therefore 4y^2 + 10y - 45 = 0 \therefore 8y + 10 = \pm\sqrt{820}$

$$\therefore y = \pm\frac{\sqrt{820} - 10}{8} = \frac{-5 \mp \sqrt{205}}{4}$$

$$\therefore x = \frac{15 \mp \sqrt{205}}{2}$$

(10) $\left.\begin{aligned} x+y &= 5 \\ 3x^2+y^2 &= 19 \end{aligned}\right\}$ $\therefore 3x^2+y^2 = (x+y)(3x-3y)+4y$

$= 15x - 15y + 4y^2 = 19$

$\therefore 75 - 15y - 15y + 4y^2 = 19$

$\therefore 4y^2 - 30y + 56 = 0$

$\therefore 8y - 30 = \pm\sqrt{4} = \pm 2 \therefore 8y = 32 \text{ or } 28$ $\quad \left.\begin{aligned} \therefore\ y &= 4 \text{ or } 3\tfrac{1}{2} \\ \text{and } x &= 1 \text{ or } 1\tfrac{1}{2} \end{aligned}\right\}$

(11) $\left.\begin{aligned} x^2+3x-2y &= 4 \\ 2x^2-5x+3y &= -2 \end{aligned}\right\}$ $\left.\begin{aligned} &\therefore 7x^2 - x - 8 = 0 \\ &\therefore (x+1)(7x-8) = 0 \end{aligned}\right\}$ $\left.\begin{aligned} \therefore x &= -1 \\ y &= -3 \end{aligned}\right\}$ or $\begin{aligned} &8/7 \\ &18/49 \end{aligned}$

(12) $\left.\begin{aligned} x + y &= 5 \\ x^2 - 2y^2 &= 1 \end{aligned}\right\}$ $\therefore 5x - 5y - y^2 = 1 \ \therefore 25 - 10y - y^2 = 1$

$\therefore y^2 + 10y - 24 = 0 \therefore \left.\begin{aligned} y &= 2 \\ x &= 3 \end{aligned}\right\}$ $\left.\begin{aligned} &\text{or } -12 \\ &\text{or } +17 \end{aligned}\right\}$

(13) $\left.\begin{aligned} 2x + y &= 3 \\ x^2 + 2xy &= 3 \end{aligned}\right\}$ $\left.\begin{aligned} &\therefore 1\tfrac{1}{2}x + 2\tfrac{1}{4}y - \tfrac{3}{4}y^2 = 3 \\ &\therefore \quad 6x + 9y - 3y^2 = 12 \end{aligned}\right\}$ $\left.\begin{aligned} &\therefore y^2 - 2y + 1 = 0 \\ &\therefore y = 1 \text{ and } x = 1 \end{aligned}\right\}$

or (*ii*) $\left.\begin{aligned} 4x^2 + 2xy &= 6x \\ x^2 + 2xy &= 3 \end{aligned}\right\}$ $\therefore 3x^2 - 6x + 3 = 0 \therefore \left.\begin{aligned} x &= 1 \\ \text{and } y &= 1 \end{aligned}\right\}$

(14) $\left.\begin{aligned} x + y &= 2 \\ x^2 + y^2 + 2x + 3y &= 7 \end{aligned}\right\}$ $\therefore 4x + y + 2y^2 = 7 \therefore 8 - 3y + 2y^2 = 7$

$\therefore 2y^2 - 3y + 1 = 0 \therefore \left.\begin{aligned} y &= 1 \\ x &= 1 \end{aligned}\right\}$ or $\left.\begin{aligned} &\tfrac{1}{2} \\ &1\tfrac{1}{2} \end{aligned}\right\}$

(15) $\left.\begin{aligned} 2x^2 + xy + y^2 &= 8 \\ 3x^2 - xy + 4y^2 &= 17 \end{aligned}\right\}$ $\therefore x^2 + y^2 = 5$

And by cross-multiplication

$34x^2 + 17xy + 17y^2 = 24x^2 - 8xy + 32y^2$

$\therefore 10x^2 + 25xy - 15y^2 = 0 \therefore 2x^2 + 5xy - 3y^2 = 0$

$\therefore (x+3y)(2x-y) = 0 \therefore x = -3y \text{ or } \tfrac{1}{2}y$

Substituting in $x^2 + y^2 = 5$, we have

$9y^2 + y^2 = 5 \text{ or } \tfrac{5}{4}y^2 = 5 \therefore y^2 = \tfrac{1}{2} \text{ or } 4$

$\therefore y = \pm\frac{1}{\sqrt{2}} \text{ or } \pm 2.$

and $\therefore x = \pm 3\sqrt{\frac{1}{2}}$ or $\pm\frac{1}{2}\sqrt{\frac{1}{2}}$ or $\pm 6$ or $\pm 1$

*N.B.*: Test for the correct sign plus or minus.

(16) $\left.\begin{array}{l} 2x^2 + xy + y^2 = 77 \\ 2x^2 + 3xy \quad = 92 \end{array}\right\}$ $\therefore 184x^2 + 92xy + 92y^2 = 154x^2 + 231xy$
$\therefore 30x^2 - 139xy + 92y^2 = 0$

$\therefore (5x - 4y)(6x - 23y) = 0 \therefore x = \frac{4}{5}y$ or $\frac{23}{6}y$

$\therefore$ By substitution,

$\left.\begin{array}{l} y = \pm 5 \\ x = \pm 4 \end{array}\right\}$ or $\begin{array}{l} 2 \pm \sqrt{6/7} \\ \frac{23}{3} \pm \sqrt{7} \end{array}$

(17) $\left.\begin{array}{l} 3x^2 - 4xy + 2y^2 = 1 \\ y^2 - x^2 \quad = -15 \end{array}\right\}$ $\therefore$ By subtraction, $4x^2 - 4xy + y^2 = 16$
$\therefore 2x - y = \pm 4$

$\therefore$ By substitution, $4x^2 \mp 16x + 16 - x^2 = -15$
$\therefore 3x^2 \mp 16x + 31 = 0$ & so on.

(18) $\left.\begin{array}{l} 2x^2 - 7xy + 3y^2 = 0 \\ x^2 + xy + y^2 = 13 \end{array}\right\}$ $\therefore x = 3y$ or $\frac{1}{2}y$

$\left.\begin{array}{l} \therefore y = \pm 1 \\ \text{and } x = \pm 3 \end{array}\right\}$ or $\left.\frac{\pm\sqrt{\frac{13}{7}}}{\pm 2\sqrt{\frac{13}{7}}}\right\}$

(19) $\left.\begin{array}{l} 3x^2 - 4xy + 2y^2 = 1 \\ y^2 - x^2 \quad = 0 \end{array}\right\}$ $\therefore x = \pm y.$
$\therefore 3x^2 - 4x^2 + 2x^2 = 1$ $\left.\begin{array}{l} \therefore x = \pm 1 \\ \therefore y = \pm 1 \end{array}\right\}$

or $3y^2 + 4y^2 + 2y^2 = 1$ $\left.\begin{array}{l} \therefore y = \pm\frac{1}{3} \\ \text{and } x = \pm\frac{1}{3} \end{array}\right\}$

(20) $\left.\begin{array}{l} x^2 - xy = 12y^2 \\ x^2 + y^2 = 68 \end{array}\right\}$ $\therefore x = 4y$ or $-3y$

$\therefore$ By substitution, $17y^2 = 68$ or $10y^2 = 68$

$\therefore y = \pm 2$ or $\pm\sqrt{34/5}$

and $x = \pm 8$ or $\pm 3\sqrt{34/5}$

(21) $\left.\begin{array}{l} x^2 - 2xy + y^2 = 2x - 2y + 3 \\ x^2 + xy + 2y^2 = 2x - y + 3 \end{array}\right\}$

(*i*) By *Śūnyaṃ Anyat* $\therefore y = 0$

$\therefore$ Let $x - y = a \therefore a^2 - 2a - 3 = 0 \therefore a = 3$ or $-1$

$\therefore x - y = 3$ or $-1$.

Now, substitute and solve.

or (*ii*) By subtraction, $3xy + y^2 = y$

$\therefore y = 0$ or $3x + y = 1$

Substitute and solve.

*N.B.*: The *Śūnyaṃ Anyat* method is the best.

(22) $\left.\begin{aligned} 3x^2 + 2xy - y^2 &= 0 \\ x^2 + y^2 &= 2x(y + 2x) \end{aligned}\right\} \therefore x = -y \text{ or } \frac{1}{3}y$

$\therefore$ Substitute and solve

or (*ii*) By transposition,

$$-3x^2 - 2xy + y^2 = 0$$

This means that the two equations are not independent; and therefore, any value may be given to y and a corresponding set of values will emerge for x!

"Harder" simultaneous quadratics will be taken up at a later stage.

# 22

# Factorisation and Differential Calculus

In this chapter the relevant *Sūtras* (*Guṇaka-Samuccaya* etc.), dealing with successive differentiations, covering Leibniz's theorem, Maclaurin's theorem, Taylor's theorem etc., and given a lot of other material which is yet to be studied and decided on by the great mathematicians of the present-day western world, is also given.

Without going into the more abstruse details connected herewith, we shall, for the time-being, contend ourselves with a very brief sketch of the general and basic principles involved and a few pertinent sample-specimens by way of illustration.

The basic principle is, of course, elucidated by the very nomenclature, i.e. the *Guṇaka-Samuccaya* which postulates that, if and when a quadratic expression is the product of the binomials $(x+a)$ and $(x+b)$, its first differential is the sum of the said two factors and so on as already explained in the chapter on quadratic equations.

It need hardly be pointed out that the well-known rule of differentiation of a product, i.e. that if $y = uv$, when u and v be the functions of x, $\frac{dy}{dx} = v\frac{du}{dx} + u\frac{dv}{dx}$ and the *Guṇaka-Samuccaya Sūtras* denote, connote and imply the same mathematical truth.

Let us start with very simple instances:

(1) $x^2 + 3x + 2 = \overset{a}{(x+1)}\,\overset{b}{(x+2)}$

$\therefore D_1$ (the first differential) $= 2x + 3 = (x+2)(x+1) = \Sigma a$

(2) $x^3 + 6x^2 + 11x + 6 = \overset{a}{(x+1)}\,\overset{b}{(x+2)}\,\overset{c}{(x+3)}$

$\therefore D_1 = 3x^2 + 12x + 11 = (x^2 + 3x + 2) + (x^2 + 5x + 6)$

$+ (x^2 + 4x + 3) = ab + bc + ac = \Sigma ab.$

$\therefore D_2 = 6x + 12 = 2(3x + 6) = 2[(x + 1) + (x + 2) + (x + 3)]$
$= 2(a + b + c) = 2\Sigma a = \underline{|2}\ \Sigma a.$

(3) $x^4 + 10x^3 + 35x^2 + 50x + 24 = (x + 1)(x + 2)(x + 3)(x + 4)$
$\therefore D_1 = 4x^3 + 30x^2 + 70x + 50 = \Sigma abc$
$D_2 = 12x^2 + 60x + 70 = 2\Sigma ab = \underline{|2}\ \Sigma ab$
$D_3 = 24x + 60 = 6(4x + 10) \underline{|3}\ \Sigma a$

(4) $x^5 + 15x^4 + 85x^3 + 225x^2 + 274x + 120$
$= (x + 1)\ (x + 2)\ (x + 3)\ (x + 4)\ (x + 5)$
$\therefore D_1 = 5x^4 + 60x^3 + 255x^2 + 450x + 274 = \Sigma abcd$
$\therefore D_2 = 20x^3 + 180x^2 + 510x + 450 = \underline{|2}\ \Sigma abc$
$\therefore D_3 = 60x^2 + 360x + 510 = \underline{|3}\ \Sigma ab$
$\therefore D_4 = 120x + 360 = 24(5x + 15) = \underline{|4}\ \Sigma a$

(5) $x^4 + 19x^3 + 116x^2 + 284x + 240 = (x + 2)\ (x + 3)\ (x + 4)\ (x + 10)$
$\therefore D_1 = 4x^3 + 57x^2 + 232x + 284 = \Sigma abc$
$\therefore D_2 = 12x^2 + 114x + 232 = \underline{|2}\ \Sigma ab$
$\therefore D_3 = 24x + 114 = 6(4x + 19) = \underline{|3}\ \Sigma a$

These examples will suffice to show the internal relationship subsisting between the factors of a polynomial and the successive differentials of that polynomial; and to show how easily, on knowing the former, we can derive the latter and vice versa.

There is another relationship too in another direction wherein factorisation and differentiation are closely connected with each other and wherein this relationship is of immense practical help to us in our mathematical work. And this is with regard to the use of successive differentials for the detection of repeated factors.

The procedure hereof is so simple that it needs no elaborate exposition at all. The following examples will serve to show the *modus operandi* in question:

(1) Factorise $x^3 - 4x^2 + 5x - 2$
$\therefore D_1 = 3x^2 - 8x + 5 = (x - 1)\ (3x - 5)$

Judging from the first and the last coefficients of E the given expression, we can rule out $(3x - 5)$ and keep our eyes on $(x - 1)$.

$\therefore D_2 = 6x - 8 = 2(3x - 4)$ $\therefore$ we have $(x - 1)^2$

$\therefore$ According to the *Ādyam Ādyena Sūtra* $E = (x-1)^2(x-2)$

(2) Factorise $4x^3 - 12x^2 - 15x - 4$

$\therefore D_1 = 12x^2 - 24x - 15 = 3(4x^2 - 8x - 5) = 3(2x-5)(2x+1)$

$\therefore D_2 = 24x - 24 = 24(x-1)$ $\therefore$ As before, we have, $(2x+1)^2$

$\therefore E = (2x+1)^2(x-4)$

(3) Factorise $x^4 - 6x^3 + 13x^2 - 24x + 36$

$\therefore D_1 = 4x^3 - 18x^2 + 26x - 24 = 2(2x^3 - 9x^2 + 13x - 12)$

$= 2(x-3)(2x^2 - 3x + 4)$

$\therefore D_2 = 12x^2 - 36x + 26$ (which has no rational factors)

$\therefore E = (x-3)^2(x^2+4)$

(4) Factorise : $2x^4 - 23x^3 + 84x^2 - 80x - 64$

$\therefore D_1 = 8x^3 - 69x^2 + 168x - 80$

$\therefore D_2 = 24x^2 - 138x + 168 = 6(4x^2 - 23x + 28) = 6(x-4)(4x-7)$

$\therefore D_3 = 48x - 138 = 6(8x - 23)$

$\therefore D_2 = 6(x-4)(4x-7)$

$\therefore D_1 = (x-4)^2(8x-5)$

$\therefore E = (x-4)^3(2x+1)$

(5) Resolve $x^4 - 5x^3 - 9x^2 + 81x - 108$ into factors.

$\therefore D_1 = 4x^3 - 15x^2 - 18x + 81$

$\therefore D_2 = 12x^2 - 30x - 18 = 6(2x^2 - 5x - 3) = 6(x-3)(2x+1)$

$\therefore D_3 = 24x - 30 = 6(4x - 5)$

$\therefore D_2 = (x-3)(12x+6)$

$\therefore D_1 = (x-3)^2(4x+9)$

$\therefore E = (x-3)^3(x+4)$

(6) Resolve $16x^4 - 24x^2 + 16x - 3$ into factors.

$\therefore D_1 = 64x^3 - 48x + 16 = 16(4x^3 - 3x + 1)$

$\therefore D_2 = 192x^2 - 48 = 48(4x^2 - 1) = 48(2x-1)(2x+1)$

$\therefore D_3 = 384x$

$\therefore D_2 = (2x-1)(96x+48)$

$\therefore D_1 = (2x-1)^2(x+1)$

$\therefore E = (2x-1)^3(2x+3)$

(7) Resolve $x^5 - 5x^4 + 10x^3 - 10x^2 + 5x - 1$ into factors.

$\therefore D_1 = 5x^4 - 20x^3 + 30x^2 - 20x + 5$

$= 5(x^4 - 4x^3 + 6x^2 - 4x + 1)$

$\therefore D_2 = 20x^3 - 60x^2 + 60x - 20 = 20(x^3 - 3x^2 + 3x - 1)$

$\therefore D_3 = 3x^2 - 6x + 3 = 3(x^2 - 2x + 1)$

$\therefore D_4 = 6x - 6 = 6(x - 1)$

$\therefore D_3 = 3(x - 1)^2$

$\therefore D_2 = 4(x - 1)^3$

$\therefore D_1 = 5(x - 1)^4$

$\therefore E = (x - 1)^5$

(8) Factorise $x^5 - 15x^3 + 10x^2 + 60x - 72$

$\therefore D_1 = 5x^4 - 45x^2 + 20x + 60 = 5(x^4 - 9x^2 + 4x - 12)$

$\therefore D_2 = 20x^3 - 90x + 20 = 10(2x^3 - 9x + 2)$

$\therefore D_3 = 60x^2 - 90 = 30(2x^2 - 3)$

$\therefore D_4 = 120x$

$\therefore D_2 = 20(x - 2)^2 (x + 1)$

$\therefore D_1 = 5(x - 2)^2 (x + 1) (x + 3)$

$\therefore E = (x - 2)^3 (x + 3)^2$

Many other such applications are obtainable from the Vedic *Sūtras* relating to चलन-कलन (*Calana-Kalana*)–Differential Calculus. They are, however, to be dealt with, later on.

# 23

# Partial Fractions

Another subject of very great importance in various mathematical operations in general and in Integral Calculus in particular is "Partial Fractions" for which the current systems have a very cumbrous procedure but which the *'Parāvartya' Sūtra* tackles very quickly with its well-known mental one-line answer process.

We shall first explain the current method; and, along-side of it, we shall demonstrate the *"Parāvartya" Sūtra* application thereto. Suppose we have to express $\frac{3x^2+12x+11}{(x+1)(x+2)(x+3)}$ in the shape of partial fractions.

The current method is as follows:

Let, $\frac{3x^2+12x+11}{(x+1)(x+2)(x+3)} = \frac{A}{x+1} + \frac{B}{x+2} + \frac{C}{x+3}$

$$\therefore \frac{3x^2+12x+11}{(x+1)(x+2)(x+3)}$$

$$= \frac{A(x^2+5x+6)+B(x^2+4x+3)+C(x^2+3x+2)}{(x+1)(x+2)(x+3)}$$

$$\therefore x^2(A+B+C)+x(5A+4B+3C)+(6A+3B+2C) = (3x^2+12x+11)$$

$\therefore$ Equating the coefficients of like powers on both sides,

$$\left.\begin{aligned} A+B+C &= 3 \\ 5A+4B+3C &= 12 \\ 6A+3B+2C &= 11 \end{aligned}\right\}$$

$\therefore$ Solving these three simultaneous equations involving three unknowns, we have, $A = 1$; $B = 1$; and $C = 1$

$$\therefore E = \frac{1}{x+1} + \frac{1}{x+2} + \frac{1}{x+3}$$

In the Vedic system, however, for getting the value of A,

(*i*) we equate its denominator to zero and thus get the *Parāvartya* value of x (i.e. –1);

(*ii*) and we mentally substitute this value –1 in the E, but without the factor which is A's denominator on the R.H.S.; and

(*iii*) we put this result down as the value of A. Similarly for B and C.

Thus, $$A = \frac{3x^2 + 12x + 11}{(x+2)(x+3)} = \frac{3 - 12 + 11}{1 \times 2} = 1;$$

$$B = \frac{3x^2 + 12x + 11}{(x+3)(x+1)} = \frac{12 - 24 + 11}{(1)(-1)} = \frac{-1}{-1} = 1;$$

and $$C = \frac{3x^2 + 12x + 11}{(x+1)(x+2)} = \frac{27 - 36 + 11}{(-2)(-1)} = \frac{2}{2} = 1$$

$$\therefore E = \frac{1}{x+1} + \frac{1}{x+2} + \frac{1}{x+3}$$

*Note*: All this work can be done *mentally*; and all the laborious work of deriving and solving three simultaneous equations is totally avoided by this method.

A few more illustrations are shown below:

(1) $$\frac{2x+3}{(x+1)(x+2)} = \frac{1}{x+1} + \frac{1}{x+2}$$ (also available by mere *Vilokanam*)

(2) $$\frac{7}{(x+1)(x+2)} = \frac{7}{x+1} - \frac{7}{x+2}$$

(3) $$\frac{2x-5}{(x-2)(x-3)} = \frac{1}{x-2} + \frac{1}{x-3}$$

(4) $$\frac{3x+13}{(x+1)(x+2)} = \frac{10}{x+1} - \frac{7}{x+2}$$

(5) $$\frac{2x+1}{x^2-5x+6} = \frac{-5}{x-2} + \frac{7}{x-3}$$

(6) $$\frac{7x-1}{1-5x+6x^2} = \frac{-5}{1-2x} + \frac{4}{1-3x}$$

(7) $$\frac{9}{x^2+x-2} = \frac{3}{x-1} - \frac{3}{x+2}$$

(8) $$\frac{x-13}{x^2-2x-15} = \frac{2}{x+3} - \frac{1}{x-5}$$

(9) $\frac{x-5}{x^2-x-2}=\frac{2}{x+1}-\frac{1}{x-2}$

(10) $\frac{x+37}{x^2+4x-21}=\frac{4}{x-3}-\frac{3}{x+7}$

(11) $\frac{5+2x-3x^2}{(x^2-1)(x+1)}=\frac{(1+x)(5-3x)}{(x+1)^2(x-1)}$

$=\frac{5-3x}{x^2-1}=\frac{1}{x-1}-\frac{4}{x+1}$

(12) $\frac{5x-18}{x^2-7x+12}=\frac{3}{x-3}+\frac{2}{x-4}$

(13) $\frac{3x^2-10x-4}{(x^2-6x+8)}=3+\frac{8x-28}{(x-2)(x-4)}=3+\frac{6}{x-2}+\frac{2}{x-4}$

(14) $\frac{x^2+x+9}{x^2+6x^2+11x+6}=\frac{9}{2(x+1)}-\frac{11}{x+2}+\frac{15}{2(x+3)}$

(15) $\frac{2x+1}{x^3-6x^2+11x-6}=\frac{3}{2(x-1)}-\frac{5}{x-2}+\frac{7}{2(x-3)}$

(16) $\frac{2x^3-11x^2+12x+1}{x^3-6x^2+11x-6}=2+\frac{x^2-10x+13}{(x-1)(x-2)(x-3)}$
$=2+\frac{2}{x-1}+\frac{3}{x-2}-\frac{4}{x-3}$

Therefore, the general formula is:

$\frac{lx^2+mx+n}{(x-a)(x-b)(x-c)}$

$\therefore A=\frac{la^2+ma+n}{(a-b)(a-c)}$; $B=\frac{lb^2+mb+n}{(b-c)(b-a)}$; and $C=\frac{lc^2+mc+n}{(c-a)(c-b)}$

If and when, however, we find one or more factors of the denominator in repetition, i.e. a square, a cube etc., a slight variation of procedure is obviously indicated. For example, let E be $\frac{3x+5}{(1-2x)^2}$

According to the current system, we say:

Let $1-2x=p\left(\text{so that } x=\frac{1-p}{2}\right)$

$\therefore E=\frac{\frac{3-3p}{2}+5}{p^2}=\frac{13-3p}{2p^2}$

$$=\frac{13}{2p^2}-\frac{3}{2p}=\frac{13}{2(1-2x)^2}-\frac{3}{2(1-2x)}$$

This is no doubt a straight and simple procedure. But even this is rather cumbrous, certainly not easy and certainly not *mental* arithmetic! And, with bigger numbers and higher numbers as will be the case in the next example, it will be still worse!

The Vedic system, however, gives us two very easy *Parāvartya* methods whereby the whole work can be done mentally, easily and speedily. They are as follows:

(i) $\dfrac{3x+5}{(1-2x)^2}=\dfrac{A}{(1-2x)^2}+\dfrac{B}{1-2x}$

$\therefore 3x+5=A+B(1-2x)\ \ldots\ \ldots\ M$

$\therefore -2Bx=3;\text{and } A+B=5$

$\therefore B=-1\frac{1}{2}\text{and } A=6\frac{1}{2}$

(*ii*) $3x+5=A+B(1-2x)\ \ldots\ \ldots\ M$

$\therefore$ By *Parāvartya* (making $1-2x=0$, i.e. $x=\frac{1}{2}$ ),

we have $A=6\frac{1}{2}$; and as this is an absolute identity, i.e. true for all values of x, let us put $x=0$

$\therefore A+B=5\ \therefore B=-1\frac{1}{2}$

Two more examples are taken by way of illustration:

(1) $\dfrac{x^3+3x+1}{(1-x)^4}$

According to the current system, we say:

let $1-x=p$ (so that $x=1-p$)

$$\therefore E=\frac{(1-p)^3+3(1-p)+1}{p^4}$$

$$=\frac{1-3p+3p^2-p^3+3-3p+1}{p^4}$$

$$=\frac{5}{p^4}-\frac{6}{p^3}+\frac{3}{p^2}-\frac{1}{p}$$

$$=\frac{5}{(1-x)^4}-\frac{6}{(1-x)^3}+\frac{3}{(1-x)^2}-\frac{1}{(1-x)}$$

But according to the Vedic procedure, we say:

(1) $A + B(1-x) + C(1-x)^2 + D(1-x)^3 = x^3 + 3x + 1$

$$\therefore (A+B+C+D) + x(-B-2C-3D) + x^2(C+3D) - Dx^2 = x^3 + 3x + 1$$

$$\left.\begin{array}{ll} \therefore -D = 1 & \therefore D = -1 \\ \therefore C + 3D = C - 3 = 0 & \therefore C = 3 \\ \therefore -B - 2C - 3D = -B - 6 + 3 = 3 & \therefore B = -6 \\ \therefore A + B + C + D = A - 6 + 3 - 1 = 1 & \therefore A = 5 \end{array}\right\}$$

or, secondly, by *Parāvartya*,

$$\begin{array}{ll} \text{Put } x = 1 & \therefore A = 5 \\ \left.\begin{array}{l}\text{Put } x = 0 \therefore A + B + C + D = 1 \\ \text{Put } x = 2 \therefore A - B + C - D = 15\end{array}\right\} & \therefore B = -6 \\ & \therefore C = 3 \\ & \therefore D = -1 \end{array}$$

all of which can be done by *mental* Arithmetic.

(2) $\dfrac{5+2x-2x^2}{(x^2-1)(x+1)} = \dfrac{A}{(x+1)^2} + \dfrac{B}{x+1} + \dfrac{C}{x-1}$

$$\left.\begin{array}{l} \therefore A(x-1) + B(x^2-1) + C(x+1)^2 \\ = (-A-B+C) + x(A+2C) + x^2(B+C) \\ = 5 + 2x - 2x^2 \end{array}\right\} \quad \left.\begin{array}{ll} \therefore A + B - C & = -5 \\ A + 2C & = 2 \\ B + B & = -2 \end{array}\right\}$$

$\therefore A = -\frac{1}{2}$; $B = -3\frac{1}{4}$; and $C = 1\frac{1}{4}$

or, secondly by *Parāvartya*,

$$\left.\begin{array}{lll} \text{Put } x = 1 & \therefore 4C = 5 & \therefore C = 1\frac{1}{4} \\ x = -1 & \therefore -2A = 1 & \therefore A = -\frac{1}{2} \\ x = 0 & & \therefore B = -3\frac{1}{4} \end{array}\right\}$$

*N.B.*: (1) It need be hardly pointed out that the current method will involve an unquestionably cumbrous and clumsy process of working, with all the attendant waste of time, energy etc.

(2) Other details of applications of *Parāvartya* and other *Sūtras* to partial fractions, will be dealt with later.

(3) Just now we take up an important part of Integral Calculus wherein, with the help of partial fractions, we can easily perform difficult integrational work.

# 24

# Integration

### By Partial Fractions

In this chapter we shall deal, briefly, with the question of integration by means of partial fractions. But, before we take it up, it will not be out of place for us to give a skeleton-sort of summary of the first principles and process of integration as dealt with by the *Ekādhika Sūtra*.

The original process of differentiation is, as is well known, a process in which we say:

Let $y = x^3$. Then $D_1\left(\text{i.e. } \frac{dy}{dx}\right) = 3x^2$;

$$D_2 = 6x; \text{ and } D_3 = 6$$

Now, in the converse process, we have:

$$\frac{dy}{dx} = 3x^2 \therefore dy = 3x^2dx$$

Integrating, $\therefore \int dy = \int 3x^2dx \therefore y = x^3$

Thus, in order to find the integral of a power of x, we add unity to the पूर्व (*Pūrva*), i.e. the original index and divide the coefficient by the new index, i.e. the original one plus unity.

A few specimen examples may be taken:

(1) Integrate $28x^3$. $\int 28x^3dx = \frac{28}{4}x^4 = 7x^4$

(2) $\int(x^4 + 3x^3 + 6x^2 + 7x - 9)\,dx$

$= \frac{1}{5}x^5 + \frac{3}{4}x^4 + 2x^3 + 3\frac{1}{2}x^2 - 9x + K$ where K is an independent term.

(3) $\int(x^a + x^{a-1} + x^{a-2} \text{ etc.})$

$$= \frac{x^{a+1}}{a+1} + \frac{x^a}{a} + \frac{x^{a-1}}{a-1} \ldots\ldots \text{ etc.}$$

(4) $\int(ax^{m+1} + bx^{m} + cx^{m-1})\,dx.$

$$= \frac{ax^{m+2}}{m+2} + \frac{bx^{m+1}}{m+1} + \frac{cx^{m}}{m} \ldots\ldots \text{etc.}$$

This is simple enough, as far as it goes. But what about complex expressions involving numerators and denominators? The following sample specimens will make the procedure by means of partial fractions clear:

(1) Integrate $\frac{7x - 1}{6x^2 - 5x + 1}$

$$\therefore \text{By } \textit{Parāvartya}, \quad \frac{7x-1}{6x^2-5x+1} = \frac{7x-1}{(2x-1)(3x-1)}$$

$$= \frac{5}{2x-1} - \frac{4}{3x-1}$$

$$\therefore \int \frac{(7x-1)dx}{6x^2-5x+1} = 5\int \frac{dx}{2x-1} - 4\int \frac{dx}{3x-1}$$

$$= \frac{5}{2}\int \frac{d(2x)}{2x-1} - \frac{4}{3}\int \frac{d(3x)}{3x-1}$$

$$= \frac{5}{2}\log(2x-1) - \frac{4}{3}\log(3x-1)$$

$$= \log\left[\frac{(2x-1)^{5/2}}{(3x-1)^{4/3}}\right]$$

(2) Integrate $\frac{x^2-7x+1}{x^3-6x^2+11x-6}$

$$\therefore \text{By } \textit{Parāvartya}, \quad \frac{x^2-7x+1}{x^3-6x^2+11x-6} = \frac{x^2-7x+1}{(x-1)(x-2)(x-3)}$$

$$= \frac{-5}{2(x-1)} + \frac{9}{x-2} - \frac{11}{2(x-3)}$$

$$\therefore \int \frac{(x^2-7x+1)\,dx}{x^3-6x^2+11x-6} = \int\left\{\frac{-5}{2(x-1)} + \frac{9}{x-2} - \frac{11}{2(x-3)}\right\}dx$$

$$= -\frac{5}{2}\int \frac{dx}{x-1} + 9\int \frac{dx}{x-2} - \frac{11}{2}\int \frac{dx}{x-3}$$

$$= -\frac{5}{2}\log(x-1) + 9\log(x-2) - \frac{11}{2}\log(x-3)$$

(3) Integrate $\dfrac{1}{x^3 - x^2 - x + 1}$

Let $\dfrac{1}{x^3 - x^2 - x + 1} = \dfrac{A}{x-1} + \dfrac{B}{(x-1)^2} + \dfrac{C}{x+1}$ ....... M

$\therefore 1 = A(x-1)(x+1) + B(x+1) + C(x-1)^2$

$= A(x^2-1) + B(x+1) + C(x-1)^2$......N

Now, let $x = 1 \therefore 1 = 2B \therefore B = \frac{1}{2}$

Differentiating (N),

$0 = 2Ax + B + 2Cx - 2C$.......P

Now put $x = 1 \therefore 2A = -\frac{1}{2} \therefore A = -\frac{1}{4}$

Differentiating (P), $2A + 2C = 0 \therefore 2C = \frac{1}{2} \therefore C = \frac{1}{4}$

$$\therefore E = \frac{-1}{4(x-1)} + \frac{1}{2(x-1)^2} + \frac{1}{4(x+1)}$$

$$\therefore \int \frac{dx}{x^3 - x^2 - x + 1} = -\frac{1}{4}\int \frac{dx}{(x-1)} + \frac{1}{2}\int \frac{dx}{(x-1)^2} + \frac{1}{4}\int \frac{dx}{(x+1)}$$

$$= -\frac{1}{4}\log(x-1) - \frac{1}{2}\frac{1}{x-1} + \frac{1}{4}\log(x+1)$$

# 25

# The Vedic Numerical Code

It is a matter of historical interest to note that, in their mathematical writings, the ancient Sanskrit writers do not use figures when big numbers are concerned in their numerical notations but prefer to use the letters of the Sanskrit Devanagari alphabet to represent the various numbers! And this they do, not in order to conceal knowledge but in order to facilitate the recording of their arguments, and the derivation conclusions etc. The more so, because, in order, to help the pupil to memorise the material studied and assimilated they made it a general rule of practice to write even the most technical and abstruse text-books in *Sūtras* or in Verse which is so mush easier—even for the children—to memorise than in prose which is so much harder to get by heart and remember. And this is why we find not only theological philosophical, medical, astronomical and other such treatises but even huge dictionaries in Sanskrit Verse! So, from this stand-point, they used verse, *Sūtras* and codes for lightening the burden and facilitating the work by versifying scientific and even mathematical material in a readily assimilable form!

The very fact that the alphabetical code as used by them for this purpose is in the natural order and can be immediately interpreted, is clear proof that the code language was resorted to not for concealment but for greater ease in verification etc., and the key has also been given in its simplest form: "कादि नव, टादि नव, पादि पञ्चक, याद्यष्टक and क्षः शून्यम्" which means:

(1) *ka* and the following *eight* letters;
(2) *ṭa* and the following *eight* letters;
(3) *pa* and the following *four* letters;
(4) *ya* and the following *seven* letters; and
(5) *kṣa* (or Kṣudra) for Zero.

Elaborated, this means:

(1) *ka, ṭa, pa* and *ya* all denote 1;
(2) *kha, tha, pha* and *ra* all represent 2;
(3) *ga, ḍa, ba* and *la* all stand for 3;
(4) *gha, dha, bha,* and *va* all denote 4;
(5) *gna, ṇa, ma* and *sa* all represent 5;
(6) *ca, ta,* and *śa* all stand for 6;
(7) *cha, tha,* and *ṣa* all denote 7;
(8) *ja, da* and *ha* all represent 8;
(9) *jha* and *dha* stand for 9; and
(10) *Kṣa* (or *Kṣudra*) means Zero!

The vowels not being included in the list make no difference; and in conjunct consonants, the last consonant is alone to be counted.

Thus *pa pa* is 11, ma ma is 55, *ṭa ṭa* is 11, *ma ra* is 52 and so on!

And it was left to the author to select the particular consonant or vowel which he would prefer at each step. And, generally, the poet availed himself of this latitude to so frame his selections as to bring about another additional meaning or meanings of his own choice. Thus, for instance, *kapa, ṭapa, papa* and *yapa* all mean 11; and the writer can by a proper selection of consonants and vowels import another meaning also into the same verse. Thus "I want *mama* and *papa*" will mean "I want 55 and 11"!

Concrete, interesting and edifying illustrations will be given later on especially in connection with recurring decimals, Trigonometry etc. wherein, over and above the mathematical matter on hand, we find historical allusions, political reflections, devotional hymns in praise of Lord Shri Krishna, Lord Shri Shankara and so on!*

This device is thus not merely a potent aid to versification for facilitating memorisation but has also a humorous side to it which adds to the fun of it!

---

*The hymn in praise of the Lord gives us the value of $\pi$ to 32 decimal places in Trigonometry.

# 26

# Recurring Decimals

It has become a sort of fashionable sign of cultural advancement for people now-a-days to talk not only grandly but also grandiosely and grandiloquently about decimal coinage, decimal weights, decimal measurements etc.; but there can be no denying or disguising of the fact that the western world as such—not excluding its mathematicians, physicists and other expert scientists—seems to have a tendency to *theorise* on the one hand on the superiority of the decimal notation and to fight shy, on the other, in actual *practice*—of decimals and positively prefer the "vulgar fractions" to them!

In fact, this deplorable state of affairs has reached such a pass that the mathematics syllabus—curricula in the schools, colleges and universities have been persistently "progressing" and "advancing" in this wrong direction to the extent of declaring that recurring decimals are not integral parts of the matriculation course in mathematics and actually instructing the pupils to convert all recurring decimals at sight into their equivalent vulgar fraction shape, complete the whole work with them and finally re-convert the fraction result back into its decimal shape!

Having invented the zero mark and the decimal notation and given them to the world as described already from the pages of Prof. Halstead and other Historians of Mathematics, the Indian Vedic system has, however, been advocating the decimal system, not on any *a priori* grounds or because of partiality but solely on its intrinsic *merits*. Its unique achievements in this direction have been of a most thrillingly wonderful character: and we have already—at the very commencement of this illustrative volume—given a few startling sample-specimens thereof. The student will doubtless remember that, at the end of that chapter, we promised to go into fuller details of this subject at a later stage. In fulfilment of that promise, we now pass on to a further exposition of the marvels of Vedic mathematics in this direction.

### PRELIMINARY NOTE

We may begin this part with a brief reference to the well-known distinction between non-recurring decimals, recurring ones and partly-recurring ones.

(*i*) A denominator containing only 2 or 5 as factors gives us an ordinary, i.e. non-recurring or non-circulating decimal fraction each 2, 5 or 10 contributing *one* significant digit to the decimal. For instance,

$$\tfrac{1}{2} = .5;\ \tfrac{1}{4} = \frac{1}{2 \times 2} = .25;\ \tfrac{1}{8} = \frac{1}{2 \times 2 \times 2} = .125;$$

$$\tfrac{1}{16} = \frac{1}{2 \times 2 \times 2 \times 2} = .0625;\ \tfrac{1}{32} = \frac{1}{2^5} = .03125;$$

$$\tfrac{1}{5} = .2;\ \tfrac{1}{10} = .1;\ \tfrac{1}{20} = .05;\ \tfrac{1}{25} = \frac{1}{5^2} = .04;$$

$$\tfrac{1}{40} = \frac{1}{10 \times 2^2} = .025;\ \tfrac{1}{50} = \frac{1}{10 \times 5} = .02;$$

$$\tfrac{1}{80} = \frac{1}{10 \times 2^3} = .0125;\ \tfrac{1}{100} = \frac{1}{10^2} = .01; \text{ and so on.}$$

(*ii*) Denominators containing only 3, 7, 11 or higher prime numbers as factors and not even a single 2 or 5 give us recurring or circulating decimals which we shall deal with in detail in this chapter and in some other later chapters too.

$$\tfrac{1}{3} = .3;\ \tfrac{1}{7} = .\dot{1}4285\dot{7};\ \tfrac{1}{9} = .\dot{1};\ \tfrac{1}{11} = .\dot{0}\dot{9};$$

$$\tfrac{1}{13} = .\dot{0}7692\dot{3};\ \tfrac{1}{17} = .\dot{0}5882352\,/\,941176 4\dot{7};$$

$$\tfrac{1}{19} = \dot{0}52631578\,/\,94736842\dot{1}; \text{ and so on}$$

(*iii*) A denominator with factors partly of the former type, i.e. 2 and 5 and partly of the latter type, i.e. 3, 7, 9 etc. gives us a mixed, i.e. partly recurring and partly non-recurring decimal, each 2, 5 or 10 contributing one non-recurring digit to the decimal.

$$\tfrac{1}{6} = \frac{1}{2 \times 3} = .1\dot{6};\ \tfrac{1}{16} = \frac{1}{3 \times 5} = .0\dot{6};$$

$$\tfrac{1}{18} = \frac{1}{2 \times 9} = .0\dot{5};\ \tfrac{1}{22} = \frac{1}{2 \times 11} = .0\dot{4}\dot{5};$$

$$\tfrac{1}{24} = \frac{1}{2^3 \times 3} = .041\dot{6}; \text{ and so on.}$$

*N.B.*: (*i*) Each 3 or 9 contributes only one recurring digit; 11 gives 2 of them; 7 gives 6; and other numbers make their own individual contribution details of which will be explained later.

(*ii*) In every non-recurring decimal with the standard numerator, i.e. 1, it will be observed that the last digit of the denominator and the last digit of the equivalent decimal, multiplied together, will always yield a product ending in zero; and

(*iii*) In every recurring decimal with the standard numerator, i.e. 1, it will be similarly observed that 9 will invariably be the last digit of the product of the last digit of the denominator and the last digit of its recurring decimal equivalent nay, the product is actually a continuous series of nines!

Thus, $\frac{1}{2}=.5$; $\frac{1}{5}=.2$; $\frac{1}{10}=.1$; $\frac{1}{4}=.25$; $\frac{1}{8}=.125$;

$\frac{1}{16}=.0625$; $\frac{1}{25}=.04$ $\frac{1}{125}=.008$; etc.

And $\frac{1}{3}=.\dot{3}$; $\frac{1}{7}=.\dot{1}42857$; $\frac{1}{9}=.\dot{1}$; $\frac{1}{11}=.\dot{0}\dot{9}$;

$\frac{1}{13}=.\dot{6}7692\dot{3}$; etc.

And this enables us to determine beforehand, the last digit of the recurring decimal equivalent of a given vulgar fraction. Thus $\frac{1}{17}$ in its decimal shape must necessarily end in 7; $\frac{1}{19}$ in 1; $\frac{1}{21}$ in 9; $\frac{1}{23}$ in 3; and so on. The immense practical utility of this rule in the conversion of vulgar fractions into their decimal shape has already been indicated in the first chapter and will be expatiated on, further ahead in this chapter and in subsequent chapters.

Let us first take the case of $\frac{1}{7}$ and its conversion from the vulgar fraction to the decimal shape. We note here:

(*i*) that the successive remainders are 3, 2, 6, 4, 5 and 1 and that, inasmuch as 1 is the original figure with which we started, the same remainders are bound to repeat themselves in the same sequence endlessly. And this is where we stop the division-process and put the usual recurring marks the dots on the first and the last digits in order to show that the decimal has begun its characteristic recurring character.

```
7)1.0(.142857
  7
  --
   30
   28
   --
    20
    14
    --
     60
     56
     --
      40
      35
      --
       50
       49
       --
        1
```

At this point, we may note that inasmuch as the first dividend 10 when divided by 7 gives us the first remainder 3, and, with a zero affixed to it, this 3 will as 30 become our second dividend and inasmuch as this process will be continuing indefinitely (until a remainder repeats itself and warns us that the recurring decimal's recurring character has begun to manifest itself) it stands to reason that there should be a uniform ratio in actual action. In other words, because the first dividend 10 gives us the first remainder 3 and the second dividend 30, therefore *Ānurūpyeṇa*, i.e. according to the ratio in question or by simple rule of three, this second dividend 3 should give us the second remainder 9! In fact, it is a "*Geometrical Progression*" that we are dealing with!

And when we begin testing the successive remainders from this standpoint, we note that the said inference about the Geometrical Progression with the common ratio 1:3 is correct. For, although, when we look for $3 \times 3 = 9$ as the second remainder, we actually find 2 there instead, yet as 9 is greater than 7 the divisor, it is but proper that, by further division of 9 by 7, we get 2 as the remainder. And then we observe that this second remainder 2 yields us the third remainder 6, and thereby keeps up the Geometrical Progression with the same ratio 1:3. In the same way, this 6 gives us 18 which being greater than the divisor and being divided by it gives us 4 as the fourth remainder. And 4 gives us 12 which after division by 7 gives us 5 as the fifth remainder! And, by the same ratio, this 5 gives us 15 which when divided by 7 gives us 1 as the sixth remainder. And as this was the dividend which we began with, we stop the division-process here!

The fun of the Geometrical Progression is no doubt there; but it is not for the mere fun of it, but also for the practical utility of it, that we have called the student's attention to it. For, in the actual result, it means that, once we know the ratio between the first dividend and the first remainder 1:3 in the present case, we can—without actual further division—automatically put down all the remainders by maintaining the 1:3 Geometrical Progression. For example, in the present case, since the ratio is uniformly 1:3, therefore the second remainder is 9 which after deducting the divisor, we set down as 2; and so on until we reach 1.

$$7)1.0(\text{G.P.} 1, 3, 2, 6, 4, 5$$
$$\frac{7}{3}$$

Thus our chart reads as follows:

1, 3, 2, 6, 4, 5

Yes, but what do we gain by knowing the remainders before-hand without actual division? The answer is that, as soon as we get the first remainder, our whole work is practically over. For, since each remainder with a zero affixed automatically becomes the next dividend, we can *mentally* do this affixing *at sight*, mentally work out the division at each step and put down the quotient automatically without worrying about the remainder! For, the remainder is already there in front of us!

Thus the remainders 1, 3, 2, 6, 4 and 5 give us the successive dividends 10, 30, 20, 60, 40 and 50; and, dividing these mentally, by 7, we can go forward or backward and obtain all the Quotient-digits, 1, 4, 2, 8, 5, and 7. And, as it is a pure circulating decimal, our answer is 142857 !

$$\begin{array}{c} \text{Dividend - digits} \\ 1,\ 3,\ 2,\ 6,\ 4,\ 5 \\ \hline 1\ \ 4\ \ 2\ \ 8\ \ 5\ \ 7 \\ \text{Quotient - digits} \end{array}$$

There is, however, a still more wonderful Vedic method by which, without doing even this little division-work, we can put down the quotient-digits automatically forward or backward, from any point whatsoever! The relevant *Sūtra* hereon says: शेषाण्यङ्केन चरमेण (*Śeṣāṇyaṅkena Carameṇa*) and means: The remainders *by* the last digit.

As explained in another context in the very first chapter of this volume, the word *by* indicates that the operation is not one of addition or of subtraction but of division and of multiplication!

The division-process whereby we affix a zero to 1, 3, 2 etc., divide the product by 7 and set down the quotient has been shown just above. We now show the reverse process of multiplication, which is still easier.

In so doing, we put down not the dividend—nucleus digits but the remainders themselves in order: 3, 2, 6, 4, 5, 1.

And, as we know from a previous paragraph that 7 is the last digit, we multiply the above given remainders by 7 and put the last, i.e. the right-hand-most digit down under each of the remainders totally ignoring the other digit or digits, if any, of the product. And lot the answer is there in front of us again, really looking more like magic than like mathematics! Thus,

$$\begin{array}{c} \text{(Remainders)} \\ 3,\ 2,\ 6,\ 4,\ 5,\ 1 \\ \hline 1\ \ 4\ \ 2\ \ 8\ \ 5\ \ 7 \\ \text{(Quotient - digit)} \end{array}$$

3 multiplied by 7 gives us 21; and we put down only 1;

$2 \times 7$ gives us 14; and we put down only 4;

$6 \times 7$ gives us 42; and we put down only 2;

$4 \times 7$ gives us 28; and we put down only 8;

$5 \times 7$ gives us 35; and we put down only 5; and

$1 \times 7$ gives us 7; and we put down 7. And the answer is $.\dot{1}4285\dot{7}$

At this point, we may remind the student of a very important point which we have already explained in Chapter One regarding the conversion of $\frac{1}{19}$, $\frac{1}{29}$ and $\frac{1}{49}$ to their recurring decimal shape. This is in connection with the facts that the two halves of these decimals together total a series of nines; that, once

$$\begin{array}{l} 7)1.0(.\dot{1}4285\dot{7} \\ \quad\ \ 7 \\ \hline \quad\ \ 30 \\ \quad\ \ 28 \\ \hline \quad\ \ \ 20 \\ \quad\ \ \ 14 \\ \hline \end{array}$$

half the answer is known, the other half can be had by putting down the complements from nine of the digits already obtained; and that, as the ending of the first half of the result synchronises with our reaching of the difference between the numerator and the denominator as the remainder, we know when exactly we should stop the division or multiplication, as the case may be and begin the mechanical subtraction from 9 of the digits already found!

```
 60
 56
 ---
  40
  35
  ---
   50
   49
   ---
    1
```

The student can easily realise how, inasmuch as this rule is applicable to every case wherein $D \sim N$ comes up as a remainder, it therefore means an automatic reduction of even the little labour involved, by exactly one-half!

Going back to the original topic re: the conversion of vulgar fractions into their equivalent decimal shape and how the geometrical progressional ratio can give us beforehand—without actual division—all the remainders that will come up in actual division, we now take up $\frac{1}{13}$ as another illustrative example and observe how the process works out therein:

(1) (*i*) $\frac{1}{13}$ Here the successive Dividents—nucleus—digits are 1, 10, 9, 12, 3 and 4. Affixing a zero to each of them and dividing the dividends by 13, we get 0, 7, 6, 9, 2 and 3 as the first digits of the quotient in the answer.

```
13)1.00 (.076
    91( 923
    ---
     90 999
     78
     ---
     120
     117
     ---
       30
       26
       ---
        40
        39
        ---
         1
```

(*ii*) Or, secondly, re-arranging the remainders so as to start from the first actual remainder, we have: 10, 9, 12, 3, 4 and 1. And multiplying these by 3 the last digit of the answer in the present case we put down merely the right-hand-most digit of each product; and these are the successive quotient-digits! Here too, as usual, we go forward or back-ward or in any sequence which we may choose. And the answer is $.\dot{0}7692\dot{3}$

```
13) 1.00 (G.P. 1, 10, 9, 12, 3, 4
          -----------------------
           0   7  6   9  2  3

13) 1.00 (10, 9, 12, 3, 4  1
          ------------------
           0  7   6  9  2  3
```

(*iii*) And here too we observe, in operation, the rule about complements from nine! And it commences from the point at which we obtain 12 the difference between the numerator and the denominator as the remainder.

$$\begin{array}{r} 076 \\ 923 \\ \hline 999 \end{array}$$

(*iv*) In the above charts, we may avoid big numbers by using a *minus* where a big number is threatened.

Thus, instead of taking 3 as in the case of 7, we may take –3 as the common geometrical ratio and will find the geometrical progression intact; and naturally the product of each remainder-digit by the last digit remains intact too and gives us the same answer: $.\dot{0}7692\dot{3}$ !

G.P. $1, \frac{3}{10}, 9, \frac{-27}{12}$ etc.

$$\begin{array}{cccccc} -\frac{3}{10}, & 9, & -\frac{27}{12}, & -\frac{36}{3}, & -\frac{9}{4}, & -\frac{12}{1} \\ \hline 0 & 7 & 6 & 9 & 2 & 3 \end{array}$$

We pass now on to still another and easier method which comes under the *Ekādhika Sūtra* which we have expounded and explained at sufficient length already in the first chapter and which therefore we need only summarise and supplement here but need not elaborate again.

The *Ekādhika Sūtra* which means 'by the preceding one increased by one' has already been shown at work in a number of ways and in a number of directions and on a number of occasions and will similarly come into operation still further, in many more ways and in many more contexts.

## Numbers Ending in Nine

(*i*) If and when the last digit of the denominator is 9, we know beforehand that the equivalent recurring decimal ends in 1.

(*ii*) In the case of $\frac{1}{19}$, the last but one digit is 1; we increase it by 1 and make it 2. In $\frac{1}{29}$ we work with $2+1=3$. In $\frac{1}{39}$ and in $\frac{1}{49}$, we operate with 4 and 5 respectively and so on.

(*iii*) In the multiplication-process by *Ekādhika Pūrva*, in all these cases, we put 1 down as the last digit, i.e. in the right-hand-most place; and we go on multiplying that last digit 1 from the right towards the left by 2, 3, 4 and 5 respectively; and when there is more than one digit in that product, we set the last of those digits down there and carry the rest of it over to the next immediately preceding digit towards the left.

(*iv*) When we get D~N as the product, we know we have done half the work; we stop the multiplication there; and mechanically put down the remaining half of the answer by merely taking down the complements from nine.

(*v*) The division-process by *Ekādhika Sūtra* follows the same rules *vide supra*.

(1) We may first consider the fraction $\frac{1}{19}$ as our first illustration of the method described:

(*i*) Putting 1 as the last digit and continually multiplying by 2 towards the left, we get the last four digits towards the left without the least difficulty.

$$\ldots\ldots\ 9\ 4\ 7\ 3\ 6\ 8\ 4\ 2\ 1$$
$$1\quad 1\ 1$$

(*ii*) $8 \times 2 = 16.$ Therefore put 6 down immediately to the left of 8 with 1 to carry over. $6 \times 2 +$ the 1 carried over $= 13.$ Put the 3 to the left of the 6 with 1 to carry over. $3 \times 2 + 1 = 7.$ Set it down before the 3 with nothing to carry over. 7 × 2 = 14. Therefore put the 4 before the 7 with 1 to carry over, 4 × 2 + the 1 carried over = 9.

(*iii*) We have thus got 9 digits by continual multiplication from the right towards the left. And now $9 \times 2 = 18$ (which is D~N). This means that half the work is over and that the earlier 9 digits are obtainable by putting down the complements from nine of the digits already determined. So, we have

$$\frac{1}{19} = 0.\dot{0}\,5\,2\,63\,1\,5\,7\,8\,/\,9\,4\,7\,3\,6\,8\,4\,2\,1$$

(2) Let us now examine the case of $\frac{1}{29}$ :

Begin with 1 as usual at the extreme right and go on multiplying by 3 each time, "carrying over" the surplus digit or digits if any to the left, i.e. to be added to the next product to be determined. Thus, when we have obtained 14 digits, i.e.

. . . . . . 9 6 5 5 1 7 2 4 1 3 7 9 3 1

we find that we have reached 28; we know we have done half the work; and we get the first 14 digits by simply subtracting each of the above digits from nine.

.0 3 4 4 8 2 7 5 8 6 2 0 6 8 /

$$\therefore \frac{1}{29} = 0.\dot{0}\,3\,4\,4\,8\,2\,7\,5\,8\,6\,2\,0\,6\,8\,/\,9\,6\,5\,5\,1\,7\,2\,4\,1\,3\,7\,9\,3\,\dot{1}$$

(3) Next let us take $\frac{1}{39}$

Take 1 again at the extreme right end and continually multiply by 4 from the right to the left. Thus, we have:

$$\frac{1}{39} = .\dot{0}2564\dot{1}.$$

Note in this case that ∴ 39 is a multiple of 3 and 13 and not a prime number like 19 and 29 and ∵ 3 and 13 give only 1 and 6 recurring decimals, there is a difference in its behaviour, i.e. that the two halves are not complementary with regard to 9 but only in relation to 6! In fact, D~N, i.e. 38 does not come up at all as an interim product as 18 and 28 did. And so, the question of complements from 9 does not arise at all; and the decimal equivalent has only 6 figures and not 38!

The reason for this is very simple. As $\frac{1}{39} = \frac{1}{13} \div 3$, as $\frac{1}{13}$ has only 6 recurring decimals in its decimal equivalent and because, for reasons to be explained a little later, this decimal equivalent of $\frac{1}{13}$ is exactly divisible by 9, much more therefore is it divisible by 3. And, consequently, when we divide it by 3 and exhaust the six digits, we find that there is no remainder left. In other words, $\frac{1}{39}$ has only 6 digits in its recurring decimal 1 shape.

These have been obtained by the self-same *Ekādhika* process as served our purpose in the case of $\frac{1}{19}$ and $\frac{1}{29}$.

We next take up and examine the case of $\frac{1}{49}$ which, besides following the rules hereinabove explained, has the additional merit of giving us the clue to a still easier process for the conversion of vulgar fractions into their recurring decimal shape:

(1) $\frac{1}{49} = \frac{1}{7 \times 7} = \frac{.142857}{7}$

(*i*) If ∴ we go on dividing 1 by 49 or $.\dot{1}4285\dot{7}$ by 7 until the decimal begins to recur, we shall doubtless get our answer. But this will mean 42 steps of labourious working and is therefore undesirable.

(*ii*) We therefore adopt either of the *Ekādhika* methods and go on multiplying from right to left by 5 or dividing from left to right by 5.

(*iii*) On completing 21 digits, we find 48, i.e. D~N coming up and standing up before us; and we mechanically put down the other 21 digits as usual by the subtraction, from 9, of the digits already obtained. And the answer is:

$$\frac{1}{49} = \left.\begin{matrix} 0.0\,2\,0\,4\,0\,8\,1\,6\,3\,2\,6\,5\,3\,0\,6\,1\,2\,2\,4\,4\,8 \\ 9\,7\,9\,5\,9\,1\,8\,3\,6\,7\,3\,4\,6\,9\,3\,8\,7\,7\,5\,5\,1 \end{matrix}\right\}$$

(*iv*) And this gives us the clue just above referred to about a still easier method than even the *Ekādhika* ones for the conversion of vulgar fractions into recurring decimals. And it is as follows:

```
49)1.00(.020408
     98
    200
    196
    400
```

By actual division of 1 by 49, we observe that the successive remainders are in geometrical progression with the common ratio 1:2 that the dividends are similarly related and that each set of the two digits in the quotient is also so related to its predecessor.

In other words, this connotes and implies that, after putting down 02, we can automatically put down 04, 08, 16 and 32 and so on.

But when we reach 64, we find that $2 \times 64 = 128$, i.e. it has 3 digits. All that we have to do then is to add the 1 of the 128 over to the 64 already there, turn it into 65 and then put down not 28 but the remaining part of double the *corrected* figure 65 (i.e. 30) and carry the process carefully on to the very end, i.e. until the decimal starts to recur. We therefore have:

$$\frac{1}{49} = \left.\begin{array}{l} 0.\dot{0}\,2\,0\,4\,0\,8\,1\,6\,3\,2\,6\,5\,3\,0\,6\,1\,2\,2\,4\,4\,8 \\ \quad 9\,7\,9\,5\,9\,1\,8\,3\,6\,7\,3\,4\,6\,9\,3\,8\,7\,7\,5\,5\,\dot{1} \end{array}\right\}$$

This new method does not apply to all cases but only to some special cases where the denominator of the given vulgar fraction or on integral multiple thereof is very near a power of ten and thus lends itself to this kind of treatment. In such cases, however, it is the best procedure of all.

*Note*: The rule of complements from 9 is actually at work in this case too; but, inasmuch as for reasons to be explained hereafter, the actual total number of digits is 42, the first half of it ends with the 21st digit and as we have been taking up a group of two digits at each step, we naturally by-pass the 21st digit which is concealed, so to speak, in the middle of the 11th group. But, even then, the double-digit process is so very simple that continuation thereof can present no difficulty.

## OTHER ENDINGS

So far, we have considered only vulgar fractions whose denominators end in 9. Let us now go on to and study the cases of $\frac{1}{7}, \frac{1}{13}, \frac{1}{17}, \frac{1}{23}$ and other such fractions whose denominators end not in 9 but in 1, 3 or 7.

(*i*) Here too, we first make up our minds, at sight, as regards the last digit of the decimal equivalent. Thus, denominators ending in 7, 3 and 1 must necessarily yield decimals ending in 7, 3 and 9 so that the product of the last digit of the denominator and the last digit of the decimal equivalent may end in 9. Let us start with the case of $\frac{1}{7}$.

(*ii*) Put down $\frac{1}{7}$ in the shape $\frac{7}{49}$ $\qquad \frac{1}{7} = \frac{7}{49}$

(*iii*) Take 5 one more than 4 as the *Ekādhika Pūrva* for the required multiplication or division as the case may be.

(*iv*) Thus start with 7 at the right end.

$$\begin{array}{r} 8\,5\,7 \\ 2\,3\phantom{\,7} \end{array}$$

(*v*) Multiply it by 5 and set down 35 as shown in the marginal chart.

(*vi*) Multiply 5 by 5, add 3 to the product and set 28 down in the same way.

$$142 \,\Big|\, \begin{matrix} 8\ 5\ 7 \\ 2\ 5 \end{matrix}$$

Now, $5 \times 8 + 2 = 42$. But that is D~N. Therefore put 142 down as the first half according to the complements rule $\therefore \frac{1}{7} = .\dot{1}42 / 85\dot{7}$

Or $\frac{1}{7} = \frac{7}{49}$

The *Ekādhika* being 5, divide 7 by 5 and continue the division as usual with the same rule of procedure. After getting the three quotient-digits 1, 4 and 2 you find 42 as the remainder before you. So tackle the last 3 digits according to the complements rule and say:

$$\dot{1}\ 4\ 2\ /\ 8\ 5\ \dot{7}$$
$$2\ 1\ 4\ /$$

$\frac{1}{7} = \frac{7}{49} = .\dot{1}42 / 85\dot{7}$

(2) Let us now take the case of $\frac{1}{13} = \frac{3}{39}$:

$.\dot{0}76 / 923$

(*i*) The last digit is 3 $\therefore$ the last digit in the answer will be 3.

$.\dot{0}76 / 923$
23

(*ii*) The *Ekādhika* (multiplier or divisor) is $\therefore$ 4.

(*iii*) $\frac{1}{13} = \frac{3}{39}$. After 3 digits whether by multiplication or by division, 36 (D~N) comes up. So, the other half is mechanically set down. And we say: $\frac{1}{13} = .\dot{0}76 / 92\dot{3}$

(3) Next, let us take $\frac{1}{11} = \frac{9}{99}$

(*i*) the last digit is 1. The last digit of the answer will be 9.

$\dot{0}/\dot{9}$
9 /

(*ii*) The *Ekādhika* in both ways is 10.

(*iii*) Immediately after the very first digit, we get 90 which is 99~9 before us. So, the complements rule operates.

(*iv*) And, in either case, we get $\frac{1}{11} = .\dot{0}/\dot{9}$

(4) $\frac{1}{23} = \frac{3}{69}$ giving 7 as *Ekādhika* and 3 as the last digit of the answer.

$\therefore$ By both methods, Multiplication and Division.

$$\frac{1}{23} = \left.\begin{matrix} 0.\dot{0}\ 4\ 3\ 4\ 7\ 8\ 2\ 6\ 0\ 8\ 6 \\ 9\ 5\ 6\ 5\ 2\ 1\ 7\ 3\ 9\ 1\ \dot{3} \end{matrix}\right\}$$

(5) $\frac{1}{12} = \frac{7}{119}$ giving 12 as *Ekādhika* and 7 as the last digit.

$\therefore$ By both the methods multiplication and division we have.

$\frac{1}{17} = \frac{7}{119} = .\dot{0}\ 5\ 8\ 8\ 2\ 3\ 5\ 2 / 9\ 4\ 1\ 1\ 7\ 6\ 4\ \dot{7}$

## THE CODE LANGUAGE AT WORK

Not only do the Vedic *Sūtras* tell us how to do all this by easy and rapid processes of mental arithmetic; but they have also tabulated the results in the shape of special sub-*Sūtras* containing merely illustrative specimens with a master-key for "unlocking other portals" too. The abstruse details and the master-key are not given here; but a few sample-specimens are given of the way in which the code and the *Ekanyūna Sūtra* explained in Chapter 2 can be utilised for the purpose of postulating mental one-line answers to the question. The three samples read as follows:

(1) केवलैः सप्तकं गुण्यात् (*Kevalaiḥ Saptakaṃ Guṇyāt*);

(2) कलौ क्षुद्रससैः (*Kalau Kṣudrasasaiḥ);* and

(3) कंसे क्षामदाहखलैर्मलैः (*Kaṃse Kṣāmadāha-khalairmalaiḥ*).

In the first of these, *Saptaka* means 'seven'; and *Kevalaiḥ* represents 143; and we are told that, in the case of seven, our multiplicand should be 143!

In the second, *Kalau* means 13 and *Kṣudrasasaiḥ* represents 077; and we are told that the multiplicand should be 077! and

In the third, *Kaṃse* means 17; and *Kṣāma-dāha-khalairmalaiḥ* means 05882353; and we are told that the multiplicand should be this number of 8 digits.

Now, if we advert to the "*Ekanyūna*" corollary of the *Nikhilaṁ* chapter on multiplication, we shall be able to remind ourselves of the operation in question and the result to be achieved thereby. Let us do the multiplications accordingly as directed and see what happens.

(1) In the case of 7 as denominator, $143 \times 999 = 142 / 857$; and lo! These are the six recurring decimal digits in the answer!

(2) In the case of 13, $077 \times 999 = 076 / 923$; and these are the six digits in the recurring decimal equivalent of $\frac{1}{13}$! and

(3) In the case of 17, $05882353 \times 99999999 = 05882352/94117647$; and these are the 16 recurring digits in the recurring decimal equivalent of $\frac{1}{17}$.

In all the 3 cases we observe the rule of complements from 9 at work. And the sub-*Sūtra* merely gives us the necessary clue to the first half of the decimal and also a simple device *Ekanyūnena* for arriving at the whole answer! And all this is achieved with the help of the easy alphabet-code!

These results may therefore be formulated as follows:

$$\frac{1}{7} = \frac{143 \times 999}{999999} = \frac{142857}{999999} = .\dot{1}\,4285\dot{7};$$

$\frac{1}{13} = \frac{077 \times 999}{999999} = \frac{076923}{999999} = .\dot{0}7692\dot{3}$; and

$\frac{1}{17} = \frac{05882353 \times 99999999}{9999999999999999}$

$= .\dot{0}5882352 / 9411764\dot{7}$!

And, by cross-multiplication, we get from the above the following results:

(1) $7 \times 142857 = 999999$;

(2) $13 \times 076923 = 999999$; and

(3) $17 \times 05882352 / 94117647$
$= 9999999999999999$ 16 digits in all!

And, just in passing, we may note that this is the reason why, in the case of all these vulgar fractions, the last digit of the denominator 9, 3, 7 or 1, as the case may be gives, 1, 3, 7 or 9 before-hand as the last digit of the equivalent recurring decimal fraction!

## The Remainder-Quotient Complements-Cycles

We have already, again and again, noted the fact that, in the various typical cases observed and analysed by us, the two halves of the quotients added together give us a series of nines. We shall now proceed a little bit further and try to see if there be any such or similar rule governing the remainders. For the purpose of the necessary experimentation and investigation, let us take up a more detailed consideration of the remainders obtained in each case by successive divisions of the numerator by the denominator; and let us start with $\frac{1}{7}$.

We know that successive remainders are 3, 2, 6, 4, 5 and 1. We know already that, on reaching 6, i.e. D~N as the remainder, half the work has been completed and that the complementary half is about to begin. Putting the above 6 figures, therefore, into two rows of three figures each, we have:

| 3 | 2 | 6 |
|---|---|---|
| 4 | 5 | 1 |

and we observe that each vertical column of one digit from the upper row and of one from the lower one gives us the same total, i.e. 71.

| 3 | 2 | 6 |
|---|---|---|
| 4 | 5 | 1 |
| 7 | 7 | 7 |

*N.B.*: As our divisor is 7, it is but natural that no remainder higher than 6 is permissible, i.e. that the only possible remainders are 1, 2, 3, 4, 5 and 6. And these are the ones we actually find.

Let us now take up the case of $\frac{1}{13}$ and note what happens. The successive remainders are 10, 9, 12, 3, 4 and 1 the highest of which is 12. And when they are placed in two rows, we find here too, that the last three remainders are complements—from 13—of the first three remainders.

10 9 12 / 3 4 1

| 10 | 9 | 12 |
|---|---|---|
| 3 | 4 | 1 |
| 13 | 13 | 13 |

In the case of $\frac{1}{17}$ the successive remainders are:

10 15 14 4 6 9 5 16/7 2 3 13 11 8 12 1

| 7 | 2 | 3 | 13 | 11 | 8 | 12 | 1 |
|---|---|---|---|---|---|---|---|
| 17 | 17 | 17 | 17 | 17 | 17 | 17 | 17 |

The last 8 remainders are thus complements—from 17—of the first eight ones!

In the case of 19, the remainders are:

| 10 | 5 | 12 | 6 | 3 | 11 | 15 | 17 | 18 |
|---|---|---|---|---|---|---|---|---|
| 9 | 14 | 7 | 13 | 16 | 8 | 4 | 2 | 1 |
| 19 | 19 | 19 | 19 | 19 | 19 | 19 | 19 | 19 |

Here again the first nine remainders, when added successively to the next nine, give 19 each time.

Thus, it is clear that, whereas the quotient-halves are uniformly complements from nine, the remainder-halves are complements from the individual divisor in each case. And this further reduces our labour in making out a list of the remainders.

### Multiples of the Basic Fractions

Thus far, we have dealt with vulgar fractions whose numerator is unity. But what about fractions which have some other numerator? And the answer is: "There are several simple and easy methods by which, with a tabulated list before us of the results obtained by one or more of the processes above expounded, or even independently, we can readily put down the recurring decimal equivalents of the vulgar fractions of the type just under discussion."

Let us, as usual, start with $\frac{1}{7}$ and frame a chart as follows:

$\frac{1}{7} = .\dot{1}4285\dot{7}$

$\frac{2}{7} = .\dot{2}8571\dot{4}$

$\frac{3}{7} = .\dot{4}2857\dot{1}$

$\frac{4}{7} = .\dot{5}7142\dot{8}$

$\frac{5}{7} = .\dot{7}1428\dot{5}$; and

$\frac{6}{7} = .\dot{8}5714\dot{2}$

In this chart, we observe that, in all the "proper" fractions having 7 as their denominator.

(*i*) The same six digits are found as in the case of $\frac{1}{7}$;

(*ii*) they come up in the same sequence and in the same direction as in the case of $\frac{1}{7}$;

(*iii*) they, however, start from a different starting-point but travel in "Cyclic" order in what is well-known as the "clock-wise" order.

(*iv*) and with the aid of these rules, one can very easily obtain the recurring decimal equivalent of a vulgar fraction whose numerator is higher than 1.

In fact, a person who is actually looking at a statement on a board, a piece of paper, a slate etc., to the effect that $\frac{1}{7} = .\dot{1}4285\dot{7}$, has several easy alternative processes to choose from, for determining the decimal equivalent of all the other five possible fractions having the same denominator, i.e. 7. They are as follows:

### The First Method

1. The various digits can be numbered and marked in ascending order of magnitude, thus:

(*i*) Unity being the least of them, the cycle for $\frac{1}{7}$ starts with one as its starting point, travels in clock-wise cyclic order and reads : $.\dot{1}4285\dot{7}$;

(1)(3)(2)(6)(4)(5)
1 4 2 8 5 7

(*ii*) 2 being the second, $\frac{2}{7}$ starts with 2 and gives us the answer $.\dot{2}8571\dot{4}$;

(*iii*) There being no 3 at all, the third digit in ascending order is 4. So $\frac{3}{7}$ begins from 4 and reads: $.\dot{4}2857\dot{1}$;

(*iv*) The next digit, i.e. the 4th in ascending order actually being 5, 4/7 begins with 5 and gives: $.\dot{5}7142\dot{8}$;

(*v*) The fifth digit actually being 7, 5/7 commences with 7 and reads: $.\dot{7}1428\dot{5}$; and

(*vi*) The 6th and last digit being 8, the sixth and last fraction, i.e. $\frac{6}{7}$ starts with 8 and says: $.\dot{8}5714\dot{2}$

This is the first method.

Yes, but what about those cases in which the number of decimal places is more than 10 and thus, in the tabulated answer before us, some digits are found more than once?

Yes, it is perfectly true that, just as some digits are found absent as in the case of $\frac{1}{7}$ just seen, there are other cases where the same digits are found more than once. In fact, in every case wherein the number of decimal places is more than 10, this is bound to happen; and there provision too must be against it. In fact, the remedy is very simple, i.e.

that, even where digits occur more than once, there still are gradations; and, if these are taken into account, the cyclic order and the ascending order of magnitude will still operate and serve their purpose.

For example, in the case of $\frac{1}{17}$, we have $.05882...$ at the very commencement; and there are two eights before us. Yes, but 88 is greater than 82; and therefore we should take 82 first and 88 afterwards and do our numbering accordingly:

$\frac{1}{17}, = .\dot{0}$ 5 8 8 2 3 5 2 9 4 1 1 7 6 4 7

(1) (10) (15) (14) (4) (6) (9) (5) (16) (7) (2) (3) (13) (11) (8) (12). Thus, $\frac{1}{17}$ starts with zero; $\frac{2}{17}$ with 11; $\frac{3}{17}$ with 17; $\frac{4}{17}$ with 23; $\frac{5}{17}$ with 29; $\frac{6}{17}$ with 3; $\frac{7}{17}$ with 41; $\frac{8}{17}$ with 47; $\frac{9}{17}$ with 52; $\frac{10}{17}$ with 58; $\frac{11}{17}$ with 6; $\frac{12}{17}$ with 70; $\frac{13}{17}$ with 76; $\frac{14}{17}$ with 82; $\frac{15}{17}$ with 88; and $\frac{15}{17}$ with 9. The arranging in ascending order of magnitude has, of course, to be done carefully and correctly. But it must be admitted that, although the procedure of counting and numbering is quite reasonable and scientifically correct, yet it is rather cumbrous, clumsy and tiring. Hence the need for other methods.

Yes, but what about the cases wherein the number of digits in the decimal equivalent is much less than the denominator of the vulgar fraction in question and has thus no scope for meeting all the possible demands?

Yes, $\frac{1}{13}$ is such a case. The number of possible multiples is 12; and the number of digits in the decimal equivalent is only 6. (for $\frac{1}{13} = .\dot{0}76/92\dot{3}$. What is the remedy?

The remedial provision is that a multiple or two will do the trick quite satisfactorily and neatly.

Now, $\frac{1}{13} = .\dot{0}7692\dot{3}$

$\therefore$ By simple multiplication by 2,

$\frac{2}{13} = .\dot{1}5384\dot{6}$

And now, there are twelve digits in all; and these can meet the needs of all the possible multiples.

Thus, $\frac{1}{13} = \dot{0}7692\dot{3}$; and $\frac{2}{13} = .\dot{1}5384\dot{6}$

$\therefore \frac{3}{13} = .\dot{2}3076\dot{9}$; and $\frac{4}{13} = .\dot{3}0769\dot{2}$

$\therefore \frac{5}{13} = .\dot{3}8461\dot{5}$; and $\frac{6}{13} = .\dot{4}6153\dot{8}$

$\therefore \frac{7}{13} = .\dot{5}3846\dot{1}$; and $\frac{8}{13} = .\dot{6}1538\dot{4}$

$\therefore \frac{9}{13} = .\dot{6}9230\dot{7}$; and $\frac{10}{13} = .\dot{7}6923\dot{0}$

$\therefore \frac{11}{13} = .\dot{8}4615\dot{3}$; and $\frac{12}{13} = .\dot{9}2307\dot{6}$

The procedure is there and is quite correct. But, after all, one must confess that, even with this device, this counting and numbering procedure is still a cumbrous, clumsy and tiring process. Hence, let us repeat, the need for other methods.

### The Second Method

The second method is one wherein we avoid even this numbering and marking etc., and in accordance with the *Ādyam Ādyena* rule, multiply the opening digit or digits of the basic decimal fraction $.14285\dot{7}$ and determine, therefrom the starting point for the multiple in question. Thus, $.\dot{1}4285\dot{7}$

$\therefore \frac{1}{7}$ starts with .14 . . . $\therefore \frac{2}{7}$ should start with .28 etc., and in clockwise cyclic order give $.\dot{2}8571\dot{4}$;

$\frac{3}{7}$ ought to start with .42 etc., and give $.\dot{4}2857\dot{1}$;

$\frac{4}{7}$ must start with .56; but there is no .56 but only .57 before us; and so making allowance for a possible nay, the actual carrying over of a surplus digit from the right leftward, we start with 57 and say: $\frac{4}{7} = .\dot{5}7142\dot{8}$.

Similarly $\frac{5}{7}$ should start with .70; but for the same reason as in the immediately preceding case, it actually starts with .71 and gives us: $.\dot{7}1428\dot{5}$; and $\frac{6}{7}$ should start with .84, actually starts with .85 and yields the answer: $.\dot{8}5714\dot{2}$!

This is the Second method.

### The Third Method

The third process is very similar; but it bases itself not on *Ādyam Ādyena* but on *Antyam Antyena*. In other words, it deals not with the opening digit but with the closing one. Thus,

$\because \frac{1}{7}$ ends with 7, $.\dot{1}4285\dot{7}$

$\therefore \frac{2}{7}$ must end with 4 $\therefore$ It is $.\dot{2}8571\dot{4}$

$\therefore \frac{3}{7}$ should end with 1 $\quad\therefore$ It is $.\dot{4}2857\dot{1}$

$\therefore \frac{4}{7}$ ought to end with 8 $\quad\therefore$ It is $.\dot{5}7142\dot{8}$

$\therefore \frac{5}{7}$ should end with 5 $\quad\therefore$ It is $.\dot{7}1428\dot{5}$

and $\therefore \frac{6}{7}$ must end with 2 $\quad\therefore$ It is $.\dot{8}5714\dot{2}$

This is the third method and the easiest and therefore the best of the lot.

### Independent Method

The above described methods are all for the utilisation of our knowledge of the decimal shape of a fraction whose numerator is unity, for deriving the corresponding decimal form of any multiple of that fraction. This is all right, so far as it goes. But what about a person who has not got such a ready-to-hand table to refer to ? In such a case, should one newly prepare the basic chart and then manipulate it—cyclically—in one of the ways just explained, for getting the required result?

That would, of course, be *absurd.* For use by such persons, we have a totally independent method, by which, without resorting to any such previously prepared or newly prepared table, one can readily deal with the particular fraction on hand! And the whole *modus operandi* is exactly the same as has been already explained in respect of the basic fraction and without the slightest difference or deviation in any particular whatsoever therefrom.

For example, suppose you have to decimalise $\frac{3}{7}$. Your last digit will be 1 and as $\frac{3}{7} = \frac{21}{49}$, your *Ekādhika Pūrva* will be 5. Now, go on dividing by 5, in the usual manner; and you get the chart, as explained in the margin: $\frac{3}{7} = \frac{21}{49} = .\dot{4}\ 2\ 8 / 5\ 7\ \dot{1}$

$\qquad\qquad 1\ 4\ 2 /$

After you get the first there digits 4, 2 and 8, you find that your dividend is 28; but this is D~N, i.e. 49—21. So you may stop here and put the last three quotient-digits down as 5, 7 and 1 the complements, from nine, of the digits already found.

Or you may continue the division till you get 21 as the dividend; and as this was your starting-point, you may put the 6 digits down as a "recurring" decimal.

Thus $\frac{3}{7} = .\dot{4}2857\dot{1}$

Try this with $\frac{7}{13}$, $\frac{13}{17}$, $\frac{5}{19}$ and so on, with any number of cases. And you will always find the same thing happening right through all of them. Thus, for those who do not have a tabulated schedule before them, this absolutely *independent* method is also there; and you can make full use of it.

*Note*: 1. In this *independent* method, it should also be noted that if we have to decimalise $\frac{1}{7}, \frac{2}{7}, \frac{3}{7}, \frac{4}{7}, \frac{5}{7}$ etc., we have merely to divide 10, 20, 30, 40, 50 etc., by 7 and put down that remainder as the first remainder in each particular case and that the work can be done automatically thereafter.

2. or, we may pre-decide the last digit in each case by taking the last digits of 7, (1) 4, (2), 1, (2) 8, (3) 5, (4) 2 as the last digits of the decimal equivalent of $\frac{1}{7}, \frac{2}{7}, \frac{3}{7}, \frac{4}{7}, \frac{5}{7}$ and $\frac{6}{7}$!

## Recapitulation (and Supplementation)

Over and above the ones expounded and explained hereinabove, there are several other very instructive and interesting principles, features characterising this question of conversion of vulgar fractions into decimal ones in respect of the remainders, the quotients etc. For the benefit of the students, we propose now to recapitulate, summarise, supplement and conclude this portion of the subject:

(1) As regards the remainders, we have noted that, as soon as D~N comes up before us as a remainder, the remaining remainders are all complements—from the divisor, i.e. the denominator —of the remainders already obtained;

(2) This automatically means that the quotient-digits already obtained and the quotient-digits still to be found, are complements from nine!

(3) If we take any remainder and multiply it by the *Caramāṅka* the last digit, the last digit of the product is actually the quotient at that step. The formula here is शेषाणि अंकेन चरमेण (*Śeṣaṇi Aṅkena Carameṇa*) which is therefore of the utmost significance and practical utility in mathematical computations. For instance,

(1) $\frac{1}{7}$ The remainders are 3, 2, 6, 4, 5 and 1. Multiplied by 7 (the *Caramāṅka*) these remainders give successively 21, 14, 42, 28, 35 and 7. Ignoring the left-hand side digits, we simply put down the last digit (*Caramāṅka*) of each product; and lo! We get $\frac{1}{7} = .\dot{1}4285\dot{7}$!

(2) $\frac{1}{13}$ The remainders are 10, 9, 12, 3, 4 and 1. Multiplied successively by 3 the last digit, these remainders give 30, 27, 36, 9, 12 and 3. Ignoring the previous digits, we write down merely the *Caramāṅka* the last digit of each product; and lo! $\frac{1}{13} = .\dot{0}7692\dot{3}$!

(3) $\frac{1}{17}$ The remainders are 10, 15, 14, 4, 6, 9, 5, 16/7, 2, 3, 13, 11, 8, 12 and 1. Multiplied by 7, they give us successively : 70, 105, 98, 28, 42, 63, 35, 112, 49, 14, 21, 91, 77, 56, 84 and 7. Dropping the surplus, i.e. left-side digits and putting down only the *Caramāṅka* the right-hand most digits, we have $\frac{1}{17} = .\dot{0}5882352 / 9411764\dot{7}$

In fact, the position is so simple and clear that we need not multiply the whole digit, write down the product and then drop the surplus digit or digits. We need only put down the *Caramāṅka* the right-hand most digits at the very outset as each step and be done with it!

(4) The geometrical progression character of the remainders gives us a clue to the internal relationship between each remainder and its successor or its predecessor! Thus, as we know one remainder, we practically know all the rest of them. Thus,

In the case of $\frac{1}{7}$: As we know the first remainder is 3, we can multiply any remainder by 3, cast out the sevens if any and immediately spot out and announce the next remainder.

$3 \times 3 = 9; 9 - 7 = 2; \therefore 2$ is the second remainder.

$2 \times 3 = 6$. This is the third remainder

As 6 is D~N, we may stop here and by the rule of complements from the denominator, we may put down 4, 5 and 1 as the remaining three remainders.

Or, if we overlook the D~N rule or prefer to go on with our multiplication by 3 the geometrical ratio, we get:

$6 \times 3 = 18; 18 - 14 = 4;$ and this is the 4th remainder.

$4 \times 3 = 12; 12 - 7 = 5;$ and this is the 5th remainder.

$5 \times 3 = 15; 15 - 14 = 1;$ and this is the 6th and last remainder.

We have thus obtained from the first remainder, all the remainders: 3, 2, 6, 4, 5 and 1.

And from these, by multiplication by the *Caramāṅka* 7, we get all the 6 quotient-digits as explained above:

$.\dot{1}$, 4, 2, 8, 5 and $\dot{7}$

This is not all. Instead of using the first remainder 3 as our geometrical ratio, we may take the second one 2, multiply each preceding group of 2 remainders by 2 and get 32, 64 and 51 for, by casting out the sevens, $6 \times 2 - 7 = 5$; and $4 \times 2 - 7 = 1$. And multiplying these 6 digits by 7, we again get the *Caramāṅkas* 1 4 2 8 5 7 as before.

Or we may take help from the third remainder, i.e. 6, multiply the preceding group of 3 remainders and get 3 2 6, 4 5 1 for, by casting out the sevens, $3 \times 6 - 14 = 4; 2 \times 6 - 7 = 5;$ and $6 \times 6 - 35 = 1$. And, multiplying these same 6 digits by 7, we again obtain the *Caramāṅkas* 142857 as before.

This procedure is, of course, equally applicable to the fourth and fifth remainders, i.e. 4 and 5 and can get us the same result as before. This is doubtless purely academical and of no practical utility. But we are discussing a principle, nay a universally operating mathematical law and must therefore demonstrate its actual universality of application.

So, if we take the 4th remainder, i.e. 4 and multiply the preceding group of four remainders by 4, we again get 3264/51. For, $4 \times 3 - 7 = 5; 4 \times 2 - 7 = 1; 4 \times 6 - 21 = 3; 4 \times 4 - 14 = 2;$ and the only difference is that the first two digits are found to have already started repeating themselves!

If we now take the 5th remainder, i.e. 5 and multiply the preceding group of 5 remainders by 5, we again get 32645/1 ... for $3 \times 5 - 14 = 1; 2 \times 5 - 7 = 3; 5 \times 6 - 28 = 2; 5 \times 4 - 14 = 6; 5 \times 5 - 21 = 4.$

And, if we follow the same procedure with the 6th remainder, i.e. 1 and multiply the group of preceding remainders by 1, we will, of course, get the same preceding remainders over again!

(5) In the case of 17, the first four remainders are : 10, 15, 14 and 4. As 4 is a manageable multiplier, we may make use of it as a convenient and suitable remainder for this purpose. Let us therefore multiply the group of four remainders already found by 4 and cast out the seventeens wherever necessary. And then we find:

$4 \times 10 - 34 = 6$     10, 15, 14, 4
$4 \times 15 - 51 = 9$     10, 15, 14, 4, 6
$4 \times 14 - 51 = 5$     10, 15, 14, 4/6, 9, 5, 16/

$4 \times 4 = 16$. But as D~N = 16, we can stop here and set down all the other remainders by subtracting each of the above digits from 17: 7, 2, 3, 13/11, 8, 12 and 1. And, multiplying each of these 16 remainders or rather their *Caramāṅka*, i.e. units digits by 7, we get:

$\frac{1}{17}$: $.\dot{0}\,5\,8\,8\,2\,3\,5\,2\,/\,9\,4\,1\,1\,7\,6\,4\,\dot{7}$

## STILL ANOTHER METHOD

Besides (1) the corollary-*Sūtra* (2) each remainder × the last digit method, (3) the *Ekādhika* process from right to left and (4) the *Ekādhika* method from left to right, there is still another method whereby we can utilise the geometrical progression relationship and deduce the same result by a simple and easy process. And it is this, namely, that as soon as we come across a clear ratio between one remainder or dividend and another, we can take that ratio for granted as being of universal application and work it out all through. For example,

In the case of 19, we have 10 and 5 as the first two remainders and we not that 5 is just one-half of ten. Keeping this ratio in view, we can deduce that the next remainder should be one-half of 5. But, as 5 is not exactly divisible by 2, we add 19 to it, make it 24 and put down its half, i.e. 12 as the next remainder. The 12 gives 6, 6 gives 3, 3+19 gives 11, 11+19 gives 15, 15+19 gives 17 and 17+19 gives 18. And we stop there and put down the remaining half of the remainders by subtractions from 19. Having thus got the remainders, we multiply the *Caramāṅkas* by 1 the last digit of the answer and we get the quotient-digits automatically.

| 10, | 5, | 12, | 6, | 3, | 11, | 15, | 17, | 18 |
|---|---|---|---|---|---|---|---|---|
| 9, | 14, | 7, | 13, | 16, | 8, | 4, | 2, | 1 |

*N.B.*: The ratio in question may be noticed at any stage of the work and made use of at any point thereof.

In the case of $\frac{1}{17}$, we have the remainders 10 and 15 at the very start. We can make use of this ratio immediately and throughout, with the proviso that, if and when a fractional product is threatened, we can take the denominator or as many multiples thereof as may be necessary for making the digit on hand exactly divisible by the divisor on hand.

Thus, in the case of $\frac{1}{17}$, we have the remainders 10 and 15 to start with the ratio being 1 to $1\frac{1}{2}$. So, whenever one odd number crops up, its successor will be fractional. And we get over this difficulty in the way just explained.

And when we get a remainder which is numerically greater than the divisor, we cast off the divisor and put down the remainder. Thus, 10 gives us 15; 15+17 gives us 48, i.e. 14; 14 gives us 21, i.e. 4; 4 gives us 6; 6 gives us 9; 9+17 gives us 39, i.e. 5; 5+17 gives us 33, i.e. 16. And there we can stop.

| 10, | 15, | 14, | 4, | 6, | 9, | 5, | 16 |
|---|---|---|---|---|---|---|---|
| 7, | 2, | 3, | 13, | 11, | 8, | 12, | 1 |

## Number of Decimal Places

Students generally feel puzzled and non-plussed as to how to know beforehand the number of decimal places which, on division, the decimal equivalent of a given vulgar fraction will actually consist of. In answer hereto, we must point out that, having—in the immediately preceding sub-section on this subject—made a detailed, analytical study of the successive remainders, we have, in every case before us, practically a tabulated statement from which without actual division to the very end we can postulate beforehand all the forthcoming remainders. And the tabulated statement has the further merit that it can be prepared, at any time, at a moment's notice!

All this means, in effect, that,

(*i*) As soon as 1 or other starting point is reached in our *mental* analysis, we will have completed the whole work of decimalisation and therefore know the actual number of decimal places coming ahead. The cases $\frac{1}{7}, \frac{1}{13}, \frac{1}{17}, \frac{1}{19}$, etc., have all proved this.

(*ii*) As soon as we reach the difference between the numerator and denominator, we know we have done half the work and that the other half is yet to come. The cases of $\frac{1}{7}$, etc., which we have dealt with *in extenso* have proved this too.

(*iii*) As soon as we reach a fairly small and manageable remainder in our *mental* calculation, we know how many more steps we should expect.

Let us again take the case of $\frac{1}{7}$ by way of illustration. The first remainder is 3; and used as a successive multiplier with the provision for the casting out of the sevens, that first remainder multiplier brings us on to 1.

When we have done two steps and got 1 and 4 as the first two quotient-digits, we find 2 as the remainder. Multiplying the first group of two digits 14 by 2, we get 28 as the second-group with the remainder also doubled, i.e. $2 \times 2 = 4$. 14 / 28/.

Multiplying 28 by 2, we get $28 \times 2 = 56$ as the third group and $4 \times 2 = 8$ as the remainder. And then, by casting out the sevens, we obtain 57 as the quotient-group and 1 as the remainder! And as this was our starting-point, we stop further computations and decide that $\frac{1}{7}$, when decimalised, has 6 decimal places in the answer.

Going back to the case of $\frac{1}{17}$, the student will remember that, after 4 steps, we got .0588 as the quotient-digits and 4 as the remainder. Multiplying the former by the latter, we obtained 2352 as the second quotient-group and $4 \times 4 = 16$ as the remainder; and there we stopped, because we had the first 8 digits on hand and knew the other 8 digits. Thus $\frac{1}{17}$ gave us 16 digits.

As a geometrical series is of the standard form 1, r, $r^2$ and so on, we are able to utilise 2 and $2^2$ in the case of $\frac{1}{7}$, 4 and $4^2$ in the case of $\frac{1}{17}$ and so on for helping us to pre-determine the number of decimal places in the answer. This is the Algebraical principle utilised herein.

*Note*: 1. We need hardly point out that the *Ekādhika* method has the supreme and superlative merit of lightening our division and multiplication work. For instance, in the case of $\frac{1}{19}, \frac{1}{29}$ etc., we have to do our division-work, at stage after stage, by successive division, not by 19 or 29 etc., the original denominator but by 2 or 3 etc. the *Ekādhika-pūrva*. And this is the case with regard to every case, i.e. that we perform all our operations—in this system—with much smaller divisors, multipliers etc., and this rule is invariable. What a tremendous saving in effort, labour, time and cost!

2. We have purposely treated this subject at great length and in detail, because it is very essential that the whole matter should be clearly understood, thoroughly assimilated and closely followed so that, even without the help of a teacher, the student may be enabled to work out these methods independently in other similar cases and to *know*—with absolute certainty—that any and every vulgar fraction can be readily tackled and converted into the corresponding recurring decimal *whatever* may be the complexity thereof and the number of decimal places therein. In fact, inasmuch as these simple and easy processes are available—and suitable—for all possible denominators and for all possible numerators, the decimal and especially the recurring decimal should no longer be a bugbear to the student. On the contrary, they should be the most welcome of all welcome friends!

### Some Characteristic Features (General and Special)

(1) In the cases of fractions with *prime numbers* like 19, 29, 59 etc., as denominators, the maximum number of decimal places is one less than the denominator! This is self-evident and requires no elaboration.

(2) *Usually,* it, or a sub-multiple thereof, is the actual number.

(3) *Generally*, the rule of complements from nine is found in operation amongst them.

(4) For fractions like $\frac{1}{39}$, $\frac{1}{49}$, $\frac{1}{69}$ etc., where the denominators are products of prime numbers, the number of digits depends on the various respective factors in each case as will be presently elucidated.

(5) If and when the decimal-fraction obtained from one of the factors of the denominator is exactly divisible by the other factor or factors, the division by the second factor leaves no remainder. And therefore the number of decimals obtained by the first factor is not added to. Thus,

$$(i)\quad \frac{1}{7\times 3} = \frac{.\dot{1}4285\dot{7}}{3}$$

Here, the numerator on the R.H.S. being exactly divisible by 3, it divides out and leaves no remainder. Therefore, the number of digits continues the same.

This means that, in every case wherein the complementary halves from nine are found, the numerator on the R.H.S. *must* necessarily be divisible by 3, 9 etc. and by multiplying the denominator in such a case by such factors, we cause no difference to the number of decimal places in the answer. And consequently, we have:

$$\frac{1}{63}=\frac{1}{7\times 9}=\frac{.\dot{1}4285\dot{7}}{9}=.\dot{0}1587\dot{3};\text{ and so on.}$$

Going back to the *Ekanyūnena Sūtra* as explained in connection with the Sanskrit Alphabetical code, we know the $142857=143\times 999$ $=11\times 13\times 3^3\times 37$. This means that since the numerator is divisible by 11, 13, 3, 9, 27, 37, 33, 39, 99, 117, 297, 351 and 999, the multiplication of the denominator 7 by any one of these factors will make no difference to the number of decimal places in the answer.

$$(ii)\ \frac{1}{39}=\frac{1}{13\times 3}=\frac{.\dot{0}7692\dot{3}}{3}$$

Here too, all the above considerations apply. And, since $76923=77\times 999=3^3\times 7\times 11\times 37$, therefore these factors and combinations of factors will, by multiplying the denominator, make no difference to the number of decimal places. Note, $999999=999\times 1001$ $=999\times 7\times 11\times 13$.

(6) $\frac{1}{69}=\frac{1}{23\times 3}$ and comes under the same category with 22 digits in the answer just like $\frac{1}{23}$.

(7) $\frac{1}{49}$ is a special case and stands by itself. Naturally it should have been expected to provide for 48 places. But, as a matter of fact, it gives only 42; and for a perfectly valid and cognet reason, i.e. that, out of the 48 possible multiples, six, i.e. $\frac{7}{49}$, $\frac{14}{49}$, $\frac{21}{49}$, $\frac{28}{49}$, $\frac{35}{49}$ and $\frac{42}{49}$ go into a different family, as it were and take shape as $\frac{1}{7}$, $\frac{2}{7}$, $\frac{3}{7}$, $\frac{4}{7}$, $\frac{5}{7}$ and $\frac{6}{7}$; have there places there as $.\dot{1}4285\dot{7}$, $.\dot{2}8571\dot{4}$ and so forth and need no place in the $\frac{1}{49}$ etc., group! And thus, since 6 go out of the 48, the remaining 42 account for the 42 places actually found in the decimal equivalent of $\frac{1}{49}$! This is not a poet's mere poetic phantasy but a veritable mathematical verity!

(8) $\frac{1}{79}$ is, in a way, an exception, as it contains only 13 digits. And, as this is an odd number, the question of the two complementary halves does not arise! 13, however, *is* a sub-multiple of 78; and there is no deviation from the normal in this respect. An at-sight-one-line mental method will soon be given for $\frac{1}{79}$ in this very chapter.

(9) Similarly $\frac{1}{89}$ has 44 digits and thus conforms to the sub-multiple rule. And this implies that, like $\frac{1}{13}$, it will need another complete turn

of the wheel in one of its multiples in order to meet the needs of all the multiples. An *incredibly easy* method will be shown in this very chapter for reeling off the answer in this case.

(10) $\frac{1}{99}$ has only two recurring places $.\dot{0}\dot{1}$ ; but the whole gamut can be and has been provided for, therewith.

(11) In the case of basic fractions ending in 3, the denominator is first multiplied by 3 and gives us the *Ekādhika*, and the last digit in the answer is also 3.

(12) $\frac{1}{33}$ like $\frac{1}{99}$ has only two decimal places $.\dot{0}\dot{3}$ .

(13) $\frac{1}{43}$ has only 21 digits. 21 is a sub-multiple of 42 but is odd and gives no scope for the complementary halves.

(14) $\frac{1}{53}$ has only 13 digits a sub-multiple but odd.

(15) $\frac{1}{83}$ has only 41 digits similarly.

(16) $\frac{1}{73}$ is special. Since $73 \times 137 = 10001$ and since

$$10001 \times 9999 = 99999999 \therefore \frac{1}{73} = \frac{127}{10001} = \frac{137 \times 9999}{99999999}$$

$$=.\dot{0}136/986\dot{3} \text{ by } \textit{Ekanyūna Sūtra}.$$

(17) And, conversely, $\frac{1}{137} = \frac{73}{10001} = \frac{73 \times 9999}{99999999} = .\dot{0}072992\dot{7}$

(18) $\frac{1}{93}$ will be discussed a little later.

(19) In the case of fractions whose denominators end in 7, the last digit is also 7; and the *Ekādhika* is obtained from the denominator multiplied by 7.

(20) $\frac{1}{7}$ and $\frac{1}{17}$ have been dealt with in detail already.

(21) $\frac{1}{27}$ and $\frac{1}{37}$ are special because $27 \times 37 = 999$. And their decimal forms are $.\dot{0}3\dot{7}$ and $.\dot{0}2\dot{7}$ .

(22) $\frac{1}{47}$ has 46 digits.

(23) $\frac{1}{57} = \frac{1}{19 \times 3}$ and has only 18 digits.

(24) $\frac{1}{67}$ has 33 digits (odd).

(25) $\frac{1}{77}$ has been discussed already (number $\frac{1}{7}$ ).

(26) $\frac{1}{87} = \frac{1}{29 \times 3}$ and has 28 digits.

(27) $\frac{1}{97}$ has its full quota of 96 digits.

(28) In the case of fractions with denominators ending in 1, the *Ekādhika* comes from the denominator multiplied by 9, and the last digit is 9.

(29) $\frac{1}{11} = .\dot{0}\dot{9}$

(30) $\frac{1}{21} = \frac{1}{7 \times 3}$ and has been discussed under 7.

(31) $\frac{1}{31}$ will come up a little later.

(32) $\frac{1}{41}$ is special $\because$ 41 × 271 = 11111

$\therefore \frac{1}{41} = \frac{271}{11111} = \frac{271 \times 9}{99999} = .\dot{0}243\dot{9}$ (odd)

(33) And, conversely, $\frac{1}{271} = \frac{41}{11111} = \frac{41 \times 9}{99999} = \frac{00369}{99999} = .\dot{0}036\dot{9}$

(34) $\frac{1}{51} = \frac{1}{17 \times 3}$ and has 16 digits.

(35) $\frac{1}{61}$ has 60 digits.

(36) $\frac{1}{71}$ has 35 digits (odd).

(37) $\frac{1}{81} = \frac{1}{27 \times 3} = \frac{\dot{0}3\dot{7}}{3} = .\dot{0}1234567\dot{9}$ (a very interesting number).

(38) $\frac{1}{91} = \frac{1}{13 \times 7}$ and has already been discussed under 7 and under 13. And besides, $\frac{1}{91} = \frac{11}{1001} = \frac{11 \times 999}{999999} = .\dot{0}10/98\dot{9}$

### BUT

But here a big but which butts in and exclaims: "Yes, all this is all right in its own way and so far as it goes. But, as our denominators go on increasing, we note that, although the last digit of the decimal fraction is 1, 3, 7, or at the most 9 and no more, yet, the *Ekādhika Pūrva* goes on increasing steadily all the time and we have to multiply or divide successively by bigger and bigger *Ekādhikas*, until, at last, with only two-digit denominators like 61, 71 and 81 and so on, we have now to deal with 55, 64, 73 etc., as our multipliers and divisors, and surely this is *not* such an easy process.

The objection is unobjectionable, nay, it is perfectly correct. But we meet it with quite a variety of sound and valid answers which will be found very cogent and reasonable. They are as follows:

(*i*) Even the biggest of our *Ekādhikas* are nowhere—in respect of bigness—near the original divisor. In every case, they are smaller. But this is only a theoretical and dialectical answer from the *comparative* standpoint and does not really meet the intrinsic objection about the Vedic methods being not only *relatively* better but also being free from all such flaws altogether! We therefore go on and give a satisfactory answer from the positive and constructive stand-point.

(*ii*) Even though the *Ekādhika* is found to be increasingly unmanageably big, yet the remainders give us a simple and easy device for getting over this difficulty. This we shall demonstrate presently.

(*iii*) The *Ekādhika* so far explained and applied is not the whole armoury. There are other auxiliaries too, wherein no such difficulty can crop up. These we shall expound and explain in a subsequent chapter of this very volume; and they will be found capable of solving the problem *in toto*; and

(*iv*) Above all, there is the crowning gem of all coming up in a near chapter and unfolding before our eyes a formula whereby, however big the denominator may be, we can—by mere mental one-line Vedic Arithmetic—read off the quotient and the remainder, digit by digit! This process of "Straight Division", we have already referred to and shall explain and demonstrate, in a later chapter, under this very caption "Straight (or Instantaneous) Division".

In the meantime, we take up and explain the way in which the remainders come to our rescue and solve this particular problem for us.

Let us take first the case of $\frac{1}{23}$. We know immediately that the last digit of the decimal is 3 and that the *Ekādhika* is 7. And then we work as follows:

(*i*) Multiplying digit after digit as usual by 7, we have:

$\frac{1}{23} = \frac{3}{69} = .04347826086$ / 9 5 6 5 2 1 7 3 9 1 3
6 / 3 4 3 1 1 5 2 6 2

or (*ii*) dividing digit by digit as usual by 7, we have:

$\frac{1}{23} = \frac{3}{69} = .0\ 4\ 3\ 4\ 7\ 8\ 2\ 6\ 0\ 8\ 6$ / 9 5 6 5 2 1 7 3 9 1 3
2 3 5 5 1 4 6 4 6

(*iii*) These are the usual *Ekādhika Pūrva* methods. But we observe in the first chart, after two digits 1 and 3 have been obtained, the next leftward group 39 is exactly three times the extreme-right-end one and we can immediately profit by it. Thus 39 gives us 117, out of which we put down 17 and keep 1 to carry over; 17 gives us 51+1=52. 52 gives us 156, out of which we set down 56 and keep 1 to carry over. 56 gives us

168 + 1 = 169. Of these, we put 69 down and keep 1 to be carried over; and so on. In fact, the whole procedure is exactly like the one which we followed from left to right in respect of $\frac{1}{49}$ (=.020408 16 32. . .). Thus we have:

$\frac{1}{23} = \frac{3}{69} = .\dot{0}434782608, 69, 56, 52, 17, 39, 1\dot{3}$

or (*iv*) if we wish to start from the left end, go on to the right, that too is easy enough.

We note that, the first digits being completed, we get 8 as the remainder. We can immediately work out this process by multiplying each two-digit group by 8 as we did in the case of $\frac{1}{49}$ by 2 and frame the following chart:

```
.04 : 32 : 72 : 24 : 08 : 64 : and so on
    :  2 :  6 :  2 :    :  5 :
-------------------------------
    : 34 : 78 : 26 : 08 : 69 :
```

These multiplications by 3 to the left and by 8 to the right are easy enough. Aren't they?

Let us now take up and try $\frac{1}{47}$ as promised at an earlier stage. Obviously, the last digit is 7 and the *Ekādhika* is 33. This is rather unwieldly as a multiplier or as divisor. We should therefore try and see what we can get from the remainders. We find them to be 10, 6 etc. We can immediately pounce upon this 6 for our purpose and work in this way:

.02 being the first two digits of the quotient and 6 being our ratio, the next two digits are obviously 12. These × 6 should give us $\dot{7}2$; but as 4 will be coming over from the right, we add the 4 and put down 76. This should give us 456, of which the first digit has already been taken over to the left. So 56 remains. But this will be increased by 3 coming from the right and will become 59—This gives us 57, 44 and 68 for the next three 2-digit groups and 08 for the one thereafter. The 08 group of two digits gives us 48 which, with the carried digit becomes 51. This gives us 06 and 36 which becomes 38. And then we have 28 turning into 29, then 74 which becomes 78 and so forth. Thus we have:

$\frac{1}{47}$ =.02 12 76 59 57 44 68 08 51 06 382 /
97 8

Here we notice that, exactly after 23 digits, the complements from nine have begun. So, we can complete the second half and say:

$\frac{1}{47}$ =.$\dot{0}$2 12 76 59 57 44 68 08 51 06 382
97 87 23 40 42 55 31 91 48 93 61$\dot{7}$

We have thus avoided the complicated divisions by the original divisor 47 and also the divisions and multiplications by the unmanageable *Ekādhika* 33; and, with the easy remainder 6 as our multiplier, we have been able to obtain all the 46 digits of the answer!

This merely shows that these are not cut-and-dried mechanical processes but only rules capable of being applied to the special kind of cases which they are particularly designed to meet and fit into.

And, as for a cut-and-dried formula capable of universal application, that too is forthcoming as already indicated and will be dealt with, very soon.

Let us now take up $\frac{1}{31}$ which, a little earlier, we promised to deal. In this case, the last digit is 9; and the *Ekādhika* is 28 which is *nearly* as big as the original denominator itself! We should therefore sift the remainders and find a suitable *auxiliary* therefrom.

In this case, we find 7 is the first significant remainder. So, leaving the *Ekādhika* process out of account for the moment, we may use the geometrical progression principle and achieve our purpose thereby as we did with 6 in the case of $\frac{1}{47}$. But let us proceed further and see whether a still more easily manageable remainder is available further up.

Well, we observe:

$\frac{1}{31}$ =.032258 with remainder 2. The actual remainders in order are:

10, 7, 8, 18, 25 and 2!

This suits us most admirably, and we proceed further with the help of 2. Thus:

.032258 / 064516 / 129,032 / 258064. . .

But this means that, after only 15 digits an odd number, the decimal has already begun to *recur*! So, we simply say:

$$\frac{1}{31} = .\dot{0}32258064516\ 12\dot{9}!$$

What a simple and easy device!

Let us now take up $\frac{1}{97}$. The last digit is 7; but the *Ekādhika* will be 68! So, we seek help from the remainders. They are: 10, 3 etc., and the quotient-digits are .0103 . . .

So, multiplying each quotient-group of 2 digits each by 3 as we did, by 2, in the case of $\frac{1}{49}$, we get:

$\frac{1}{97}$ =.010309 27 81

2

83 etc.

Let us take one more example (i.e. $\frac{1}{997}$) and conclude. The last digit is 7; but the *Ekādhika* will be 698! It will surely not be an enviable task for even the most practised and experienced statistician to multiply or divide, at each step, by such a big figure! We therefore again seek help from the remainders and the geometrical progression rule.

The quotient-digits are .001 etc., and the successive remainders are 10, 100, 3 etc.! This means that we should multiply each group of three quotient-digits by 3 and get our answer to any number of decimal-places. We thus have:

$$\frac{1}{997} = .001 : 003 : 009 : 027 : 081 : 243 : 729$$

$$: \quad : \quad : \quad : \quad : \quad : \frac{\quad 3}{732 \text{ etc.}}$$

### The Converse Operation

Having dealt, *in extenso*, with the conversion of vulgar fractions into their equivalent recurring decimals, we now take up the converse process, i.e. the conversion of decimals into the equivalent vulgar fractions. We do not, however, propose to go into such a detailed and exhaustive analytical study thereof as we have done in the other case but only to point out and explain *one* particular principle, which will be found very useful in this particular operation and in many subsequent ones.

The principle is based on the simple proposition that $.\dot{9} = \frac{9}{9} = 1; .\dot{9}\dot{9} = \frac{99}{99} = 1; .\dot{9}9\dot{9} = \frac{999}{999} = 1$; and so forth *ad infinitum*. It therefore follows that all recurring decimals whose digits are all nines are *ipso facto* equal to unity; and if a given decimal can be multiplied by a multiplier in such a manner as to produce a product consisting of only nines as its digits, the operation desired becomes automatically complete.

(1) For instance, let us first start with the now familiar decimal $.\dot{0}7692\dot{3}$. In order to get 9 as the last digit, we should multiply this by 3. Setting this product down .230769 we find that, in order to get 9 as the penultimate digit, we should add 3 to the 6 already there. And, in order to get that 3, we should multiply the given multiplicand by 1. On doing this, we find that the totals of two rows are all nines! So we stop there and argue that, because the given decimal ×13 = 999999 (i.e. 1), therefore the fraction should be $\frac{1}{13}$. In fact, it is like saying $13x = 1$

$$\begin{array}{r} .0\,7\,6\,9\,2\,3 \\ 1\,3 \\ \hline .2\,3\,0\,7\,6\,9 \\ \dot{0}.\,7\,6\,9\,2\,3 \\ \hline .\dot{9}\,9\,9\,9\,9\,\dot{9} \\ \hline \end{array} = 1$$

$$\therefore x = \frac{1}{13}$$

(2) Secondly, let us take the case of $.\dot{0}3\dot{7}$ and see how this works. Here as the last digit is 7, so, in order to get 9 as the last digit of the product, we should multiply it by 7. And, putting 259 down, we should add 4 to obtain 9 as the penultimate digit. And, in order to get that 4 there, we should multiply the multiplicand by 2. And, on doing so, we find that the product is $.\dot{9}9\dot{9}(=1)$. Therefore, the fraction $\times\ 27 = 1 \therefore x = \frac{1}{27}$

$$\begin{array}{r} \dot{0}\ 3\ 7 \\ 2\ 7 \\ \hline 2\ 5\ 9 \\ 0\ 7\ 4 \\ \hline 0\ .\dot{9}\ 9\ \dot{9} = 1 \end{array}$$

(3) When we try the case of $142857\dot{}$, we find that multiplication by 7 gives us the all-nine product. $.\dot{9}9999\dot{9}(=1)$; and therefore we say $.14285\dot{7} = \frac{1}{7}$

(4) $.04761\dot{9}$. We first multiply by 1, see 1 in the penultimate place, have to add 8 thereto, multiply by 2 for getting that 8 and thus find that the required answer is $\frac{1}{21}$.

$$\begin{array}{r} .4\ 7\ 6\ 1\ \dot{9} \\ 2\ 1 \\ \hline .0\ 4\ 7\ 6\ 1\ 9 \\ 0.9\ 5\ 2\ 3\ 8\ \ \\ \hline .9\ 9\ 9\ 9\ 9\ 9 = 1 \end{array}$$

(5) Similarly, we may take up various other decimals including the long big ones like the equivalents of $\frac{1}{19}, \frac{1}{29}, \frac{1}{49}, \frac{1}{63}, \frac{1}{117}, \frac{1}{17}, \frac{1}{19}, \frac{1}{23}, \frac{1}{69}$ etc., and invariably we find our purpose achieved.

(6) But, what about decimals ending in even numbers or 5? Well, no integral multiplier can possibly get us 9 as the last digit in the product. And what we do in such a case is to divide off by the powers of 2 and 5 involved and use this new method with the final quotient thus obtained. Thus, if we have to deal with $.\dot{2}8571\dot{4}$ we divide it off by 2, get $.\dot{1}42857$ as the quotient and find that multiplication thereof by 7 gives us the product $.\dot{9}9999\dot{9} = 1$. And therefore we say:

$$\begin{array}{r} .2)\ ,\dot{2}8571\dot{4} \\ \hline .142857 \end{array}$$

$$\frac{x}{2} \times 7 = 1 \therefore x = \frac{2}{7}$$

(7) Let us now try the interesting decimal $.\dot{0}123456\dot{9}$. On applying this new method, we find that multiplication by 81 gives us 1 as the product $\therefore x = \frac{1}{81}$

$$\begin{array}{r} .\dot{0}\ 1\ 2\ 3\ 4\ 5\ 6\ 7\ 9 \\ 8\ 1 \\ \hline .0\ 1\ 2\ 3\ 4\ 5\ 6\ 7\ 9 \\ .9\ 8\ 7\ 6\ 5\ 4\ 3\ 2\ \ \\ \hline .\dot{9}\ 9\ 9\ 9\ 9\ 9\ 9\ 9\ \dot{9} = 1 \end{array}$$

*N.B.* 1. The student should also make use of the *Ekanyūna* formula. This is readily applicable in every case of "Complementary halves" including $\frac{1}{7}, \frac{1}{13}, \frac{1}{19}$ etc.

Thus, $\dot{1}4285\dot{7} = \frac{143 \times 999}{999\ 999} = \frac{11 \times 13}{1001} = \frac{11 \times 13}{11 \times 13 \times 7} = \frac{1}{7}$

Similarly, $.\dot{0}7692\dot{3} = \frac{77 \times 999}{1001 \times 999} = \frac{1}{13}$; and so on.

2. Similarly, with regard to other factors too, it goes without saying that the removal, in general, of common factors from the decimal and the denominator facilitates and expedites the work.

3. The subsequent chapters on "auxiliary fractions" and "divisibility" etc., will expound and explain certain very simple and easy processes by which this work of arithmetical factorisation can be rendered splendidly simple and easy.

4. Above all, the forthcoming "Straight Division" method will not merely render the whole thing simple and easy but also turn it into a pleasure and delight even to the children.

### Some Salient Points and Additional Traits

Thus, the *Ekādhika* process forwards and backwards and the geometrical progression relationship between the remainders have given us the following three main principles:

(*i*) The quotient-complements from 9;

(*ii*) The remainder-complements from the denominator; and

(*iii*) The multiplication of the *Caramāṅka* (last digit) of the remainders by the *Caramāṅka* (last digit) of the decimal, for obtaining each digit of the quotient.

Now, *apropos of* and in connection with this fact, the following few important and additional traits should also be observed and will be found interesting and helpful:

(1) In the case of $\frac{1}{19}$, the remainders are (1), 10, 5, 12, 6, 3, 11, 15, 17, 18/9, 14, 7, 13, 16, 8, 4, 2, 1 and the quotient digits are: 0 5 2 6 3 1 5 7 8/9 4 7 3 6 8 4 2 1.

(*i*) Each remainder by itself, is even and with the addition of the denominator, if odd, is double the next remainder. This follows from the *Ekādhika* being 2.

(*ii*) Each quotient-digit is the last digit of its corresponding remainder. This is because 1 is the last digit of the decimal.

(2) In the case of $\frac{1}{29}$, the remainders are: (1) 10, 13, 14, 24, 8, 22, 17, 25, 18, 6, 2, 0, 26, 28/19, 16, 15, 5, 21, 7, 12, 4, 11, 23, 27, 9, 3 and 1.

(*i*) The quotient-digits are the last digits thereof for the same reason as above;

(*ii*) Each remainder by itself or in conjunction with the denominator or double of it = three times its successor; and

(*iii*) Each remainder *plus* its successor's successor = the next remainder thereafter. Thus

$10+14=24$; $13+24-29=8$; $14+8=22$; $24+22-29=17$; $8+17=25$; $22+25-29=18$; $17+18-29=6$; $25+6-29=2$; $18+2=20$; $6+20=26$; and so on!

*N.B.* : Note the casting off of the denominator all through.

(3) In the case of $\frac{1}{89}$, the quotient-digits are:

0 112 3595 5056 1797 752 808
9887 6404 4943 8202 247 191

and the remainders are (1), 10, 11, 21, 32, 53, 85, 49, 45, 5, 50, 55, 16, 71, 87, 69, 67, 47, 25, 72, 8, 80, 88, 79, 78, 68, 57, 36, 4, 40, 44, 84, 39, 34, 73, 18, 2, 20, 22, 42, 64, 17, 81, 9 and 1. (Note the Ratio 9:1).

The remarkable thing here is that the numerator+the first remainder =the second remainder and that all through, the sum of any two consecutive remainders is the next remainder thereafter! Thus $1+10=11$; $10+11=21$; $11+21=32$; and so on to the very end.

The general form herefor is a, d, a+d, a+2d, 2a+3d, 3a+5d, 5a+8d etc. The student who knows this secret relationship between each remainder and its successor can reel the 44 digits of the answer off, at sight, by simple addition!

(4) In the case of $\frac{1}{79}$ the remainders are:

(1) 10, 21, 52, 46, 65, 18, 22, 62, 67, 38, 64, 8 and 1.

The general form herefor is a, d, a+2d, 2a+5d etc. Knowledge of this relationship will be of splendid practical utility in this case. Note the Ratio 8:1.

(5) In the case of $\frac{1}{69}$ the remainders are:

(1) 10, 31, 34, 64, 19, 52, 37, 25, 43, 16, 22, 13, 61, 58, 28, 4, 40, 55, 67, 49, 7 and 1. Note the Ratio 7:1.

The general form herefor is obviously a, d, a+3d, 3a+10d, 10a+33d etc.

(6) In the case of 59, the remainders are:

(1) 10, 41, 56, 29, 54, 9, 31, 15, 32, 25, 14, 22, 43, 17, 52, 48, 8, 21, 33, 35, 55, 19, 13, 12, 2, 20, 23, 53, 58/49, 18, 3, 30, 5, 50, 28, 44, 27, 34, 45, 37, 16, 42, 7, 11, 51, 38, 26, 24, 4, 40, 46, 47, 57, 39, 36, 6 and 1. Note the Ratio 6:1.

Here the general form is a, d, a+4d, 4a+17d etc.

## Inductive Conclusion

Having thus examined the cases of $\frac{1}{89}$, $\frac{1}{79}$, $\frac{1}{69}$ and $\frac{1}{59}$ we note the following:

(*i*) In every case, we start with 1 the basic numerator as a sort of pre-natal remainder which is perfectly justified because we are dealing with a recurring decimal; and we call it a;

(*ii*) In every case, the first actual remainder is 10; and we call it d;

(*iii*) And then the successive remainders are a, d, a+d, a+2d, a+3d, a+4d respectively wherein the coefficient of d is obviously the deficit of the penultimate digit from 9!

Thus for $\frac{1}{89}$, we have a + 1d;

for $\frac{1}{79}$, we have a + 2d;

for $\frac{1}{69}$, we have a + 3d;

for $\frac{1}{59}$; we have a + 4d; and so on.

(*iv*) And this relationship is maintained systematically all through. In other words, each remainder+the next one or double that or three times that etc.=the further subsequent remainder. Arguing thus, let us try $\frac{1}{39}$. As 3 is 6 less than 9, ∴ the general form should be a + 6d. This means 1, 10, 61, (i.e. 9), 64 (i.e. 12), 3, 30 (i.e. 4) and 27 (i.e. 1). And we find this to be actually correct;

(*v*) And, in case the penultimate digit is more than 9, we should react by subtracting d and not add to it at the rate of 1 for each surplus. Thus, our chart will now read – a, d, a – d, d – (a – d), i.e. 2d – a, and so on. For instance, for $\frac{1}{109}$, the remainders will be (1) 10, –9, 19, –28, 47 and so on;

(*vi*) And, over and above all these details which are different for different numbers as explained above, there is one multiplier, namely 10 which is applicable to all cases! 'And thus, whatever fraction we may be dealing with 2, 4, 5, 8 or any remainder whatsoever can be safely put in into the next place with a zero added! The student will observe that, in *all* the examples dealt with hereinabove not only in this particular sub-section, every such remainder of two digits ending in zero has been invariably preceded by the same number without the zero!

With the help of this rule applicable in all cases and the special rules about d, 2d, 3d, 4d etc., enjoined for the different individual cases, the student should easily be in a position to make a list of the successive remainders in each case and therefrom, by *Caramāṅka* multiplication, put down the successive quotient-digits without further special labour!.

These and many more interesting features are there in the Vedic decimal system, which can turn mathematics for the children, from its present excruciatingly painful character to the exhilaratingly pleasant and delightful character it really bears!

We have, however, already gone into very great details; and this chapter has already become very long. We therefore conclude this chapter here and hold the other things over for a later stage in the student's progress.

# 27

# Straight Division

We now go on, to the long-promised Vedic process of straight (at sight) division which is a simple and easy application of the *Ūrdhva-tiryak Sūtra* which is capable of *immediate* application to *all* cases and which we have repeatedly been describing as the "crowning gem of all" for the very simple reason that over and above the universality of its application, it is the most supreme and superlative manifestation of the Vedic ideal of the at-sight mental-one-line method of mathematical computation.

### Connecting Link

In order to obtain a correct idea of the background, let us go back, very briefly to the methods which we employed in the earlier chapters on division; and let us start with the case of $\frac{38982}{73}$.

According to the first method under the *Nikhilaṁ* etc., *Sūtra*, our chart will read as follows:

$$\begin{array}{l|lll|lll|l}
\frac{73}{27} & 3 & 8 & 9 & 8 & 2 & & \\
 & & 6 & 21 & & & & \\
 & & & 28 & 98 & & & \\
 & & & & 116 & 406 & & \therefore Q = 534 \\
\hline
 & 3 & 14 & 58 & (26) & 28 & & \text{and } R = 0 \\
\hline
 & 5 & 3 & 4 & (36) & 00 & & \\
\hline
\end{array}$$

According to the second method by *Parāvartya* formula, we say:

$$\begin{array}{l|lll|ll|}
\frac{73}{1\bar{3}3} & 3 & 8 & 9 & 8 & 2 \\
3\bar{3} & & 9 & -9 & & \\
 & & & 51 & -51 & \\
 & & & & 153 & -153 \\
\hline
 & 3 & 17 & 51 & 23 & 949 \\
\hline
 & 5 & 3 & 4 & 0 & 0 \\
\hline
\end{array}$$

We have felt, and still feel, that even these comparatively short, intellectual and interesting methods are cumbrous and clumsy from the idealistic Vedic standpoint. And hence the clamant need for a method which is free from all such flaws and which fulfils the highest ideal of the Vedic *Sūtras*.

And that is as follows:

| | | | | |
|---|---|---|---|---|
| 3 : | 38 | 9 | 8 : | 2 : |
| 7 : | | 3 | 3 : | 1 : |
| : | 5 | 3 | 4 : | 0 : |

Out of the divisor 73, we put down only the first digit, i.e. 7 in the divisor-column and put the other digit, i.e. 3 "on top of the flag" by the *Dhvajāṅka Sūtra*, as shown in the chart alongside.

The entire division is to be by 7; and the procedure is as explained below:

As one digit has been put on top, we allot one place at the right end of the dividend to the remainder portion of the answer and mark it off from the digits by a vertical line.

(*i*) We divide 38 by 7 and get 5, as the quotient and 3 as the remainder. We put 5 down as the first quotient-digit and just prefix the remainder 3 up before the 9 of the dividend. In other words, our actual second-step gross dividend is 39. From this, we however, deduct the product of the indexed 3 and the first quotient-digit 5, i.e. $3 \times 5 = 15$. The remainder 24 is our actual net-dividend. It is then divided by 7 and gives us 3 as the second quotient-digit and 3 as the remainder, to be placed in their respective places as was done in the first step. From 38 the gross dividend thus formed, we subtract $3 \times$ the second quotient-digit 3, i.e. 9, get the remainder 29 as our next actual dividend and divide that by 7. We get 4 as the quotient and 1 as the remainder. This means our next gross dividend is 12 from which, as before, we deduct $3 \times$ the third quotient-digit 4, i.e. 12 and obtain 0 as the remainder. Thus we say : Q is 534 and R is zero. And this finishes the whole procedure; and all of it is one-line mental Arithmetic in which all the actual division is done by the simple-digit Divisor 7.

The Algebraical Proof hereof is very simple and is based on the very elementary fact that all arithmetical numbers are merely Algebraical expressions wherein x stands for ten. For instance, $3x^2 + 5x + 1$ is merely the algebraical general expression of which with x standing for 10 the arithmetical value is 351.

Remembering this, let us try to understand the steps by means of which 3 8 9 8 2 is sought to be divided by 73. Algebraically put with x standing for 10, this dividend is $38x^3 + 9x^2 + 8x + 2$; and this divisor is $7x + 3$. Now, let us proceed with the division in the usual manner.

When we try to divide $38x^3$ by $7x$, our first quotient-digit is $5x^2$; and, in the first step of the multiplication of the divisor by $5x^2$, we get the product $35x^3 + 15x^2$; and this gives us the remainder $3x^3 + 9x^2 - 15x^2$, which really means $30x^2+9x^2 - 7x+3)\ 38x^3+9x^2+8x+2\ (5x^2+3x+4$ $15x^2=24x^2$. This plus 8x being our second-step dividend, we multiply the divisor by the second quotient-digit 3x and subtract the product $21x^2+9x$ there-from and thus get $3x^2-x$ as the remainder. But this $3x^2$ is really equal to 30x which with $-x+2$ gives us $29x+2$ as the last-step dividend. Again multiplying the divisor by 4, we get the product $28x+12$; we subtract this $28x+12$, thereby getting $x-10$ as the remainder. But x being 10, this remainder vanishes! And there you have the whole thing in a nut-shell.

$$\begin{array}{r} 35x^3+15x^2 \\ \hline 3x^3-\ 6x^2 \\ =24x^2+8x \\ 21x^2+9x \\ \hline 3x^2-\ x \\ =29x+2 \\ 28x+12 \\ \hline x-10 \\ \hline =10-10=0 \\ \hline \end{array}$$

It will be noted that the arithmetical example just hereabove dealt with $\left(\text{i.e. } \frac{38982}{73}\right)$ is merely the arithmetical form of $\frac{38x^3+9x^2+8x+2}{7x+3}$ and the arithmetical chart has merely shown the above given algebraical operation in its arithmetical shape wherein $x=10$ and that, whatever the algebraical working has taken a remainder-digit over to the right with a zero added, the arithmetical chart shows that particular remainder prefixed to the digit already there.

$$\begin{array}{c|l} 3 & 38\ 9\ 8 : 2 \\ 7 & \quad\ 3\ 3 : 1 \\ \hline & \quad 5\ 3\ 4 : 0 \end{array}$$

Thus, where $3x^2$ has been counted as $30x^2$ and added to the $9x^2$ already there and produced $39x^2$ as the result, this algebraical operation has been graphically pictured as the prefixing of 3 to 9 and making it 39! And similarly, in the next step of the division, the remainder 3 is prefixed to the 8 already there: and we have to deal with 38; and similarly, at least, the 1 prefixed to the 2 gives us 12 which the $3\times4$ subtracted therefrom cancels out.

In other words the given expression $38x^3+9x^2+8x+2$ with 10 substituted for x is actually the same as $35x^3+36x^2+37x+12$. And we say:

$$\frac{38x^3+9x^2+8x+2}{7x+3} = \frac{35x^3+36x^2+37x+12}{7x+3} = 5x^2+3x+4.$$ And, graphically, this algebraical operation is demonstrated arithmetically in the manner shown in the margin.

$$\begin{array}{c|l} 3 & 38\ 9\ 8 : 2 \\ 7 & \quad\ 3\ 3 : 1 \\ \hline & \quad 5\ 3\ 4 : 0 \end{array}$$

The procedure is very simple and needs no further exposition and explanation. A few more illustrative instances with running comments, as usual will however, be found useful and helpful and are therefore given below:

(1) Divide 529 by 23.
The procedure is exactly the same and is simple and easy.

```
  3 : 5   2 : 9
2   :     1 : 0
    : 2   3 : 0
```

(2) Divide 4096 by 64.

```
  4 : 40  9 : 6 :
6   :     4 :1 :
    : 9   4 : 0 :
```

(3) Divide 16384 by 128.

```
   8 : 16  3   8 : 4
12   :     4 11  :6
     : 1   2   8 : 0
```

(4) Divide 7632 by 94.
(*ii*) *New Nikhilaṁ* method or (*ii*) Newest Vedic method

```
94 : 7  6 : 3 2  :
06 :      0 : 42  :
   :        : 036 :
   :    76 :  4 88 :
   :    81 : 18   :
```

```
  4 : 7 6 3 : 2 :
9   :     4 :2 :      ∴ Q = 81
    : 8 1   :18 :       R = 18
```

(5) Divide 601325 by 76.
Here, in the first division by 7, we can put 8 down as the first quotient-digit; but the remainder then left will be too small for the subtraction expected at the next step. So, we take 7 as the quotient-digit and prefix the remainder 11 to the next dividend-digit. N.B. For purposes of reference and verification, it will be a good plan to *underline* such a quotient-digit because the chart offers itself for verification at every step and any reconsideration necessary at any stage need not involve ourgoing back to the beginning and starting the whole thing over again.

```
  6 : 60   1  3 2 : 5
7   :      11 6 2 : 2
    : 7    9  1 2 : 13
```

(6) Divide 3100 by 25.

```
  5 : 3   1   0 : 0 :
2   :     1   2 : 2 :
    : 1   2 4   : 0 :
```

*Note*: In algebraic terminology, $3100 = 3x^3 + x^2 = 2x^3 + 9x^2 + 8x + 20$ and the above example is the arithmetical way of stating that $2x^3 + 9x^2y + 8xy^2 + 29y^3 = (2x + 5y)(x^2 + 2xy + 4a^2)$ (i.e. $25 \times 124 = 3100$).

(7) Similar is the case with regard to the division of $38x^3 + 9x^2 + 8x + 2$ by $(x - 1)$, wherein $Q = 38x^2 + 47x + 55$ and $R = 57$.

(8) Divide 695432 by 57.

```
 7 : 6 9 5 4 3 : 2  :
5  :   1 2 1 0 : 3  :
   : 1 2 2 0 0 : 32 :
```

(9) Divide 3279421 by 53.

```
 3 : 32 7 9 4 2 : 1  :
5  :    2 4 6 5 : 6  :
   :  6 1 8 7 5 : 46 :
```

(10) Divide 7777777 by 38.

```
 8 : 7 7 7 7 7 7 : 7  :
3  :   1 1 5 7 8 : 7  :
   : 2 0 4 6 7 8 : 13 :
```

(11) Divide 500001 by 89.

```
 9 : 50 0  0 0 : 1
8  :    10 7 8 : 15
   : 5  6  1 7 : 88
```

(12) Divide 37941 by 47.

```
 7 : 37 9 4 : 1  :
4  :    5 3 : 6  :
   : 8  0 7 : 12 :
```

(13) Divide 745623 by 79.

```
 9 : 74 5 6 2 : 3  :
7  :    11 6 9 : 9  :
   :  9 4 3 8 : 21 :
```

(14) Divide 7453 by 79.
(to 3 places of decimals)

```
 9 : 74 5 3 .0 0 :
7  :    11 6 6 5 0 :
   :  9 4 .3 4 2 2 :
```

(15) Divide 710.014 by 39
(to 3 places of decimals)

```
 9 : 7 1 0. 0 1 4 :
3  :   4 8 2 2 6 4 :
   : 1 8. 2 0 5 5 :
```

(16) Divide 220 by 52
(to 3 places of decimals)

```
 2 : 2 2 0 .0 0 0
5  :       2 2 1 4
   :    4. 2 3 0 8 :
```

(17) Divide 7.3 by 53 (to 5 places of decimals)

```
 3 : 7 .3 0 0 0 0 :
5  :    2 5 6 4 4 :
   : 0.1 3 7 7 3 6 :
```

(18) Divide 71 by 83 (to 5 places of decimals)

```
 3 : 7 1 0 0 0 0 :
8  :     7 6 5 3 :
   : 0 .85 5 4 2... :
```

(19) Divide 1337 by 79.

(*i*) By the *New Nikhilaṁ* method

```
79 : 13 : 3    7 :
21 :  2 : 1      :
   :    : 10   5 :
   : 15 : 1 : 52 :
   :    :   : 21 :
   : 16 :   : 73 :
```

(*ii*) By the newest Vedic method

```
 9 : 13 3 : 7  :
7  :    6 : 12 :
   :  1 6 : 73 :
```

(20) Divide 1681 by 41.

```
  1 : 16  8   1 :
4   :     0   0 :
    :  4  1   0 :
```

*N.B.*: The Algebraical form is:

$$\frac{16x^2+8x+1}{4x+1}=4x+1$$

(21) Divide 115491 by 137.

```
   7 : 115   4  9 : 1 :
13   :      11  6 : 2 :
     :   8   4  3 : 0 :
```

or in Algebraical form:

$$13x+7 : 115x^3+4x^2+9x+1(8x^2+4x+3$$

$$: 104x^3+56x^2$$

$$11x^3-52x^2+9x = 58x^2+9x$$

$$\therefore E = 104x^3+108x^2+67x+21 \quad 52x^2+28x$$

$$6x^2-19x+1$$

$$=41x+1$$

$$39x+21$$

$$2x-20=0$$

$$\therefore Q = 8x^2+4x+3 \text{ (i.e. 843) } \& R = 0$$

(22) Divide 7458 by 127. (to 3 places of decimals)

```
   7 : 74   5  8 0 0 0 0
12   :     14 14 8 7 8 4
     :  5   8 .7 2 4 4...
```

(23) Divide 3517 by 127. (to 3 places of decimals)

```
   7 : 35  1  7  0  0  0
12   :    11 13 16 10 13
     :  2  7 .6  9  2  9...
```

(24) Divide 7031985 by 823.

Here, the divisor is of 3 digits. All the difference which this makes to us is that, instead of putting one extra digit on top, we put both the extra digits 23 there; and we adopt a slightly different *modus operandi* on the *Ūrdhva-tiryak* lines in respect of the subtraction-portion of the work.

```
  23 : 70  3  1  9 : 8 5
8    :     6  7  5 : 4 4 3      and
     :  8  5  4  4   3 3        R = 273
```

In this instance, we divide 70 by 8 and set 8 and 6 down in their proper places as usual. Thus, our second gross dividend is now 63. From that, we subtract 16 the product of the first of the flag-digits, i.e. 2 and the first quotient-digit, i.e. 8 and get the remainder $63-16=47$ as the actual dividend. And, dividing it by 8, we have 5 and 7 as Q and R respectively and put them down at their proper places. So now, our gross dividend is 71; and we deduct, by the *Ūrdhva-Tiryak* rule, the cross-products of the two flag-digits 23, and the two quotient-digits 8,

5, i.e. 10+24=34; and our remainder is 71–34=37. We then continue to divide again by 8 and subtract etc., in the same manner by cross-multiplication as just now explained by the *Ūrdhva-Tiryak* method until the last digit of the dividend is reached. And that finishes the task.

And, in other divisions too, irrespective of the number of digits in the divisor, we follow the same method. And, in every case, our actual divisor is *of one digit only* or at the most, a small two-digit one like 12, 16 and so on which one can easily divide by. And all the rest of the digits of the divisor are hoisted on the flag-top. And this is the whole secret of the "Straight Division" formula.

*Note*: If instead of the decimal places in the quotient, you want the remainder, you can have it in the usual way. In this case, 23 and **44** by cross-multiplication give us 20, which when taken to the right means 200; $3 \times 4$ the last flag-digit $\times$ the last obtained quotient-**digit=12**. Subtracting the total of these two, i.e. 212 from 485, we have **R=273**, i.e. R = 485 – 200 – 12 = 273.

Some more instances of division by three-digit divisions etc. are cited below:

(1) Divide 1064321 by 743 (to 4 places of decimals)

```
  43 : 10 6 4 3 2 1 0 0 :
7    :    3 4 4 5 7 7 6 5:
     :   14 3 2 .4 6 4 3 :
```

∴ R = 521 – 170 – 6 = 345

(2) Divide 222220 by **735** (to 3 places of **decimals**)

```
  35 : 22   2 2 2 0 0  :
7    :      1 3 3 5 3 3:
     : 3 0 2.3 4 0     :
```

R = 320 – 60 – 10 = 250

(3) Divide 888 by 672 (to 3 places of decimals)

```
  72 :8  8 8 0 0 :
6    :   2 3 3 4 2:
     :1 .3 2 1 5 :
```

```
or  3̄2 : 8 : 8     .8 :
   7   :   : 1        :
       : 1 : R = 216  :
```

*or* by mere *Vilokanaṁ* (Inspection)

(4)

```
  28 : 6 3 8 1 8 :  2   7 :
5    :   1 1 6 5 :        :
     : 1 2 0 8 6 : R = 419 :
```

```
or  32̄ : 6 3 8 1 8 :  2   7 :
   5   :   1 0 4 5 :  4  12 :
       : 1 2 0 8 6 : R = 419:
```

(5) Divide 13579 by 975

```
  75 : 13 5 :   7 9 :
9    :    4 : 11    :
     : 1  3 :       :
```

R = 1179 – 260 – 15 = 904

(6) Divide 513579 by 939

```
  39 : 51  3  5 :  7  9:
9    :     6 12 : 14   :
     : 5   4  6 :      :
```

R = 1479 – 540 – 54 = 885

(7) Divide 7143 by 1171

(*i*) By the new *Parāvartya* method

$$\begin{array}{r|l} 1171 & :7:\ 1\quad 4\quad 3: \\ -1-7-1 & :\quad :-7-49-7: \\ \hline & :7:6\quad 45\quad 4: \\ \hline & :\ 7:-10-5-4: \\ \hline & :\ 6:117\qquad : \end{array}$$

(*ii*) By the *new Parāvartya* (*Vinculum*) method

$$\begin{array}{r|l} 1171 & :7:\ 1\quad 4\quad 3: \\ -1-7-1 & :\quad :-14+21-7: \\ -2+3-1 & :7:-13+25-4: \\ & :7:-1054 \\ \hline & :6:\ 1\quad 1\quad 7\ : \end{array}$$

(*iii*) By the newest *Vedic* method

$$\begin{array}{rl} 71:71:4\ 3: & \therefore Q=6 \\ 11\quad :\qquad :5\ 1 & :\text{and } R=543-426=117 \\ \hline :\ 6:1\ 0: & \end{array}$$

(8) Divide 4213 by 1234 (to 4 places of decimals)

$$\begin{array}{l} 34:42:\ 1\ 3\ 0\ 0\ 0 \\ 12\quad :\qquad :6\ 4\ 7\ 3\ 2 \\ \hline \quad :\ 3:4\ 1\ 4\ 1\ 0 \end{array}$$

(9) Divide 46781 by 1483 (to 3 places of decimals)

$$\begin{array}{l} 83:46:\ 7\ 8\ 1\ 0 \\ 14\quad :\qquad :4\ 9\ 11\ 12\ 3 \\ \hline \quad :\ 3:\ 1\ .5\ 4\ 5 \end{array}$$

(10) Divide 3124 by 1532 (to 3 places of decimals)

$$\begin{array}{l} 32:31:\ 2\ 4\ 0 \\ 15\quad :\qquad :1\ 6\ 0 \\ \hline \quad :\ 2.:0\ 4\ 0 \end{array}$$

(11) Divide 333333 by 1782 (to 3 places of decimals)

$$\begin{array}{l} 82:33:\ 3\ 3\ 3\ 3 \\ 17\quad :\qquad :16\ 19\ 8\ 11 \\ \hline \quad :1:\ 8\ 7\ .0\ 5\ 6 \end{array}$$

(12) Divide 46315 by 1054 (to 3 places of decimals)

$$\begin{array}{l} 54:46:\ 3\ 1\ 5\ 0 \\ 10\quad :\qquad :6\ 13\ 10\ 8 \\ \hline \quad :\ 4:\ 3.9\ 4\ 2 \end{array}$$

$R=1315-310-12=993$

(13) Divide 75313579 by 1213

$$\begin{array}{l} 13:75:\ 3\ 1\ 3\ 5\ :79 \\ 12\quad :\qquad :3\ 3\ 11\ 11\ 11 \\ \hline \quad :6:2\ 0\ 8\ 8 \end{array}$$

$R=1179-344=835$

(14) Divide 135791 by 1245

$$\begin{array}{l} 45:13\ 5\ 7:91 \\ 12\quad :\quad 1\ 11\ :4 \\ \hline \quad :1\ 0\ 9\ : \end{array}$$

$R=491-405$ or $130-44=86$

(15) Divide 13579 by 1616

$$\begin{array}{l} 16:135:\ 79 \\ 16\quad :\qquad :7 \\ \hline \quad :\ 8\quad : \end{array}$$

$R=779-128$ or $690-39=651$

(16) Divide 135791 by 1632

$$\begin{array}{l|l}
\ \ 32 : 135\ \ 7 : \ 91 & \\
16\ \ : \qquad 7 : 5 & R = 591 - 250 - 6 = 335 \\
\hline
\quad\ : \ \ 8\ 3\ : &
\end{array}$$

(17) Divide 97531 by 1627

$$\begin{array}{l}
\ \ 27 : 97\ \ \ 5 : \ \ 31 \\
16\ \ : \ \ 17\ \ : 21 \\
\hline
\quad\ : \ 5\ \ 9\ :
\end{array}
\qquad or \qquad
\begin{array}{l}
\ \ 27 : 97\ \ 5 : \ \ 31 \\
16\ \ : \ \ 1\ \ : 3 \\
\hline
\quad\ : \ 6\ \ 0 :
\end{array}$$

$R = 2131 - 593 = 1538$ and $R = 331 - 420 = -89$

i.e. $Q = 59$ and $R = 1538$

(18) Divide 97531 by 1818

$$\begin{array}{l}
\ \ 18 : 97\ \ 5 : \ \ 31 \\
18\ \ : \ \ 7\ \ : 16 \\
\hline
\quad\ : \ 5\ \ 3\ :
\end{array}
\qquad R = 1631 - 454 (or\ 1200 - 23) = 1177$$

(19) Divide 13579 by 2145

$$\begin{array}{l}
\ \ 45 : 135 : \ 79 \\
21\ \ : \qquad : 9 \\
\hline
\quad\ : \ 6\ \ :
\end{array}
\qquad And\ R = 979 - 270\ or\ 930 - 221 = 709$$

(20) Divide 135791 by 2525

$$\begin{array}{l}
\ \ 25 : 135\ \ 7 : \ 91 \\
25\ \ : \qquad 10\ : 22 \\
\hline
\quad\ : \ 5\ \ 3\ :
\end{array}
\qquad And\ R = 2291 - 325\ or\ 1980 - 14 = 1966$$

(21) Divide 5011 by 439 (to three places of decimals)

$$\begin{array}{l}
\ 39 : 5\ 0\ 1\ 1\ 0 : \\
4\ \ : \ 1\ 3\ 3\ 6\ : \\
\hline
\quad : 11\ .415\quad :
\end{array}
\qquad or \qquad
\begin{array}{l}
\ 4\bar{1} : 50 : 110\ : \\
4\ \ : \ 1 : .222 : \\
\hline
\quad : 11 : .415\ :
\end{array}
\qquad
\begin{array}{l}
R = 19\bar{8} = 182 \\
or\ 311 - 129 \\
= 182
\end{array}$$

(22) Divide 1561 by 349 (to three places of decimals)

$$\begin{array}{l}
\ 49 : 15\ \ 6\ \ 1\ \ 0 \\
3\ \ : \ \ 3\ \ 8\ \ 85\ \ 10 \\
\hline
\quad : 4\ .475\ \ 277
\end{array}
\qquad or \qquad
\begin{array}{l}
\ 5\bar{1} : 15\ .6\ \ 1\ \ 0 : \\
3\ \ : \ \ 3\ \ 4\ \ 4\ : \\
\hline
\quad : 4\ .47277\ \ :
\end{array}$$

And $R = 361 - 196$ or $200 - 35 = 165$ And $R = 361 - 196 = 165$

(23) Divide 47 by 798 (to five places of decimals)

```
 98 : .4 7 0 0 0  0        or   02 : .4 7 0 0 0 0
7   :  4    12 19 15           8   :  4 7 6 6 4
    : .0 5 8 9  0                  : .0 5 8 8 9 7
```

(24) Divide 1111 by 839

```
 39 : 11 : 1 1 :            or
8   :     :3 4  :
    :  1 :  3   : By mere Vilokanaṁ (Inspection)
```

And R = 311 − 39 = 272

We now extend the jurisdiction of the *Sūtra* and apply it to divisors consisting of a large number of digits. The principle involved being the same, the procedure is also identically the same as in the foregoing examples. And the division by a single digit or a small two-digit divisor continues exactly the same. A few illustrative instances are given hereunder:

(1) Divide 7031.95 by 8231 (to 5 decimal places)

```
 231 : 70  3 1 9 5 0 0
8    :     6 7.5 4 6 3
     : .8 5 4  32
```

(2) Divide 995 311 by 16123

```
  123 : 99  5 :  3 1 1
16    :     3 :13
      : 6 1   :
```

R = 13311 − 1503 (or $12000 + \overline{1}\,\overline{9}\,\overline{0} + \overline{2}$) = 11808

(3) Divide 975 311 by 16321

```
  231 : 97  5 :3   1  1
16    :     1 :3
      :  6  0 : R = 1500
```

R = 1500 + $\overline{5}\,\overline{0}$ + 1 or 3311 − 1860 = 1451

(4) Divide 975 311 by 16 333

```
  333 : 97  5 :3   1  1
16    :  17   :16
      : 5  9  :
```

R = 16311 − 4627 = 11664
or $12100 + \overline{4}\,\overline{1}\,\overline{0} + 2\overline{6}$ = 11664

(5) Divide 975 311 by 18123

```
  123 : 97  5 :   311
18    :     7 :16
      : 5   3 :
```

R = $15000 + \overline{2}\,\overline{0}\,\overline{0} + \overline{8}$
(or 16311 − 1519) = 14792

(6) Divide 995311 by 20321

```
   321 : 99  5 :  311
20     :   19  : 23
       : 4  8  :
```

R = 20100 − 190 − 7
(= 23311 − 3408) = 19903

(7) Divide 997531 by 30321

```
   321 : 99  7  531
30     :     9  28
       : 3 2
```

R = 27300 − 40 − 1
(or 28531 − 1272) = 27259

(8) Divide 137294 by 4794 (to 6 places of decimals).

```
   749 : 13 7 : 2  9  4   0   0  0
5      :    3 : 8 13 13  14  14  8
       :   23.  8  8  1 371
```

*or*

```
   351 : 13 7. 2 9 4 0 0 0
6      :    1  5 3 2 5 3 0
       : 23 . 88   1371
```

(9) Divide 53247 by 4999 (to five places of decimals)

```
   999 : 53. 2 4 7  0  0 0
4      :  1  4 9 11 14 12 9
       : 10. 6 5 1  5  3 . . .
```

*or*

$00\bar{1}$ : 5 3. 2 4 7 0 0 0

```
5      : 0 3 2 0 2 1 0
       : 1 0.6 5 1 5 3 . . .
```

*N.B.* Better to divide by 50

(10) Divide 138462 by 39898 (to 3 places of decimals)

```
  9838 : 13 : 8 4  6  2  0  0
3      :    :4 9 13 14 17 21
       :  3.  4 7  0  9  3
```

*or*

$0\bar{1}0\bar{2}$ : 13 : 8 4 6 2 :

```
4      :    :1 2 3 4 :
       :  3.  4 7 03 :
```

*N.B.* Better divide by 40

(11) Divide 131 by 19799 (to 5 places of decimals)

9799 : 1 3 1 0 0 0
1 : 1 7 11 13 19
: 00 6 6 1 6 . . .

*or*

0$\bar{2}$0$\bar{1}$ : 1 3 1000
2 : 1 1$\bar{1}$
: 00 6 6 16

*N.B.* Better to divide by 20

(12) Divide 76432 by 67998 (to 5 places of decimals)

7998 : 7 6 4 3 2 0
6 : 1 3 6 7 9 10
: 1. 1 2 4 0 3

*or*

$\bar{2}$00$\bar{2}$ : 7 6 4 3 2 0 0
7 : 0 1 2 6 7 18
: 1. 1 2 4 0 3 ...

(13) Divide 2537 by 48329

8329 : 2 5 3 7 0
4 : 2 5 5 10 16
: .0 5 2 4 9

*or*

$\bar{2}$33$\bar{1}$ : .2 5 3 7 0 0
5 : 2 0 3 11
: .0 5 2 4 9 . . .

(14) Divide 371628.112 by 12734 (to 5 decimal places)

2734 : 371628 112
1 : 14 8 7 11 8 12 8 11 11 11
: 29.1839 25868

*N. B.* Better divide by 12

*or*

$\bar{3}$34 : 3 7 1 6 2 8. 112 00
13 : 11 0 14 3 14 5 8 6
: 2 9 1 8 3 9 2 5 8 6 8

*N.B.* Here we have divided by 13.

(15) Divide 41326 by 31046 (to 5 decimal places)

$$\begin{array}{l|l} 1046 & 4\ 1\ 3\ 2\ 6\ 0\ 0 \\ 3 & \ \ 1\ 1\ 1\ 2\ 4 \\ \hline & 1.\ 3\ 3\ 1\ 1\ 2 \ldots \end{array} \quad \text{or} \quad \begin{array}{l|l} 105\bar{4} & 4\ 1\ 3\ 2\ 6\ 0\ 0 \\ 3 & \ \ 1\ 1\ 1\ 1\ 1\ 1 \\ \hline & 1.\ 3\ 3\ 1\ 1\ 2 \ldots \end{array}$$

(16) Divide 20014 by 137608 (to 5 decimal places)

$$\begin{array}{l|l} 37608 & 2\ 0\ 0\ 1\ 4\ 0 \\ 1 & \ \ 1\ 3\ 6\ 8\ 9 \\ \hline & .1\ 4\ 5\ 4\ 4 \ldots \end{array} \quad \text{or} \quad \begin{array}{l|l} 4\bar{2}\bar{4}1\bar{2} & 2\ 0\ 0\ 1\ 4\ 0 \\ 1 & \ \ \ 1\ 2\ 1\ 0 \\ \hline & .1\ 4\ 5\ 4\ 4 \ldots \end{array}$$

*N.B.* Better divide by 13.

(17) Divide .0034147 by 81.4256321 (to 6 decimal places)

$$\begin{array}{l|l} 14256321 & 0\ 0\ 3\ 4\ 1\ 4\ 7 \\ 8 & \ \ \ \ \ 3\ 2\ 9\ 5 \\ \hline & .0\ 0\ 0\ 0\ 419 \ldots \end{array}$$

*N.B.* The *Vinculum* method is always available but will not make much difference. In fact, it may prove stiffer.

(18) Divide .200103761 by 93.71836211 (to 5 decimal places)

$$\begin{array}{l|l} 371836211 & .2\ 0\ 0\ 1\ 0\ 3\ 7\ 6\ 1 \\ 9 & \ \ \ 2\ 5\ \ \ 7\ 7 \\ \hline & .0\ \ 2\ \ 1\ \ 3\ 5 \ldots \end{array}$$

(19) Divide 74.5129 by 9.314

$$\begin{array}{l|l} 314 & 74.\ 5\ 1\ 2\ 9\ 0\ 0\ 0\ 0 \\ 9 & \ \ \ \ 2\ 1\ 3\ 0\ 9\ 9\ 9\ 9 \\ \hline & 8.\ 0\ 0\ 0\ 0\ 9\ 6\ 6 \end{array}$$

(20) Divide 71324 by 23145

$$\begin{array}{l|l|l} 3145 & 7 & 1\ 3\ 2\ 4 \\ 2 & & 1\ 2\ 4\ 4 \\ \hline & 3 & 0\ 8\ 1 \ldots \end{array} \quad \text{or} \quad \begin{array}{l|l|l} 145 & 71 & 3\ 2\ 4 \\ 23 & & 2\ 20\ 6\ 18 \\ \hline & 3 & 0\ 8\ 1 \ldots \end{array}$$

(21) Divide 137426 by $743.2 \times 1.242 \times 80.04$ (to 4 places of decimals)

| | |
|---|---|
| 432 | : 13 7 : 4 2 6 0 0 0 |
| 7 | : 6 :7 11 7 5 4 8 |
| 42 | : 184 : 9 1 11 9 |
| 12 | : 6 12 15 15 7 11 |
| 004 | : 14 8 8 8 17 |
| 8 | : 6 4 0 4 9 |
| | : 1. 8 6 0 0 9 |
| | :=1. 8 6 0 1 (approximately) |

# 28

# Auxiliary Fractions

In our exposition of vulgar fractions and decimal fractions, we have so far been making use of processes which help to give us the *exact* results in each case. And, in so doing, we have hitherto generally followed the current system whereby multiplications and divisions by powers of ten are mechanically effected by the simple device of putting the decimal point backwards or forwards as the case may be.

### Conventional Method

For instance, we manipulate the decimal point thus:

$$(1)\ \frac{1}{800}=\frac{.01}{8};\quad (2)\ \frac{39}{70}=\frac{3.9}{7};\quad (3)\ \frac{17}{130}=\frac{1.7}{13};$$

$$(4)\ \frac{3741}{110000}=\frac{.3741}{11};\ \text{and}\ (5)\ \frac{97654}{90000000}=\frac{.0097654}{9}$$

But after this has been done, the other operations—of actual division etc.—have had to be carried out in the usual manner.

### Auxiliary Fractions

There are certain Vedic processes, however, by which, with the aid of what we call *sahāyaks* (auxiliary) fractions, the burden of the subsequent operations is also considerably lightened and the work is splendidly facilitated.

### First Type

The first and commonest type thereof is a very simple and easy application of our self-same old friend the ***Ekādhika*** *Pūrva*. And the whole *modus operandi* is to replace the denominator by its *Ekādhika*, i.e. to drop the last digit and increase the penultimate one by 1 and make a consequential alteration in the division-procedure as in the case of other ***Ekādhika*** operations.

*N.B.*: The student will remember that, in these operations, the remainder at each step of a division is not prefixed to a series of zeroes from the right-hand side, but to each quotient-digit.

### Auxiliary Fractions (A.F.) (First Type)

(1) for $\frac{1}{19}$, the Auxiliary fraction is $\frac{1}{2}$

$\therefore \frac{1}{19} = .\dot{0}52631578 / 94736842\dot{1}$

(2) for $\frac{1}{29}$, the A.F. is $\frac{1}{3}$

$\therefore \frac{1}{29} = \frac{1}{3}(AF) = .\dot{0}3448275862068 / 9655172413793\dot{1}$

(3) for $\frac{37}{59}$ AF is 3.7/6

(4) for $\frac{3}{59}$, AF is $\frac{0.3}{6}$

(5) for $\frac{73}{89}$; AF is $\frac{7.3}{9}$

(6) for $\frac{1}{119}$, AF is $\frac{0.1}{12}$

(7) for $\frac{1}{149}$, AF is $\frac{0.1}{15}$

(8) for $\frac{.7}{149}$, AF $= \frac{0.07}{15}$

(9) for $\frac{172}{1299}$, AF $= \frac{1.72}{13}$

(10) for $\frac{371}{7999}$, AF is $= \frac{0.371}{8}$

(11) for $\frac{537}{89999}$, AF is $= \frac{0.0537}{9}$

(12) for $\frac{56}{15999}$, AF is $\frac{0.056}{12}$

(13) for $\frac{50}{69999}$, AF is $\frac{0.0050}{7}$

(14) for $\frac{50}{14999999}$, AF $= \frac{0.000001}{15}$

(15) for $\frac{2175}{79999999}$, AF $= \frac{0.0002175}{8}$

(16) for $\frac{21863}{49999}$, AF $= \frac{2.1863}{5}$

In the above cases, the first eight denominators end in a single nine; the remaining eight terminate in 2, 3, 4, 3, 4, 6, 7 and 4 nines respectively. The question now is: Does it stand to reason that the *Ekādhika* should be the same in $\frac{7}{299}$ and in $\frac{7}{2999}$ irrespective of the difference in the number of nines? That would be tantamount to declaring that the same significant numerator or dividend with two different denominators or divisors will yield the same quotient! And that would be palpably absurd!

Yes, the objection is perfectly valid; and the relevant *Sūtra* has surmounted this difficulty beforehand, by providing for *groups* of quotient-digits to which the remainder at each stage of the mental division should be prefixed! And that solves the whole problem.

### Modus Operandi

For instance, let us take the sixteenth example *supra*, namely, $\frac{21863}{49999}$ whose A.F. is $\frac{2.1863}{5}$ and whose denominator ends in four nines:

Here, F is $\frac{21863}{49999}$ ; and AF is $\frac{21863}{5}$; and we have to make 5 in lieu of 49999 our working divisor. As we have dropped 4 nines from the original denominator and have 5 as our *Ekādhika* in the denominator of the auxiliary fraction, we have to divide the numerator of the latter in bundles, so to say, of 4 digits each by 5; and, whatever remainder there is, has to be prefixed not to any particular quotient-digit but to the bundle just already reached.

Thus, we take up 2.1863 to start with and divide it by 5.

We get .4372 as the first $\left[\frac{5)\ 2.1863}{Q=.4372}\right]$ and $R=3$

quotient-group and 3 as the remainder. We prefix this remainder to that *group* and say:

$_{3}.4372$ and we divide this dividend, namely, 34372 by the same divisor 5; and we get:

$$\begin{array}{l|l} {}_{3}.4372 : {}_{2}6874 & {}_{4}5374 \end{array}$$

i.e. 6874 is the second quotient-group; and 2 is the second remainder, which therefore we prefix to the second quotient group. And we continue this process with as many groups as we need.

Thus we have: $_{3}.4372,\ _{2}6874,\ _{4}5374,\ _{4}9074$ and so on to any number, or tens, or hundreds or thousands etc., of decimal places!

The proof hereof is very simple:

$$\begin{array}{ll} & 000\bar{1} : 2.1863000 \\ 5 & \quad\ \ : 1313432 \\ \hline & \quad\ \ : .43726874 \end{array}$$

$\therefore F = .4372, 6874, 5374, 9074\ldots$

*N.B.*: The prefixed remainders are not parts of the quotient but only prefixes to the quotient-group in question and are therefore to be dropped out of the answer!

This is a simple method by which we avoid divisions by long big divisors and have small and easy denominators to deal with.

The student will note that division by big denominators with a continuous series of *zeroes* on the right-hand side and division by the *Ekādhika* with the *prefixing* of the remainder at each step yield the same result! And this is why the auxiliary fraction scheme has been incorporated for lightening the burden of long big divisions.

A few more examples are given below:

(1) Express $\frac{6}{29}$ in its decimal shape.

Here, $F = \frac{6}{29} \therefore A.F. = \frac{.6}{3}$

$\therefore F = .20689655172413\dot{7}9310344827586\dot{}$ (the two rows of digits joined by a brace)

(2) $F = \frac{71}{89} \therefore AF = \frac{7.1}{9} = \therefore F = .79775280898$ etc.

(3) $F = \frac{17}{139} \therefore AF = \frac{1.7}{14} = \therefore F = .1\ 2\ 2\ 3\ 0\ 215827\ldots$
3 3 4 0 3

(4) $F = \frac{98}{179} \therefore AF = \frac{9.8}{18} = \therefore F = .54748603351955\ldots$

(5) $F = \frac{1}{43} = \frac{3}{129} \therefore AF = \frac{.3}{13} = \therefore F = .023255813953488\ldots$

(6) $F = 17/43 = 51/129; \therefore AF = \frac{5.1}{13} \therefore F = .395\ 348\ 837\ 209\ldots$

(7) $F = \frac{18}{73} = \frac{54}{219} \therefore AF = \frac{5.4}{22} \therefore F = .246\ 57534\ldots$

(8) $F = \frac{53}{799} \therefore AF = \frac{.53}{8} \therefore F = .06\ 63\ 32\ 91\ 61\ 45\ 18\ldots$
5 2 7 4 3 1 1

*N.B.*: The upper row .06633291614518. . . is the answer and the lower one 5 2 7 4 3 1 . . . is a mere scaffolding and goes out.

(9) $F = \frac{15}{899} \therefore AF = \frac{.15}{9} \therefore F = 01\ 66\ 85\ 20\ 57$
6 7 1 5 7

(10) $F = \frac{2}{1799} \therefore AF = \frac{.02}{18} \therefore F = 00\ 11\ 11\ 72\ 87\ldots$
2 2 13 15 6

(11) $F = \frac{100}{233} = \frac{300}{699} \therefore AF = \frac{3}{7} \therefore F = 42\ 91\ 84\ 54\ 93\ 56\ 22$
6 5 3 6 3 1 2
31. . .
5

(12) $F = \frac{444}{13999} \therefore AF = \frac{.444}{14} \therefore F = 031\ 716\ 551\ 182\ldots$
10 7 2 3

$$(13)\ F = \frac{97017}{29999999} \therefore AF = \frac{.0097017}{3}$$

F = 0032339/0010779/6670259 etc.
0 2 2

The student will have noted that the denominators in all the above cases ended in 9 or 3 which could be so multiplied as to yield an easy multiple ending in 9. But what about those ending in 1 which would have to be multiplied by 9 for this purpose and would, therefore, as already pointed out in the chapter on recurring decimals yield a rather unmanageable *Ekādhika*? Is there any provision for this kind of fractions?

Yes, there *is*. And this takes us on to the second type of auxiliary fractions.

### Auxiliary Fractions (Second Type)

If and when F has a denominator ending in 1, drop the 1 and decrease the numerator by unity. This is the required second type of auxiliary fractions. Thus,

(1) for 3/61, AF = 2/60 = $\frac{.2}{6}$
(2) for 36/61, AF = 35/60 = 3.5/6
(3) for 28/71, AF = 27/70 = 2.7/7
(4) for 73/91, AF = 72/90 = 7.2/9
(5) for 2/121, AF = 1/120 = .1/12
(6) for 14/131, AF = 13/130 = 1.3/13
(7) for 1/301, AF = 0/300 = .00/3
(8) for 1/901, AF = 0/900 = .00/9
(9) for 172/1301, AF = 171/1300 = 1.71/13
(10) for 2743/7001, AF = 2742/7000 = 2.742/7
(11) for 6163/8001, AF = 6162/8000 = 6.162/8
(12) for 1768/9001, AF = 1767/9000 = 1.767/9
(13) for 56/16001, AF = 55/16000 = .055/16
(14) for 50/700001, AF = 49/700000 = .00049/7
(15) for 2175/80000001, AF = 2174/80000000 = .0002174/8
(16) for 1/900000001, AF = 0/900000000 = .00000000/9

### Modus Operandi

The principles, the prefixing to the individual quotient-digits or to groups of quotient-digits etc., and other details are the same as in the *Ekādhika* auxiliary fraction. But the *procedure* is different, in a very important nay, vital particular. And this is that after the first division or group-division is over we prefix the remainder not to each quotient-digit but to its complement from nine and carry on the division in this way all through.

An illustrative instance will clarify this:

Let F be $\frac{13}{31}$ $\therefore$ AF $= \frac{12}{30} = \frac{1.2}{3}$

(*i*) We divide 1.2 by 3 and set 4 down as the first quotient-digit and 0 as the first remainder. $\begin{array}{l} .4 \\ 0 \end{array}$

(*ii*) We then divide not 04 but 05 the complement of 4 from 9 by 3 and put 1 and 2 as the second quotient-digit and the second remainder respectively. Therefore we have $\begin{array}{ll} .4 & 1 \\ 0 & 2 \end{array}$

(*iii*) We take now, not 21 but 28 as our dividend, divide it by 3 and get: $\begin{array}{lll} .4 & 1 & 9 \\ 0 & 2 & 1 \end{array}$

(*iv*) Thus, dividing 10 by 3, we have:

9 0. 2 1 1 and

so on, until finally our chart reads: 4 1 9 3

$$\begin{array}{l} F\left(\text{i.e. } \frac{13}{31}\right) = .4\ 1\ 9\ 3\ 5\ 4\ 8\ 3\ 8\ 7\ 0\ 9\ 6 \text{ etc.} \\ \qquad\qquad\quad\ \ 0\ 2\ 1\ 1\ 1\ 2\ 1\ 2\ 2\ 0\ 2\ 2\ 2 \end{array}$$

Always, therefore, remember to take the complement from 9 of each quotient-digit and not the quotient-digit itself for the purpose of further division, subtraction etc. This is the whole secret of the second type of auxiliary fraction.

Some more illustrative examples are given hereunder:

(1) $F = \frac{1}{41}$ $\therefore$ AF $= \frac{.0}{4}$

$$\left.\begin{array}{l} \therefore F = .0\ 2\ 4\ 3\ 9/0 \\ \qquad\quad 0\ 1\ 1\ \ 3\ 0/0 \end{array}\right\} \text{So, this is a definite recurring decimal.}$$

(2) $F = \frac{70}{71}$ $\therefore$ AF $= \frac{6.9}{7}$

$$\begin{array}{l} \therefore F = .9\ 8\ 5\ 9\ 1\ 5\ 4\ 9\ 2\ 9\ 5\ 7\ 7\ 4\ 6\ 4\ 7\ 8\ 8\ 7\ 3\ 2\ 3 \ldots \\ \qquad\quad 6\ 4\ 6\ 1\ 3\ 3\ 6\ 2\ 6\ 4\ 5\ 5\ 3\ 4\ 3\ 5\ 6\ 6\ 5\ 2\ 1\ 2\ 6 \end{array}$$

(3) $F = \frac{91}{171}$ $\therefore$ AF $= \frac{9.0}{17}$ $\therefore$ $\begin{array}{l} F = .5\ \ 3\ 2\ \ \ 1\ 6\ \ \ 3\ 7\ 4\ \ 2\ \ 6 \\ \qquad\ \ 5\ 3\ 2\ 10\ 6\ 12\ 7\ 4\ 11\ 15 \end{array}$

(4) $F=\frac{10}{27}=\frac{30}{81}\therefore AF=\frac{2.9}{8}$ $\therefore .\dot{3}\ 7\ \dot{0}/3$ Evidently a recurring
$\quad 5\ 0\ 2/5$ decimal.

(5) $F=\frac{131}{701}\therefore AF=\frac{1.30}{7}$ (with groups of 2 digits)

$\therefore F=.18\ 68\ 75\ 89\ 15\ 83\ldots$
$\quad 4\ 5\ 6\ 1\ 5\ 3$

(6) $F=\frac{1400}{1401}\therefore AF=\frac{13.99}{14}$ (with two-digit groups)

$F\therefore=.99\ 92\ 86\ 22\ 41$
$\quad 13\ 12\ 3\ 5\ 3$

(7) $F=\frac{243}{1601}\therefore AF=\frac{2.42}{16}$ (with groups of two digits)

$\therefore F=.15\ 17\ 80\ 13\ 74\ 14\ 11\ 61\ 79\ldots$
$\quad 2\ 12\ 2\ 11\ 2\ 1\ 9\ 12\ 4$

(8) $F=\frac{5}{67}=\frac{15}{201}\therefore AF=\frac{.14}{2}$ (with groups of two digits)

$\therefore F=.07\ 46\ 26\ 86\ 56$
$\quad 0\ 0\ 1\ 1\ 1$

(9) $F=\frac{2743}{7001}\therefore AF=\frac{2.742}{7}$ (with 3-digit groups)

$\therefore F=.391\ 801\ 171\ 261\ 248\ 393\ 086$
$\quad 5\ 1\ 1\ 1\ 2\ 0\ 4$

*Proof* : 
```
      001 : 2. 742 999
   7      :          5
          ------------
          :  .5391
          ------------
```

(10) $F=\frac{31}{77}=\frac{93}{231}\therefore AF=\frac{9.2}{23}\therefore F=402/597$
$\quad 0$

*or* $F=\frac{31}{77}=\frac{403}{1001}\therefore\frac{.402}{1}$ (with three-digit groups)

$\therefore F=.\dot{4}02/59\dot{7}$ (evidently a recurring decimal)

(11) $F=\frac{29}{15001}\therefore AF=\frac{.028}{15}$ (with three-digit groups)

$\therefore$ F = .001 933 204 453 036 etc.
13 3 6 0 6

(12) $F = \frac{137}{13000001}$ $\therefore$ $AF = \frac{.000136}{13}$ (with 6-digit groups)

$\therefore$ F =.000010: 538460 : 727810 : 713245 etc.
6 :9 :0 :4

### Other Astounding Applications

Yes, but what about still other numbers which are neither immediately below nor immediately above a ten-power base or a multiple of ten etc., as in the above cases but a bit remoter therefrom? Well, these too have been grandly catered for, in the shape of a simple application of the *Ānurūpya Sūtra*, whereby, after the pre-fixing of each remainder to the quotient-digit in question, we have to add to or subtract from the dividend at every step, as many times the quotient-digit as the divisor, i.e. the denominator is below or above the normal which, in the case of all these auxiliary fractions, is counted as ending, not in zero or a number of zeroes but in 9 or a series of nines!

For example, let F be $\frac{15}{68}$ and suppose we have to express this vulgar fraction in its decimal shape to, say, 16 places of decimals.

Let the student should have, in the course of these peregrinations into such very simple and easy methods of work, forgotten the tremendous difference between the current method and the Vedic method and thereby deprived himself of the requisite material for the purpose of comparison and contrast, let us, for a brief while, picture the two methods to ourselves side by side and see what the exact position is.

According to the Vedic method, the process *wholly mental* is as follows:

$F = \frac{15}{68}$, $\therefore$ A.F. $\frac{1.5}{7}$. But 68 being one less than 69 the normal ending in 9 we shall have to add to each dividend, the quotient-digit in question. Thus

(*i*) when we divide 1.5 by 7, we get 2 and 1 as our first quotient-digit and our first remainder.

(*ii*) our second dividend will not be 12 but 12 + 2 = 14; and by division of that by 7, our second Q and R are 2 and 0.

$\frac{1.5}{7}$
.2
1

(*iii*) our next dividend is 02 + 2 = 04; and this gives us 0 and 4 as Q and R.

.2 2
1 0

(*iv*) our fourth dividend is $40+0$, giving us 5 and 5 as our fourth Q and R.

```
.2 2 0
 1 0 4
```

(*v*) So, our next dividend is $55+5=60$; and our Q and R are 8 and 4.

```
.2 2 0 5
 1 0 4 5
```

We can proceed on these lines to as many places of decimals as we may need. And, in the present case wherein 16 decimal-places have been asked for, we toss off digit after digit mentally and say:

```
.2 2 0 5 8
 1 0 4 5 4
```

```
F =.2 2 0 5 8 8 2 3 5 2 9 4 1 1 7 6 etc.
    1 0 4 5 4 0 2 3 1 6 1 0 1 5 3 2
```

Over against this, let us remind ourselves of the current method for answering this question:

```
68) 15.0 (.2205882352941176 etc.
    136
    ---
     140
     136
     ---
       400
       340
       ---
        600
        544
        ---
         560
         544
         ---
          160
          136
          ---
           240
           204
           ---
            360
            340
            ---
             200
             136
             ---
              640
              612
              ---
               280
               272
               ---
                 80
                 68
                 ---
                 120
                  68
                  ---
                  520
                  476
                  ---
                   440
                   408
                   ---
```

Alongside of this cumbrous 16-step process, let us once again put down the *whole working* by the Vedic method and say:

$F\left(\frac{15}{68}\right) = .2\ 2\ 0\ 5\ 8\ 8\ 2\ 3\ 5\ 2\ 9\ 4\ 117\ 6\ 4$ etc.

$\quad\quad\quad 1\ 0\ 4\ 5\ 4\ 0\ 2\ 3\ 1\ 6\ 1\ 0\ 1\ 5\ 3\ 2\ 4$

A few more illustrative examples are given hereunder:

(1) Express 101/138 in its decimal shape (20 places).

(*i*) *Routine Method*:

$138)101.0\ (.\dot{7}3188405797\dot{1}01449275$

```
 966
 ----
  440
  414
  ----
   260
   138
   ----
   1220
   1104
   -----
    1160
    1104
    -----
      560
      552
      ---
       800
       690
       ----
       1100
        966
       ----
       1340
       1242
       ----
         980
         966
         ---
          140
          138
          ---
            200
            138
            ---
             620
             552
             ---
              680
              552
              ----
              1280
              1242
              ----
                380
                276
                ----
               1040
                966
                ---
                 740
                 690
                 ---
                  50
                 ---
```

(*ii*) *Vedic Method:*

A.F. $= \frac{101}{14}$ (with one below the normal 139)

$\therefore$ F = .73 1 8 8 4 0 5 7 9 7 1 0 1 4 4 9 2 7 5 etc.
3 2 12 10 4 0 8 10 128 0 0 2 6 6 122 10 64

*Note*: More-than-one-digit quotient if any, should be carried over (as usual) to the left.

(2) (*i*) *Conventional method:*

```
97)73.0(.7525773195876288659 7 etc.
   679
   ---
    510
    485
    ---
     250
     194
     ---
      560
      485
      ---
       750
       679
       ---
        710
        679
        ---
         310
         291
         ---
          190
           97
          ---
          930
          873
          ---
           570
           485
           ---
            850
            776
            ---

740
679
---
 610
 582
 ---
  280
  194
  ---
   860
   776
   ---
    840
    776
    ---
     640
     582
     ---
      580
      485
      ---
       950
       873
       ---
        770
        679
        ---
         91
```

(*ii*) *Vedic at-sight method:*

*Data* : F $= \frac{73}{97}$ $\therefore$ AF $= \frac{7.3}{10}$ (but with 2 below the normal 99).

$\therefore$ Add twice the Q-digit at each step.

*Actual Working:*

$\therefore$ F = .7 5 2 5 7 7 3 1 9 5 8 7 6 2 8 8 6 5 9 7 etc.
3 1 5 6 5 1 1 9 3 7 5 4 1 8 6 4 4 8 5 7

(3) Express $\frac{17}{127}$ as a decimal (20 places).

(*i*) *Current method*—

127)17.0(.13385826771653543306 etc.

127
430
381
490
381
1090
1016
740
635
1050
1016
340
254
860
762
980
889
910
889
210
127
830
762
680
635

450
381
690
635
550
508
420
381
390
381
900
882
18

(*ii*) *Vedic at-sight method*—

$\because F = \frac{17}{127} \therefore AF = \frac{1.7}{13}$ (but with 2 below the normal 129)

$\therefore$ Double the Q-digit to be added at every step.

$\therefore$ F = .1 3 3 8 5 8 2 6 7 7 1 6 5 3 5 4 3 3 0 6 etc.
4 4 10 5 9 1 8 8 7 0 8 5 3 6 4 3 3 0 9 12

(4) Express $\frac{5236}{8997}$ in decimal form (21 decimal places)

(*i*) *Usual method:*

```
8997)5236.0(.581/971/768/367/233/522/285 etc.
     44985
     -----
      73750
      71976
      -----
       17740
        8997
       -----
        87430          21010
        80973          17994
        -----          -----
         64570          30160
         62979          26991
         -----          -----
          15910          31690
           8997          26991
          -----          -----
           69130          46990
           62979          44985
           -----          -----
            61510          20050
            53982          17994
            -----          -----
             75280          20560
             71976          17994
             -----          -----
              33040          25660
              26991          17994
              -----          -----
               60490          76660
               53982          71976
               -----          -----
                65080          46840
                62979          44985
                -----          -----
                                1855
```

What a tremendous mass andmess of multiplications, subtractions etc.!

(*ii*) *Vedic at-sight method:* (but with 2 below the normal 8999 and also with groups of 3 digits at a time).

$\therefore F = \frac{5236}{8997} \therefore AF = \frac{5.236}{9}$

$\therefore$ Add twice the Q-digit at every step.

$\therefore$ F = .581 : 971 : 768 : 367 : 233 : 522 : 285 etc.
  7  :4  :1  :1  :4  :1  :1

(5) Express $\frac{21863}{49997}$ as a decimal (16 places)

(*i*) *Conventional method:*

```
49997)21863.0(.4372 / 8623 / 7174 / 2304 / etc.
      199988
      ------
       186420
       149991
       ------
        364290
        349979
        ------
         143110
          99994
         ------
          431160
          399976
          ------
           311840
           299982
           ------
            118580          What a horrible mess?
             99994
            ------
             185860
             149991
             ------
              358690
              349979
              ------
                87110
               49997
               ------
               371130
               349979
               ------
                211510
                199998
                ------
                 115120
                  99994
                 ------
                  151260
                  149991
                  ------
                   126900
                    99994
                   ------
                    26906
                   ------
```

(*ii*) *Vedic at-sight method*:

$\therefore F = \dfrac{21863}{49997} \therefore AF = \dfrac{2.1863}{5}$

(with 2 below the normal 49999 and with groups of four digits each).

∴ Add double the Q-digit at every step)

$$\therefore F = 4372 : \frac{62}{8\,5(12)3} : \frac{71}{6(11)}\frac{4}{73} : \frac{2}{(12)}\frac{30}{2(10)4}$$

3 : 1 : 4 : 4

*N.B.*: Very carefully that the extra or surplus, i.e. left-hand side parts of Q-digit have been "carried over" to the left.

This excess is due to the additional multiplication and can be got over in the manner just indicated. A method for *avoiding* this difficulty altogether is also available but will be dealt with at a later stage.

(6) Express $\frac{17}{76}$ as a decimal (eight places)

(*i*) *Current method*:

```
76)17.0(.22368421 etc.
   152
   ---
    180
    152
    ---
     280
     228
     ---
      520
      456
      ---
       640
       608
       ---
        320
        304
        ---
         160
         152
         ---
           80
           76
           --
            4
```

*N. B.*: Note 84 : 21 :: 4 : 1.

*Even this is bad enough.*

(*ii*) *Vedic at-sight method*:

$\because F = \frac{17}{76} \therefore AF = \frac{1.7}{8}$ (but with 3 less than the normal 79)

Thrice the Q-digit is to be added at every step.

∵ F = .2 2 3 6 8 4 2 1 etc.
  1 2 4 4 0 0 0 0

(7) Express $\frac{17125}{59998}$ as a decimal (12 places)

(*i*) *Usual method*:

```
59998)17125.0(.2854 / 2618 / 0872
       119996
      -------
        512540
        479984
       -------
         325560
         299990
        -------
          255700
          239992
         -------
           157080
           119996
          -------
            370840
            359988
           -------
             108520
              59998
            -------
               485220
               479984
              -------
                523600
                479984
               -------
                 436160
                 419986
                -------
                  161740
                  119996
                 -------
                    41744
                  -------
```

(*ii*) *Vedic at-sight method*:

$\because F = \frac{17125}{59998} \therefore AF = \frac{1.7125}{6}$ (with 1 less the normal and with 4-digit groups)

$\therefore$ only one Q-digit is to be added.

$\therefore F = .2854 : 2618 : 0872$ etc.
$\quad 1 \quad : 0 \quad : 4$

These examples should suffice to bring vividly home to the student the extent and magnitude of the difference between the current cumbrous methods and the Vedic at-sight one-line process in question.

Yes, but what about other numbers, in general, which are no where near any power or multiple of ten or a "normal" denominator-divisor ending in 9 or a series of nines? Have they been provided for, too?

Yes, they have. There are methods whereby, as explained in an earlier chapter the one dealing with recurring decimals we can easily transform any miscellaneous or non-descript denominator in question—by simple multiplication etc.—to the requisite standard form which will bring them within the jurisdiction of the auxiliary fractions hereinabove explained.

In fact, the very discovery of these auxiliary and of their wonderful utility in the transmogrification of frightful looking denominators of vulgar fractions into such simple and easy denominator-divisors must suffice to prepare the scientifically-minded seeker after knowledge, for the marvellous devices still further on in the offing.

We shall advert to this subject again and expound it still further, in the next two subsequent chapters dealing with divisibility and the application of the *Ekādhika Pūrva* etc., as positive and negative osculators in that context.

# 29

# Divisibility and Simple Osculators

We now take up the interesting and intriguing question as to how one can determine before-hand whether a certain given number however long it may be is divisible by a certain given divisor and especially as to the Vedic processes which can help us herein.

The current system deals with this subject but only in an ultra-superficial way and only in relation to what may be termed the most elementary elements thereof. We need not now enter into details of these including divisibility by 2, 5, 10, 3, 6, 9, 18, 11, 22 and so on, as they are well-known even to the mathematics-pupils at a very early stage of their mathematical study. We shall take these for granted and start with the intermediate parts and then go on to the advanced portions of the subject.

### The Osculators

As we have to utilise the "वेष्टनs" (*Veṣṭanas = Osculators*) throughout this subject of divisibility, we shall begin with a simple definition thereof and the method of their application.

Owing to the fact that our familiar old friend the *Ekādhika* is the first of these osculators, i.e. the positive osculator, the task becomes all the simpler and easier. Over and above the huge number of purposes which the *Ekādhika* has alredy been shown to fulfil, it has the further merit of helping us to readily determine the divisibility or otherwise of a certain given dividend by a certain given divisor.

Let us, for instance, start with our similar familiar old friend or experimental-subject or shall we say, "Guinea-pig" the number 7. The student need hardly be reminded that the *Ekādhika* for 7 is derived from $7 \times 7 = 49$ and is therefore 5. The *Ekādhika* is a clinching test for divisibility; and the process by which it serves this purpose is technically called *Veṣṭana* or "Osculation".

Suppose we do not know and have to determine whether 21 is divisible by 7. We multiply the last digit, i.e. 1 by the *Ekādhika* (or Positive Osculator, i.e. 5) and add the product, i.e. 5 to the previous digit, i.e. 2 and thus get 7. This process is technically called "*Osculation*". And, if the result of the osculation is the divisor itself or a repetition of a previous result), we say that the given original dividend 21 is divisible by 7.

A trial chart (for 7) will read as follows:

$14; 4\times5+1=21;$ and $1\times5+2=7$ $\therefore$ Yes.

21(already dealt with);

$28; 8\times5+2=42; 2\times5+4=14$ (already dealt with)

$35; 5\times5+3=28$ (already dealt with);

42 (already dealt with);

$49; 9\times5+4=49$ (*Repetition means divisibility*).

$56; 6\times5+5=35$ (already dealt with);

$63; 3\times5+6=21$ (already dealt with);

$70; 0\times5+7=7$ $\therefore$ Yes

$77; 7\times5+7=42$ (already done);

$84; 4\times5+8=28$ (already over);

$91; 1\times5+9=14$ (already dealt with);

$98; 8\times5+9=49$ (already done);

Now let us try and test, say, 112.

$112; 2\times5+1=11; 11\times5+1=56$ $\therefore$ Yes.

or $2\times5+11=21$ $\therefore$ Yes.

We next try and test for 13; and we find the repetitions more prominent there. The *Ekādhika* is 4. Therefore we go on multiplying leftward by 4. Thus,

$13; 3\times4+1=13$

$26; 6\times4+2=26$

$39; 9\times4+3=39$

$52; 2\times4+5=13$

$65; 5\times4+6=26$

$78; 8\times4+7=39$

The repetition etc., is uniformly there and in correct sequence too, i.e. 13, 26, 39! $\therefore$ Yes.

$91; 1 \times 4 + 9 = 13$

$104; 4 \times 4 + 10 = 26$

### Examples of the Osculation Procedure (Veṣṭana)

A few examples will elucidate the process:

(1) 7 continually osculated by 5 gives 35, 28 42, 14, 21 and 7.
(2) 5 so osculated by 7 gives 35, 38, 59, 68, 62, 20, 2 and so on.
(3) 9 (by 7) gives 63, 27, 51, 12, 15 etc.
(4) 8 (by 16) gives 128, 140, 14 etc.
(5) 15 (by 14) gives 71, 21, 16 etc.
(6) 18 (by 12) gives 97, 93, 45, 64, 54, 53, 41, 16 etc.
(7) 36 (by 9) gives 57, 68, 78, 79, 88, 80, 8 etc.
(8) 46 (by 3) gives 22, 8 etc.
(9) 49 (by 16) gives 148, 142, 46, 100, 10, 1 etc.
(10) 237 (by 8) gives 79, 79 etc., and is $\therefore$ divisible by 79.
(11) 719 (by 9) gives 152, 33, 30, 3 etc.
(12) 4321 (by 7) gives 439, 106, 25, 19, 64, 34, 31, 10, 1 etc.
(13) 7524 (by 8) gives 784, 110, 11, 9 etc.
(14) 10161 (by 5) gives 1021, 107, 45, 29, 47, 39, 48, 44, 24, 22, 12, 11, 6 etc.
(15) 35712 (by 4) gives 3579, 393, 51, 9 etc.
(16) 50720 (by 12) gives 5072, 531, 65, 66, 78, 103, 46, etc.

*N.B.*: We need not carry on this process indefinitely. We can stop as soon as we reach a comparatively small number which gives us the necessary clue as to whether the given number is divisible or not by the divisor whose *Ekādhika* we have used as our osculator! Hence the importance of the *Ekādhika*.

### Rule for Ekādhikas

(1) For 9, 19, 29, 39 etc. (all ending in 9), the *Ekādhikas* are 1, 2, 3, 4 etc.

(2) For 3, 13, 23, 33 etc. (all ending in 3) multiply them by 3; and you get 1, 4, 7, 10 etc. as the *Ekādhikas*.

(3) For 7, 17, 27, 37 etc. (all ending in 7) multiply them by 7; and you obtain 5, 12, 19, 26, etc. as the *Ekādhikas*.

(4) For 1, 11, 21, 31, etc. (all ending in 1), multiply them by 9; and you get 1, 10, 19, 28 etc. as the *Ekādhikas*.

### Osculation by own Ekādhika

*Note* that the osculation of any number by its own *Ekādhika* will as in the case of 7 and 13 go on giving that very number or a multiple thereof. Thus,

(1) 23 osculated by 7(its *Ekādhika*) gives $7 \times 3 + 2 = 23$;

46 (osculated by 7) gives $7 \times 6 + 4 = 46$;

69 (similarly) gives $7 \times 9 + 6 = 69$;

92 (likewise) gives $2 \times 7 + 9 = 23$;

115 (similarly) gives $7 \times 5 + 11 = 46$; and so on.

Now, 276 osculated by 7 by way of testing for divisibility by 23 gives $7 \times 6 + 27 = 69$ which again gives 69! ∴ Yes. Thus, all the multiples of 23 fulfil this test, i.e. of osculation by its *Ekādhika* 7. And this is the whole secret of the *Veṣṭana* sub-*Sūtra*.

### Modus Operandi of Osculation

Whenever a question of divisibility comes up, we can adopt the following procedure. Suppose, we wish to know—without actual division—whether 2774 is divisible by 19 or not. We put down the digits in order as shown below. And we know that the *Ekādhika* osculator is 2.

(*i*) We multiply the last digit 4 by 2, add the product 8 to the previous digit 7 and put the total 15 down under the second right-hand digit.

$$\begin{array}{cccc} 2 & 7 & 7 & 4 \\ & & 15 & \end{array}$$

(*ii*) We multiple that 15 by 2, add that 30 to the 7 on the upper row, cast out the nineteens from that 37 and put down the remainder 18 underneath that 7.

$$\begin{array}{cccc} 2 & 7 & 7 & 4 \\ & 37 & 15 & \\ \hline & (18) & & \end{array}$$

*N.B.*: This casting out of the nineteens may be more easily and speedily achieved by first osculating the 15 itself getting 11, adding it to the 7 to the left-hand on the top-row and putting the 18 down thereunder.

$$\begin{array}{cccc} 2 & 7 & 7 & 4 \\ & 18 & 15 & \end{array}$$

(*iii*) We then osculate that 18 with the 2 to the left on the upper row and get 38; or we may osculate the 18 itself, obtain 17, add the 2 and get 19 as the final osculated result. And, as 19 is divisible by 19, we say the given number 2774 is also divisible thereby.

$$\begin{pmatrix} 2 & 7 & 7 & 4 \\ 19 & 18 & 15 & \end{pmatrix}$$

This is the whole process; and our chart says:

By 19?

∴ The osculator is 2 $\begin{Bmatrix} 2 & 7 & 7 & 4 \\ 19 & 18 & 15 & \end{Bmatrix}$ ∴ Yes.

*Or* secondly, we may arrive at the same result as effectively but less spectacularly by means of a continuous series of osculations of the given number 2774 by the osculator 2 as hereinbefore explained. And we can say:

$\because$ 2774 osculated by the osculator 2 gives us 285, 38 and 19,
$\therefore$ 2774 is divisible by 19.

*N.B.*: The latter method is the shorter but more mechanical and cumbrous of the two; and the former procedure looks neater and more pictorially graphic, nay, spectacular. And one can follow one's own choice as to which procedure should be preferred.

*Note*: Whenever, at any stage, a bigger number than the divisor comes up, the same osculation-operation can always be performed.

Some more specimen examples are given below:

(1) By 29? $\therefore$ The osculator is 3. $\left\{\begin{matrix}3 & 2 & 8 & 9 & 6\\ 27 & 8 & 31 & 27 & \end{matrix}\right\}$ $\therefore$ No.

*Or* E osculated by 3 gives 3307, 351, 38, 27, etc. $\therefore$ No.

(2) By 29? $\therefore$ The osculator is 3. $\left\{\begin{matrix}9 & 3 & 1 & 4 & 8\\ 29 & 26 & 27 & 28 & \end{matrix}\right\}$ $\therefore$ Yes

*Or* The osculation-results are 9338, 957, 116 and 29 $\therefore$ Yes

(3) By 29? $\therefore$ The osculator is 3. $\left\{\begin{matrix}2 & 4 & 3 & 4 & 5 & 2 & 1\\ 29 & 9 & 21 & 6 & 20 & 5 & \end{matrix}\right\}$ $\therefore$ Yes

*Or* the osculation-results are 243455, 24360, 2436, 261 and 29 $\therefore$ Yes

(4) By 39? $\therefore$ The osculator is 4. $\left\{\begin{matrix}4 & 9 & 1 & 4\\ 39 & 38 & 17 & \end{matrix}\right\}$ $\therefore$ Yes

*Or* The osculation-results are 507, 78 and 39 $\therefore$ Yes

(5) By 49? $\therefore$ The osculator is 5. $\left\{\begin{matrix}5 & 3 & 3 & 2\\ 51 & 19 & 13 & \\ \overline{(10)} & & & \end{matrix}\right\}$ ...No

*Or* The osculation-results are 543, 69, 51, and 10 $\therefore$ Yes

(6) By 59? $\therefore$ The osculator is 6. $\left\{\begin{matrix}1 & 9 & 1 & 5 & 7 & 3\\ 59 & 49 & 46 & 37 & 25 & \end{matrix}\right\}$ $\therefore$ Yes

*Or* The osculation-results are 19175, 1947, 236 and 59 $\therefore$ Yes

(7) By 59? $\therefore$ The osculator is 6. $\left\{\begin{matrix}1 & 2 & 5 & 6 & 7\\ 59 & 49 & 57 & 48 & \end{matrix}\right\}$ $\therefore$ Yes

*Or* The osculation-results are 1298, 177 and 59 $\therefore$ Yes

(8) By 59? ∴ The osculator is 6. $\left\{\begin{matrix} 4 & 0 & 1 & 7 & 9 & 1 \\ 47 & 17 & 52 & 38 & 15 & \end{matrix}\right\}$ ∴ Yes

*Or* The osculation-results are 40185, 4048, 452 and 57 ∴ No

(9) By 79? ∴ The osculator is 8. $\left\{\begin{matrix} 6 & 3 & 0 & 9 & 4 & 8 & 2 & 1 \\ 13 & 70 & 38 & 64 & 76 & 9 & 10 & \end{matrix}\right\}$ ∴ No

*Or* The osculation-results are 6309490, 630949, 63166,
6364, 668, 130 and 13 ∴ No

(10) By 43? ∴ The osculator is 13. $\left\{\begin{matrix} 1 & 4 & 0 & 6 & 1 \\ 129 & 119 & 118 & 19 & \end{matrix}\right\}$ ∴ Yes

*Or* The osculation-results are 1419, 258 and 129 ∴ Yes

(11) By 53? ∴ The osculator is 16. $\left\{\begin{matrix} 2 & 1 & 9 & 5 & 3 \\ 149 & 39 & 62 & 53 & \end{matrix}\right\}$ ∴ No

*Or* The osculation-results are 2243, 272 and 59 ∴ No

(12) By 179? ∴ The osculator is 18. $\left\{\begin{matrix} 7 & 1 & 4 & 5 & 5 & 0 & 1 \\ 179 & 109 & 6 & 20 & 150 & 18 & \end{matrix}\right\}$ ∴ Yes

*Or* The osculation-results are 714568, 71600, 7160, 716 and 179 ∴ Yes

(13) Determine whether 5293240096 is divisible by 139 or not

(A) *By the current method* (just by way of contrast):

```
139)5293240096(38080864
    417
    ----
    1123
    1112
    -----
      1124
      1112
      -----
        1200
        1112
        -----
          889
          834
          ---
           556
           556
           ---
             0
```

∴ Yes

(B) *By the Vedic method*:

By 139? ∴ The *Ekādhika* (osculator) is 14.

$\left\{\begin{matrix} 5 & 2 & 9 & 3 & 2 & 4 & 0 & 0 & 9 & 6 \\ 139 & 89 & 36 & 131 & 29 & 131 & 19 & 51 & 93 & \end{matrix}\right\}$ Yes

*Or* The osculation-results are 529324093, 52932451, 5293259, 529451, 52959, 5421, 556 and 139 ∴ Yes

*Note*: In all the above cases, the divisor either actually ended in 9 or could—by suitable multiplicaton—be made to yield a product ending in 9 for the determination of the required *Ekādhika* or Osculator in each case. But what about the numbers ending in 3, 7 and 1 whose *Ekādhika* may generally be expected, to be a bigger number? Is there a suitable provision for such numbers being dealt with without involving bigger *Ekādhika* multipliers?

Yes, there is, and we proceed to deal with this.

## THE NEGATIVE OSCULATOR

This is an application of the *Parāvartya Sūtra* and is called the negative osculator because it is a process not of *addition* as in the case of the *Ekādhika* but of *Subtraction* leftward. And this actually means a consequent alternation of *plus* and *minus*.

## EXAMPLES OF THE NEGATIVE OSCULATION PROCESS

(1) 36 thus osculated by 9, gives $3-54=-51$

(2) 7 osculated by 5 gives $0-35=-35$

(3) 35712 osculated by 4 will yield $8-3571=-3563$

## HOW TO DETERMINE THE NEGATIVE OSCULATOR

Just as the *Ekādhika* the positive *Veṣṭana* has been duly defined and can be correctly ascertained, similarly the negative osculator will also require to be determined by means of a proper definition and has been so defined with a view to proper recognition.

It consists of two clauses:

(*i*) In the case of all divisors ending in 1, simply drop the one; and

(*ii*) In the other cases, multiply so as to get 1 as the last digit of the product, i.e. 3 by 7, 7 by 3 and 9 by 9; and then apply the previous sub-clause, i.e. drop the 1.

*Note*: For facility of symbolisation, the positive and the negative osculators will be represented by P and Q respectively.

## EXAMPLES OF NEGATIVE OSCULATORS

(1) For 11, 21, 31, 41, 51 and other numbers ending in 1, Q is 1, 2, 3, 4, 5 and so on. Note that, by this second type of osculators, we avoid the big *Ekādhikas* produced by multiplying these numbers by 9.

(2) For 7, 17, 27, 37, 47, 57 etc., we have to multiply them by 3 in order to get products ending in 1. And they will be 2, 5, 8, 11, 14, 17 and so on. In these cases too, this process is generally

calculated to yield smaller multipliers than the multiplication by 7 is likely to do.

(3) For 3, 13, 23, 33, 43, 53 etc., we have to multiply them by 7; and the resultant negative osculators will be 2, 9, 16, 23, 30, 37 etc., which will *generally* be found to be bigger numbers than the *Ekādhikas*.

(4) For 9, 19, 29, 39, 49, 59 etc., we have to multiply these by 9; and the resultant negative osculators will be 8, 17, 26, 35, 44, 53 etc., all of which will be much bigger than the corresponding *Ekādhikas*.

### Important and Interesting Feature

*Note*: A very interesting and important feature about the relationship between P and Q, is that, whatever the divisor (D) may be, P+Q=D, i.e. the two osculators together invariably add up to the divisor. And this means that, if one of them is known, the other is automatically known being the complement thereof from the divisor, i.e. the denominator.

### Specimen Schedule of Osculators P and Q

| Number | Multiple for P | Multiple for Q | P | Q | Total |
|---|---|---|---|---|---|
| 1 | 9 | (1) | 1 | 0 | 1 |
| 3 | 9 | 21 | 1 | 2 | 3 |
| 7 | 49 | 21 | 5 | 2 | 7 |
| 9 | (9) | 81 | 1 | 8 | 9 |
| 11 | 99 | (11) | 10 | 1 | 11 |
| 13 | 39 | 91 | 4 | 9 | 13 |
| 17 | 119 | 51 | 12 | 5 | 17 |
| 19 | (19) | 171 | 2 | 17 | 19 |
| 21 | 189 | (21) | 19 | 2 | 21 |
| 23 | 69 | 161 | 7 | 16 | 23 |
| 27 | 189 | 81 | 19 | 8 | 27 |
| 29 | (29) | 261 | 3 | 26 | 29 |
| 31 | 279 | (31) | 28 | 3 | 31 |
| 33 | 99 | 231 | 10 | 23 | 33 |
| 37 | 259 | 111 | 26 | 11 | 37 |
| 39 | (39) | 351 | 4 | 35 | 39 |
| 41 | 369 | (41) | 37 | 4 | 41 |
| 43 | 129 | 301 | 13 | 30 | 43 |
| 47 | 329 | 141 | 33 | 14 | 47 |
| 49 | (49) | 441 | 5 | 44 | 49 |

| | | | | | |
|---|---|---|---|---|---|
| 51 | 459 | (51) | 46 | 5 | 51 |
| 53 | 159 | 371 | 16 | 37 | 53 |
| 57 | 399 | 171 | 40 | 17 | 57 |
| 59 | (59) | 531 | 6 | 53 | 59 |
| 61 | 549 | (61) | 55 | 6 | 61 |
| 63 | 189 | 441 | 19 | 44 | 63 |
| 67 | 469 | 201 | 47 | 20 | 67 |
| 69 | (69) | 621 | 7 | 62 | 69 |
| 71 | 639 | (71) | 64 | 7 | 71 |
| 73 | 219 | 511 | 22 | 51 | 73 |
| 77 | 539 | 231 | 54 | 23 | 77 |
| 79 | (79) | 711 | 8 | 71 | 79 |
| 81 | 729 | (81) | 73 | 8 | 81 |

*N.B.*: It will be noted :—

(*i*) that P + Q always equals D;
(*ii*) multiples of 2 and 5 are inadmissible for the purposes of this schedule;
(*iii*) and these will have to be dealt with by dividing off all the powers of 2 and 5 which are factors of the divisor concerned.

### A Few Sample Examples

(1) for 59, P is 6 $\therefore$ Q = 53
(2) for 47, Q is 14 $\therefore$ P = 33
(3) for 53, P is 16 $\therefore$ Q = 37
(4) for 71, Q is 7 $\therefore$ P = 64
(5) for 89, P is 9 $\therefore$ Q = 80
(6) for 83, P is 25 $\therefore$ Q = 58
(7) for 91, P is 82 $\therefore$ Q = 9
(8) for 93, Q is 65 $\therefore$ P = 28
(9) for 97, P is 68 $\therefore$ Q = 29
(10) for 99, Q is 89 $\therefore$ P = 10
(11) for 101, P is 91 $\therefore$ Q = 10
(12) for 103, Q is 72 $\therefore$ P = 31
(13) for 107, P is 75 $\therefore$ Q = 32
(14) for 131, Q is 13 $\therefore$ P = 118
(15) for 151, P is 136 $\therefore$ Q = 15
(16) for 201, Q is 20 $\therefore$ P = 181

P+Q=D throughout

*Note*: (1) If the last digit of a divisor be 3, its P<its Q;

(2) If the last digit be 7, its Q<its P; and

(3) in the actual working out of the subtractions of the osculated multiples for the negative osculators, the actual result will be an alternation of *plus* and *minus*.

*Explanation*: (1) In the removal of brackets, a series of subtractions actually materialises in an alternation of + and −. For example,

$$a-\left[b-\left\{c-\left(d-\overline{e-f}\right)\right\}\right]$$

$$=a-b+c-d+e-f.$$

Exactly similar is the case here.

(2) When we divide $a^n+b^n$ by $(a+b)$, the quotient consists of a series of terms which are alternately plus and minus. Exactly the same is the case here.

*Note*: The student will have to carefully *remember* this alternation of positives and negatives. But the better thing will be, not to rely on one's memory at each step but to *mark* the digits beforehand, alternately, say, by means of a *Vinculum* from right to left, on all the even-place digits, so that there may be an automatic safeguard against the possible playing of any pranks by one's memory.

Armed with this safeguard, let us now tackle a few illustrative instances and see how the plan works out in actual practice.

(1) By 41? ∴ The Negative Osculator is 4

$$\left\{\begin{matrix} \bar{1} & 6 & \bar{5} & 7 & \bar{6} & 3 \\ -41 & -10 & \bar{4} & 31 & 6 & \end{matrix}\right\} \therefore \text{Yes}$$

*Or*

The osculation-results are 16564, 1640, 164 and 0 ∴ Yes

(2) By 31? $\therefore Q=3 \left\{\begin{matrix} \bar{6} & 6 & \bar{0} & 3 \\ 0 & 33 & 9 & \end{matrix}\right\} \therefore \text{Yes}$

*Or*

The osculation-results are 651, 62 and 0 ∴ Yes

(3) By 41? $\therefore Q=4 \left\{\begin{matrix} 1 & \bar{1} & 2 & \bar{3} & 4 \\ 0 & 10 & 13 & 13 & \end{matrix}\right\} \therefore \text{Yes}$

*Or*

The osculation-results are 1107, 82 and 0 ∴ Yes

(4) By 47? $\therefore Q = 14 \left\{\begin{matrix} \bar{7} & 4\,\bar{2} & 1 & \bar{6} & 5 \\ 11 & 102\;7 & 51 & 64 & \end{matrix}\right\} \therefore$ No

*Or*

The osculation-results are 74146, 7330, 733 and 31 ∴ No

(5) By 51? $\therefore Q = 5 \left\{\begin{matrix} \bar{4} & 3 & \bar{7} & 3 & \bar{2} & 1 \\ -5 & 10 & 32 & 18 & 3 & \end{matrix}\right\} \therefore$ No

*Or*

The osculation-results are 43727, 4337, 398 and 1 ∴ No

(6) By 61? $\therefore Q = 6 \left\{\begin{matrix} 1 & \bar{9} & 5 & \bar{8} & 1 \\ 0 & -51 & -7 & -2 & \end{matrix}\right\} \therefore$ Yes

*Or*

The osculation-results are 1952, 183 and 0 ∴ Yes

(7) By 67? $\therefore Q = 20 \left\{\begin{matrix} \bar{1} & 0 & \bar{1} & 7 & \bar{1} & 2 & \bar{0} & 3 \\ 0 & -10 & 101 & \bar{5} & -81 & -4 & 60 & \end{matrix}\right\} \therefore$ Yes

*Or*

The osculation-results are 1017060, 101706, 10050, 1005 and 0 ∴ Yes

(8) By 91? $\therefore Q = 9 \left\{\begin{matrix} 9 & \bar{8} & 0 & \bar{4} & 5 & \bar{9} & 0 & \bar{5} & 3 \\ 84 & 69 & 49 & 56 & 37 & 44 & 16 & 22 & \end{matrix}\right\} \therefore$ No

*Or*

The osculation-results are 98045878, 9804515, 980406, 97986, 9744, 938 and 21 ∴ No

(9) By 61? $\therefore Q = 6 \left\{\begin{matrix} \bar{1} & 2 & \bar{2} & 1 & \bar{3} & 0 & \bar{5} & 4 \\ 0 & \bar{1}\bar{0} & \bar{2} & 0 & 10 & 53 & 19 & \end{matrix}\right\} \therefore$ Yes

*Or*

The osculation-results are 1221281, 122122, 12200, 1220, 122 and 0 ∴ Yes

(10) By 71? $\therefore Q = 7 \left\{\begin{matrix} \bar{8} & 0 & \bar{9} & 0 & \bar{4} & 5 \\ 0 & 62 & 19 & 4 & 31 & \end{matrix}\right\} \therefore$ Yes

*Or*

The osculation-results are 80869, 8023, 781 and 71 ∴ Yes

(11) By 131? $\therefore$ Q = 13 $\left\{\begin{matrix} 1 & \bar{3} & 3 & \bar{7} & 9 & \bar{0} & 3 \\ 0 & 10 & 1 & 20 & 123 & 39 & \end{matrix}\right\}$ $\therefore$ Yes

*Or*

The osculation-results are 133751, 13362, 1310 and 131 $\therefore$ Yes

(12) By 141? $\therefore$ Q = 14 $\left\{\begin{matrix} 4 & \bar{8} & 9 & \bar{8} & 8 & \bar{5} & 7 \\ 94 & 87 & 37 & 2 & 41 & 93 & \end{matrix}\right\}$ $\therefore$ No

*N.B.* But this dividend yielding the same results is divisble by 47 whose Q is also 14. ($94 = 47 \times 2$)

# 30

# Divisibility and Complex Multiplex Osculators

The cases so far dealt with are of a simple type, involving only small divisors and consequently small osculators. What then about those wherein bigger numbers being the divisors, the osculators are bound to be correspondingly larger?

The student-inquirer's requirements in this direction form the subject-matter of this chapter. It meets the needs in question by formulating a scheme of *groups of digits* which can be osculated, not as individual digits but in a lump, so to say.

### Examples of Multiplex Veṣṭana, i.e. Osculation

(1) 371 osculated by 4 for 2 digits at a time, gives $3+71\times4(=287)$ and $81+256\ (=337)$ for plus oscillation and minus oscillation respectively.

(2) 1572 osculated by 8 for 2 digits gives $15+576(=591)$ and $15-576(=-561)$ respectively.

(3) 8132 osculated by 8 (P and Q) for 2 digits gives $81+256\ (=337)$ and $81-256(=-175)$ respectively.

(4) 75621 osculated by 5 (P and Q) for 3 digits gives $75+3105$ $(=3180)$ and $75-3105\ (=-3030)$ respectively.

(5) 61845 osculated by 7 (P and Q) for 3-digit groups gives $61+5915\ (=5976)$ and $61-5915(=-5854)$ respectively.

(6) 615740 osculated by 8 (P and Q) for 3-digit packets gives $615+5920\ (=6535)$ and $615-5920\ (=-5305)$ respectively.

(7) 518 osculated by 8 (P and Q) for 4-digit bundles gives $0+4144\ (=4144)$ and $0-4144\ (=-4144)$ respectively.

(8) 73 osculated by 8 (P and Q) for five-digit groups yields $0+584(=584)$ and $0-584(=-584)$ respectively.

(9) 210074 osculated by 8 (P and Q) for five-digit bundles give $2+80592$ $(=80594)$ and $2-80592$ $(=-80590)$ respectively.

(10) 7531 osculated by 2 (P) for 3 digits gives $7+1062=1069$

(11) 90145 osculated by 5 (Q) for 3 gives $-725+90=-635$

(12) 5014112 osculated by 7 (Q) for 4 gives $501-28784=-28283$

(13) 7008942 osculated by 3 (P) for 2 gives $126+70089=70215$

(14) 7348515 osculated by 8 (P) for 3 gives $7348+4120=11468$

(15) 59076242 osculated by 7 (Q) for 2 gives $-590762+294=-590468$

## Categories of Divisors and their Osculators

In this context, it should be noted that, as there are various types of divisors, there are consequent differences as to the nature and type of osculators positive and/or negative which will suit them. They are generally of two categories:

(*i*) those which end in nine or a series of nines in which case they come within the jurisdiction of the *Ekādhika*, i.e. the Positive Osculator or, which terminate in or contain series of zeroes ending in 1, in which case they come within the scope of operations performable with the aid of the *Viparīta*, i.e. the negative osculator; and

(*ii*) those which, by suitable multiplication, yield a multiple of either of the two sorts described in sub-section (*i*) and can thus be tackled on that basis.

## The First Type

We shall deal, first, with the first type of divisors, namely, those ending in 9 or a series of nines or 1 or a series of zeroes ending in unity and explain a technical terminology and symbology which will facilitate our operations in this context.

(1) Let the divisor be 499. It is obvious that its osculator P is 5 and covers 2 digits. This fact can be easily expressed in symbolical language by saying : $P_2=5$

(2) In the case of 1399, it is obvious that our osculator positive is 14 and covers 2 digits $\therefore P_2=14$

(3) As for 1501, Q obviously comes into play, is 15 and covers 2 digits. In other words, $Q_2=15$

(4) For 2999, P is 3 and covers 3 digits $\therefore P_3 = 3$

(5) For 5001, $Q_3 = 5$

(6) For 7001, $Q_3 = 7$

(7) For 79999, $P_4 = 8$

(8) For 119999, $P_4 = 12$

(9) For 800001, $Q_5 = 8$

(10) For 900001, $Q_5 = 9$

(11) For 799999, $P_5 = 8$

(12) For 120000001, $Q_7 = 12$

### Correctness of the Symbology

The osculation-process invariably gives us the original number itself or a multiple thereof or zero: For example,

(*i*) 499 (with $P_2 = 5$) gives us $4 + 5\,(99) = 4 + 495 = 499$

(*ii*) 1399 (with $P_2 = 14$) gives $13 + 14(99) = 13 + 1386 = 1399$

(*iii*) 1501 (with $Q_2 = 15$) gives $15 \times 1 - 15 = 0$

(*iv*) 2999 (with $P_3 = 3$) gives $2 + 3(999) = 2 + 2997 = 2999$

(*v*) 5001 (with $Q_3 = 5$) gives $5 \times 1 - 5 = 0$

(*vi*) 7001 (with $Q_3 = 7$) gives $7 \times 1 - 7 = 0$

(*vii*) 79999 (with $P_4 = 8$) gives $7 + 8(9999) = 79999$

(*viii*) 119999 (with $P_4 = 12$) gives $11 + 12(9999) = 119999$

(*ix*) 800001 (with $Q_5 = 8$) gives $8 \times 1 - 8 = 0$

(*x*) 900001 (with $Q_5 = 9$) gives $9 \times 1 - 9 = 0$

(*xi*) 799999 (with $P_5 = 8$) gives $7 + 8(99999) = 799999$

(*xii*) 120000001 (with $Q_7 = 12$) gives $12 \times 1 - 12 = 0$

*N.B.*: The osculation-rule is strictly adhered to; and the P's and the Q's invariably yield the original dividend itself and zero respectively?

### Utility and Significance of the Symbology

The symbology has its deep significance and high practical utility in our determining of the divisibility or otherwise of a certain given number however big by a certain given divisor however large, inasmuch as it throws light on (1) the number of digits to be taken in each group and (2) the actual osculator itself in each individual case before us.

A few simple examples of each sort will clarify this:

(1) Suppose the question is: Is 106656874269 divisible by 499?

Here, at sight, $P_2 = 5$. This means that we have to split the given expression into 2-digit groups and osculate by 5. Thus,

$$\left\{\begin{array}{cccccc} 10 & 66 & 56 & 87 & 42 & 69 \\ 499 & 497 & 186 & 525 & 387 & \end{array}\right\} \therefore \text{Yes}$$

($69 \times 5 = 345$; $345 + 42 = 387$; $435 + 3 + 87 = 525$; $5 \times 25 + 5 + 56 = 186$; $5 \times 86 + 66 + 1 = 497$; $5 \times 97 + 4 + 10 = 499$!)

*Or*

The osculation-results are 1066569087, 10666125, 106786, 1497 and 499 $\therefore$ Yes

(2) Is 126143622932 divisible by 401?

Here $Q_2 = 4$

$$\therefore \left\{\begin{array}{cccccc} \bar{1}\bar{2} & 61 & \bar{4}\bar{3} & 62 & \bar{2}\bar{9} & 32 \\ -16 & 400 & 185 & 458 & 99 & \end{array}\right\} \therefore \text{No}$$

*Or*

The osculation-results are 1261436101, 12614357, 125915, 1199 and $-385$. $\therefore$ No

(3) Is 69492392 divisible by 199?

Here $P_2 = 2$

$$\therefore \left\{\begin{array}{cccc} 69 & 49 & 23 & 92 \\ 199 & 65 & 207 & \end{array}\right\} \therefore \text{Yes}$$

*Or*

The osculation-results give 695107, 6965 and 199 $\therefore$ Yes

(4) Is 1928264569 divisible by 5999?

Here $P_3 = 6$

$$\therefore \left\{\begin{array}{cccc} 1 & 928 & 264 & 569 \\ 5999 & 4999 & 3678 & \end{array}\right\} \therefore \text{Yes}$$

*Or*

The osculation-results are 1931678 and 5999 $\therefore$ Yes

(5) Is 2188 6068 313597 divisible by 7001?

Here $Q_3 = 7$

$$\therefore \left\{\begin{matrix} 21 & \overline{886} & 068 & \overline{313} & 597 \\ 0 & -3 & 6127 & 3866 & \end{matrix}\right\} \therefore \text{Yes}$$

*Or*

The osculation-results give 21 886 064 134, 21885126, 21003 and 0. $\therefore$ Yes

(6) Is 30102 1300602 divisible by 99?

Here $P_2 = 1$

As $P_2 = 1$ and continuous multiplications by 1 can make no difference to the multiplicand, the sum of the groups will suffice for our purpose: $\because 30+10+21+30+06+2 = 99 \therefore$ Yes.

The second method amounts to the same thing and need not be put down.

(7) Is 2130 1102 1143 4112 divisible by 999?

Here $P_3 = 1$

and $\because 2 + 130 + 110 + 211 + 434 + 112 = 999$

$\therefore$ (By both methods) Yes

(8) Is 7631 3787 858 divisible by 9999?

Here $P_4 = 1$

and $\because = 763 + 1378 + 7858 = 9999$

$\therefore$ (By both methods) Yes.

(9) Is 2037760003210041 divisible by 9999?

Here $P_4 = 1$

and $\because 2037 + 7600 + 0321 + 0041 = 9999$

$\therefore$ (By both methods) Yes.

(10) Is 5246 7664 0016 201452 divisible by 1001?

Here $Q_3 = 1$; and

$$\because S = \overline{524} + 676 + \overline{648} + 016 + \overline{201} + 452 = 221 = 13 \times 17$$

$\therefore$ Divisible by 13

But $1001 = 7 \times 11 \times 13$ $\therefore$ Divisible by 13 but not by 7 or by 11.

$\therefore$ No

### The Second Category

The second type is one wherein the given number is of neither of the standard types to which P and Q readily and instantaneously apply but requires a multiplication for the transformation of the given number to either or both of the standard forms and for ascertaining P and Q or both suitable for our purpose in the particular case before us.

### The Process of Transformation

In an earlier chapter on "Recurring Decimals" we have shown how to convert a given decimal fraction into its vulgar-fraction shape, by so multiplying it as to bring a series of nines in the product. For example, in the case of $.\dot{1}4285\dot{7}$, we had multiplied it by 7 and got $.\dot{9}9999\dot{9}\ (=1)$ as the product and thereupon argued that, because 7× the given decimal = 1, ∴ that decimal should be the vulgar fraction $\frac{1}{7}$

(1) $.\dot{1}4285\dot{7} \times 7 = .\dot{9}9999\dot{9} = 1$

$\therefore x = \frac{1}{7}$

Similarly, with regard to $.\dot{0}7692\dot{3}$, we had multiplied it by 3 in order to get 9 as the last digit of the product; argued that, in order to get 9 as the penultimate digit, we should add 3 to the already existing 6 there and that this 3 could be had only by multiplying the original given decimal by 1; then found that the product was now a series of nines; and then we had argued that, ∵ $13x = 1$, ∴ $x$ must be equal to $\frac{1}{13}$. And we had also given several more illustrations of the same kind for demonstrating the same principle and process.

(2)
$$\begin{array}{r} .\dot{0}7692\dot{3} \\ 13 \\ \hline .23076\dot{9} \\ 0.76923\phantom{0} \\ \hline \dot{9}9999\dot{9} = 1 \end{array}$$

$\therefore x = \frac{1}{13}$

As P in the present context requires, for osculation, numbers ending in 9 or a series of nines, we have to adopt a similar procedure for the same purpose; and, in the case of Q too, we have to apply a similar method for producing a number which will terminate in 1 or a series of zeroes ending in 1.

### The Modus Operandi

A few examples of both the kinds will elucidate the process and help the student to pick up his P and Q. And once this is done, the rest will automatically follow as explained above.

(1) Suppose the divisor is 857. ∵ 857×7=5999, we can therefore, at once say : $P_3 = 6$.

The test and proof of the correctness hereof is that any multiple of the divisor in question must necessarily fulfil this condition, i.e. on osculation by $P_3$ must yield 857 or a multiple thereof.

For instance, let us take $857 \times 13 = 11141$. As $P_3 = 6$ $\therefore$ $11 + 6\ (141) = 857$! And this proves that our osculator is the correct one.

(2) Let us now take 43. $\because 43 \times 7 = 301, \therefore Q_2 = 3$.

Taking $43 \times 3$ $(=129)$ for the test, we see 129 yields $29 \times 3 - 1 = 86$; and 86 is a multiple of 43 being exactly double of it. So, our Q is correct.

The significance of this fact consists in the natural consequence thereof, namely, that any number which is really divisible by the divisor in question must conform to this rule of divisibility by the P process or the Q process.

*N.B.*: Remember what has already been explained as regards P or Q being greater.

In this very case of 43, instead of multiplying it by 7, getting 301 as the product and ascertaining that $Q_2 = 3$ is the osculator, we could also have multiplied the 43 by 3, got 129 as the product, found $P_1 = 13$ to be the positive osculator and verified it. Thus, in the case of $43 \times 2 = 86$, $\therefore 8 + 6(13) = 86 \therefore P_1 = 13$ is the correct positive osculator.

Multiplication by 13 at every step being necessarily more cumbrous than by 3, we should naturally prefer $Q_3 = 3$ to $P_1 = 13$.

In fact, it rests with the student to choose between P and Q and in view of the bigness or otherwise of multiplier-osculator etc. decide which to prefer.

(3) Ascertain the P and the Q for 137.

$$\begin{array}{r} 137 \\ 27 \\ \hline 959 \\ 274 \\ \hline \underline{3699} \end{array} \qquad\qquad \begin{array}{r} 137 \\ 73 \\ \hline 411 \\ 959 \\ \hline \underline{10001} \end{array}$$

$$\therefore P_2 = 37 \qquad\qquad \therefore Q_3 = 14$$

Obviously $Q_4 = 1$ is preferable (to $P_2 = 37$).

(Test : $137 \times 8 = 1296$ $\because Q_4$ gives $960 - 1 = 959 \therefore$ Yes.)

(4) Determine the P and the Q for 157

$\because 157 \times 7 = 1099 \therefore P_2 = 11$ And $157 \times 93 = 14601 \therefore Q_2 = 146$.

$\therefore P_2 = 11$ is to be preferred.

(Test : $157 \times 7 = 1099 \therefore P_2$ gives $10 + 1089 = 1099$)

(5) Find out the P and Q for 229

$\because 229 \times 131 = 29999 \therefore P_4 = 3$

This Osculator being so simple, the Q need not be tried at all. But on principle, $\because 229 \times 69 = 15801$

$\therefore Q_2 = 158$ obviously a big multiplier

Test for $P_4 = 3$

$229 \times 100 = 22900 \therefore P_4 = 3$ gives $8702 = 229 \times 38$

(6) Find P and Q for 283

$\because 283 \times 53 = 14999 \therefore P_4 = 15$

and $\because 283 \times 47 = 13301 \therefore Q_3 = 133$

$\therefore P_3 = 15$ is preferable

Test $283 \times 4 = 1132; 1 + 15\ (132) = 1981 = 283 \times 7$

(7) Find P and Q for 359

$\because 359 \times 61 = 21899 \therefore P_2 = 219$

and $\because 359 \times 339 = 14001 \therefore Q_3 = 14$

Obviously $Q_3 = 14$ is to be preferred.

Test : (*i*) $319 \times 3 = 1077 \therefore Q_3 = 14$ gives $14 \times 77 - 1 = 1077$

and (*ii*) $359 \times 115 = 41285 \therefore Q_2 = 14$ gives $14 \times 285 - 41 = 3949$ $= 359 \times 11$

(8) Ascertain P and Q for 421

$\because 421 \times 19 = 7999 \therefore P_3 = 8$

and $\because 421 \times 81 = 34101 \therefore Q_2 = 341$

Obviously $P_3 = 8$ is the better one

Test: $421 \times 5 = 2105 \therefore P_3 = 8$ gives $2 + 840 = 842 = 421 \times 2$

(9) Determine P and Q for 409

$\because 409 \times 511 = 208999 \therefore P_3 = 209$

and $\because 409 \times 489 = 200001 \therefore Q_5 = 2$

Obviously the Q osculator is preferable.

Test: $409 \times 1000 = 409000$

$\therefore Q_5 = 2$ gives $18000 - 4 = 17996 = 409 \times 44$

Having thus studied the multiplex osculator technique and *modus operandi*, we now go on to and take up actual examples of divisibility which can be easily tackled by the multiplex osculatory procedure.

### Model Applications to Concrete Examples

(1) Is 79158435267 divisible by 229?

$\because 229 \times 131 = 29999 \therefore P_4 = 3$

$$\therefore \begin{Bmatrix} 791 & 5843 & 5267 \\ 5725 & 21644 & \end{Bmatrix}$$

But $5725 = 229 \times 25 \therefore$ Yes

(2) Is 6056200566 divisible by 283?

$\because 283 \times 53 = 14999 \therefore P_3 = 15$

$$\therefore \begin{Bmatrix} 6 & 056 & 200 & 566 \\ 6226 & 10414 & 8690 & \end{Bmatrix}$$

But $6226 = 283 \times 22 \therefore$ Yes

(3) Is 7392 60261 divisible by 347?

$\because 347 \times 317 = 109999 \therefore P_4 = 11$

$$\therefore \begin{Bmatrix} 7 & 3926 & 6251 \\ 73654 & 0627 & \end{Bmatrix}$$

But $73654 = 347 \times 212 \therefore$ Yes

(4) Is 867 311 7259 divisible by 359?

$\because 359 \times 39 = 14001 \therefore Q_3 = 14$

$$\therefore \begin{Bmatrix} \bar{8} & 673 & \bar{1}\bar{1}\bar{7} & 259 \\ 3590 & 257 & 278 & \end{Bmatrix} \therefore \text{Yes}$$

(5) Is 885648437 divisible by 367?

$\because 367 \times 3 = 1101 \therefore Q_2 = 11$

$$\therefore \begin{Bmatrix} 8 & \bar{8}\bar{5} & 64 & \bar{8}\bar{4} & 37 \\ 734 & 66 & 314 & 323 & \end{Bmatrix}$$

But $734 = 367 \times 2 \therefore$ Yes

(6) Is 490 222 8096 divisible by 433?

$\because 433 \times 3 = 1299 \therefore P_2 = 13$

$$\therefore \begin{Bmatrix} 49 & 02 & 22 & 80 & 96 \\ 1257 & 1292 & 399 & 1328 & \end{Bmatrix} \therefore \text{No}$$

(7) Is 51 888 888 37 divisible by 467?

$\because 467 \times 3 = 1401 \therefore Q_2 = 14$

$$\therefore \left\{ \begin{matrix} 51 & \bar{8}\bar{8} & 88 & \bar{8}\bar{8} & 37 \\ -467 & -37 & 504 & 430 & \end{matrix} \right\} \therefore \text{Yes}$$

*N.B.*: The alternative method of successive mechanical osculations is also, of course, available but will prove generally less neat and tidy and will also be more tedious.

(8) Is 789405 35994 divisible by 647?

$\because 647 \times 17 = 10999 \therefore P_3 = 11$

$$\therefore \left\{ \begin{matrix} 78 & 940 & 535 & 994 \\ 1294 & 6110 & 11469 & \end{matrix} \right\}$$

But $1294 = 647 \times 2 \therefore$ Yes

(9) Is 2093 1726 7051 0192 divisible by 991?

$\because 991 \times 111 = 110001 \therefore Q_4 = 11$

$$\therefore \left\{ \begin{matrix} \bar{2}\bar{0}\bar{9}\bar{3} & 1726 & \bar{7}\bar{0}\bar{5}\bar{1} & 0192 \\ -30721 & -52603 & -4939 & \end{matrix} \right\}$$

But $30721 = 991 \times 331 \therefore$ Yes

(10) Is 479466 54391 divisible by 421?

$\because 421 \times 19 = 7999 \therefore P_3 = 8$

$$\therefore \left\{ \begin{matrix} 47 & 946 & 654 & 391 \\ 1694 & 7205 & 3782 & \end{matrix} \right\} \text{No}$$

(11) What change should be made in the first digit of the above number in order to render it divisible by 421?

*Answer*: As 1684 is exactly $4 \times 421$, the only change needed in order to reduce the actually present 1694 into 1684 is the alternation of the first digit from 4 to 3.

# 31

# Sum and Difference of Squares

Not only with regard to questions arising in connection with and arising out of Pythagoras' Theorem which we shall shortly take up but also in respect of matters relating to the three fundamental Trigonometrical-Ratio-relationships as indicated by the three formulae $\sin^2\theta + \cos^2\theta = 1$, $1 + \tan^2\theta = \sec^2\theta$ and $1 + \cot^2\theta = \mathrm{cosec}^2\theta$ etc., we have often to deal with the difference of two square numbers, the addition of two square numbers etc. And it is desirable to have the assistance of rules governing this subject and benefit by them.

### DIFFERENCE OF TWO SQUARE NUMBERS

Of the two, this is much easier. For, *any* number can be expressed as the difference of two square numbers. The Algebraical principle involved is to be found in the elementary formula $a^2 - b^2 = (a+b)(a-b)$. This means that, if the given number can be expressed in the shape of the product of two numbers, our task is automatically over. And this "if" imposes a condition which is very easy to fulfil. For, even if the given number is a prime number, it can be correctly described as the product of itself and of unity. Thus $7 = 7 \times 1$, $17 = 17 \times 1$, $197 = 197 \times 1$ and so on.

In the next place, we have the derived formula:

$(a+b)^2 - (a-b)^2 = 4ab$; and therefore ab can always come into the picture as $\left(\frac{a+b}{2}\right)^2 - \left(\frac{a-b}{2}\right)^2$ = i.e. (half the sum)$^2$ – (half the difference)$^2$ and as *any* number can be expressed as ab, the problem is readily solved. And the larger the number of factorisations possible, the better. In fact, if we accept fractions too as permissible, the number of possible solutions will be *literally infinite*.

For example, suppose we have to express 9 as the difference of two squares. We know that—

$$9 = 9 \times 1 \therefore 9 = \left(\frac{9+1}{2}\right)^2 - \left(\frac{9-1}{2}\right)^2 = 5^2 - 4^2. \text{ Similarly}$$

(1) $13 \times 1 = \left(\frac{14}{2}\right)^2 - \left(\frac{12}{2}\right)^2 = 7^2 - 6^2$

(2) $12 = 6 \times 2 = 4^2 - 2^2$

*or* $4 \times 3 = 3\frac{1}{2}^2 - \frac{1}{2}^2$

*or* $12 \times 1 = 6\frac{1}{2}^2 - 5\frac{1}{2}^2$

(3) $48 = 8 \times 6 = 7^2 - 1^2$

*or* $12 \times 4 = 8^2 - 4^2$

*or* $16 \times 3 = 9\frac{1}{2}^2 - 6\frac{1}{2}^2$

*or* $48 \times 1 = 24\frac{1}{2}^2 - 23\frac{1}{2}^2$

*or* $24 \times 2 = 13^2 - 11^2$

The question, therefore, of expressing any number as the difference of two squares presents no difficulty at all !

## The Sum of Two Square Numbers

Inasmuch, however, as $a^2 + b^2$ has no such corresponding advantage or facilities etc., to offer, the problem of expressing any number as the sum of two square numbers is a tough one and needs very careful attention. Therefore, we now proceed to deal with this.

## A Simple Rule in Operation

We first turn our attention to a certain simple rule at work in the world of numbers, in this respect.

We need not go into the relevant original *Sūtras* and explain them especially to our non-Sanskrit-knowing readers. Suffice it for us, for our present purpose, to explain their purport and their application.

Let us take a particular series of "mixed" fractions, namely, $1\frac{1}{3}, 2\frac{2}{5}, 3\frac{3}{7}, 4\frac{4}{9}, 5\frac{5}{11}$ etc. which fulfil three conditions:

(*i*) that the integer-portion consists of the natural numbers in order;
(*ii*) that the numerators are exactly the same; and
(*iii*) that the denominators are the odd numbers, in order, commencing from 3 and going right on.

It will be observed that, when all these fractions are put into shape as "improper" fractions.

i.e. as $\frac{4}{3}, \frac{12}{5}, \frac{24}{7}, \frac{40}{9}, \frac{60}{11}$ etc.

the sum of $D^2$ and $N^2$ is invariably equal to $(N+1)^2$ the General Algebraical form being.

$$n + \frac{n}{2n+1} = \left\{\frac{2n(n+1)}{2n+1}\right\}$$

$\therefore D = 2n+1;$ and $N = 2n(n+1)$

$\therefore D^2 + N^2 = (N+1)^2$

*or* $(2n+1)^2 + 4n^2(n+1)^2 = (2n^2+2n+1)^2$

The *shape* of it is perhaps frightening; but the *thing* in itself is very simple; and the best formula is $D^2 + N^2 = (N+1)^2$.

This means that when $a^2$ (given) $+x^2$ is a perfect square, we can readily find out $x^2$. Thus, for instance,

(*i*) If the given number be 9, $2n+1 = 9 \therefore n = 4$

$\therefore 4\frac{4}{9}$ is the fraction we want. And $9^2 + 40^2 = 41^2$

(*ii*) If a be 35, $2n+1 = 35 \therefore n = 17 \therefore$ The fraction wanted is

$17\frac{17}{35} = \frac{612}{35} \therefore 35^2 + 612^2 = 613^2$

(*iii*) If $a = 57$, $2n+1 = 57 \therefore n = 28 \therefore$ The required fraction is

$28\frac{28}{57} = \frac{1624}{57} \therefore 57^2 + 1624^2 = 1625^2$

(*iv*) If $a = 141$, $2n+1 = 141 \therefore n = 70 \therefore$ The wanted fraction is

$70\frac{70}{141} = \frac{9940}{141} \therefore 141^2 + 9940^2 = 9941^2$

*Note*: Multiples and sub-multiples too behave in exactly the same manner according to *Ānurūpya*, i.e. proportionately. For instance,

Let $a = 35 \therefore 2n+1+35 \therefore n = 17$

$\therefore$ The fraction wanted is $17\frac{17}{35} = \frac{612}{35} \therefore 35^2 + 612^2 = 613^2$

$\therefore 70^2 + 1224^2 = 1226^2$

### A Simpler Method (for the Same)

This same result can also be achieved by a simpler and easier method which does not necessitate the "mixed" fractions, the transforming of them into the "improper"–fraction–shape etc., but gives us the answer immediately.

It will be observed that, in all the examples dealt with above,

Since $D^2 + N^2 = (N+1)^2$

$\therefore D^2 = (N+1)^2 - N^2 = 2N + 1 = N + (N+1)$

In other words, the square of the given number is the sum of two consecutive integers at the exact middle. For instance, if 7 be the given number, its square = 49 which can be split up into the two consecutive integers 24 and 25.

$\therefore\ 7^2 + 24^2 = 25^2$. Similarly,

(1) If $a = 9$, its square $(81) = 40 + 41\ \therefore\ 9^2 + 40^2 = 41^2$

(2) If $a = 35$, its square $(1225) = 612 + 613\ \therefore\ 35^2 + 612^2 = 613^2$

(3) If $a = 57$, its square $(3249) = 1624 + 1625$

$\therefore\ 57^2 + 1624^2 = 1625^2$

(4) If $a = 141$, its square $(19881) = 9940 + 9941$

$\therefore\ 141^2 + 9940^2 = 9941^2$

and so on. And all the answers are exactly as we obtained before by the first method.

## The Case of Even Numbers

Yes, the square of an odd number is necessarily odd and can be split up into two consecutive integers. But what about even numbers whose squares will always be even and cannot be split up into two consecutive numbers? And the answer is that such cases should be divided off by 2 and other powers of 2 until an odd number is reached and then the final result should be multiplied proportionately.

For example, if $a = 52$, we divide it by 4 and get the odd number 13. Its square $(169) = 84 + 85\ \therefore\ 13^2 + 84^2 = 85^2\ \therefore$ multiplying all the terms by $4^2$, we say : $52^2 + 336^2 = 340^2$.

There are many other simple and easy methods by which we can tackle the problem (of $a^2 + b^2 = c^2$) by means of clues and conclusions deducible from $3^2 + 4^2 = 5^2$, $5^2 + 12^2 = 13^2$, $8^2 + 15^2 = 17^2$ etc. But we do not now enter into details of these and other allied matters.

# 32

# Elementary Squaring, Cubing etc.

In some of the earliest chapters of this treatise, we have dealt, at length, with multiplication-devices of various sorts, and squaring, cubing etc., are only a particular application thereof. This is why this subject too found an integral place of its own in those earlier chapters on multiplication.

And yet it so happens that the squaring, cubing etc., of numbers have a particular entity and individuality of their own; and besides, they derive additional importance because of their intimate connection with the question of the square-root, the cuberoot etc., which we shall take up shortly. And, consequently, we shall now deal with this subject of squaring, cubing etc., mainly by way of preliminary revision and recapitulation on the one hand and also by way of presentation of some important new material on the other.

### THE YĀVADŪNAM SŪTRA (FOR SQUARING)

In the revision part of it, we may just formally remind the student of the *Yāvadūnam* formula and merely cite some examples thereof as a sort of practical memory refresher:

1. $97^2 = 94/09$;
2. $87^2 = 74_5/_1 69 = 7569$
3. $19^2 = 1_1, 8_8 1 = 28_8 1 = 361$
4. $91^2 = 82/81$
5. $965^2 = 930_1/225 = 931/225$
6. $113^2 = 126_1/69 = 12769$
7. $996^2 = 992/016$
8. $998^2 = 996/004$
9. $9997^2 = 9994/0009$
10. $1007^2 = 1014/049$

11. $9996^2 = 9992/0016$
12. $9999^2 = 9998/0001$
13. $1017^2 = 1034/289$
14. $1039^2 = 1078_1/521 = 1079/521$
15. $99991^2 = 99982/00081$
16. $99998^2 = 99996/00004$
17. $99994^2 = 99988/00036$
18. $10004^2 = 10008/0016$
19. $999978^2 = 999956/000484$
20. $999998^2 = 999996/000004$
21. $100023^2 = 100046/00529$
22. $9999873^2 = 9999746/0016129$
23. $9999999^2 = 9999998/0000001$
24. $1000012^2 = 1000024/000144$

### The Ānurūpya Sūtra (for Cubing)

This is new material. A simple example will, however, suffice to explain it:

Take the hypothetical case of one who knows only the cubes of the "first ten natural numbers", i.e. 1 to 10 and wishes to go therebeyond, with the help of an intelligent principle and procedure. And suppose he desires to begin with $11^3$.

1. The first thing one has to do is to put down the cube of the first digit in a row of 4 figures in a geometrical ratio in the exact proportion subsisting between them. Thus,

$$
\begin{array}{rcccc}
11^3 = & 1 & 1 & 1 & 1 \\
 & & 2 & 2 & \\
\hline
 & 1 & 3 & 3 & 1 \\
\hline
\end{array}
$$

(*ii*) The second step is to put down, under the second and third numbers, just two times the said numbers themselves and add up. And that is all!

A few more instances will clarify the procedure:

(1)
$$
\begin{array}{rcccc}
12^3 = & 1 & 2 & 4 & 8 \\
 & & 4 & 8 & \\
\hline
 & 1 & 6 & 2 & 8 \\
 & & 1 & & \\
\hline
 & 1 & 7 & 2 & 8 \\
\hline
\end{array}
$$

(2)
$$
\begin{array}{rcccc}
13^3 = & 1 & 3 & 9 & 27 \\
 & & 6 & 18 & \\
\hline
 & 1 & 9 & 7 & 7 \\
 & & 2 & 2 & \\
\hline
 & 2 & 1 & 9 & 7 \\
\hline
\end{array}
$$

(3) $$14^3 = \begin{array}{rrrr} 1 & 4 & 16 & 64 \\ & 8 & 32 & \\ \hline 27 & 4 & 4 & \end{array}$$

(4) $$15^3 = \begin{array}{rrrr} 1 & 5 & 25 & 125 \\ & 10 & 50 & \\ \hline 33 & 7 & 5 & \end{array}$$

(5) $$16^3 = \begin{array}{rrrr} 1 & 6 & 36 & 216 \\ & 12 & 72 & \\ \hline 40 & 9 & 6 & \end{array}$$

(6) $$17^3 = \begin{array}{rrrr} 1 & 7 & 49 & 343 \\ & 14 & 98 & \\ \hline 49 & 1 & 3 & \end{array}$$

(7) $$18^3 = \begin{array}{rrrr} 1 & 8 & 64 & 512 \\ & 16 & 128 & \\ \hline 58 & 3 & 2 & \end{array}$$

(8) $$19^3 = \begin{array}{rrrr} 1 & 9 & 81 & 729 \\ & 18 & 162 & \\ \hline 68 & 5 & 9 & \end{array}$$

(9) $$21^3 = \begin{array}{rrrr} 8 & 4 & 2 & 1 \\ & 8 & 4 & \\ \hline 9 & 2 & 6 & 1 \end{array}$$

(10) $$22^3 = \begin{array}{rrrr} 8 & 8 & 8 & 8 \\ & 16 & 16 & \\ \hline 10 & 6 & 4 & 8 \end{array}$$

(11) $$23^3 = \begin{array}{rrrr} 8 & 12 & 18 & 27 \\ & 24 & 36 & \\ \hline 12 & 1 & 6 & 7 \end{array}$$

(12) $$24^3 = \begin{array}{rrrr} 8 & 16 & 32 & 64 \\ & 32 & 64 & \\ \hline 13 & 8 & 2 & 4 \end{array}$$

(13) $$25^3 = \begin{array}{rrrr} 8 & 20 & 50 & 125 \\ & 40 & 100 & \\ \hline 15 & 6 & 2 & 5 \end{array}$$

(14) $$32^3 = \begin{array}{rrrr} 27 & 18 & 12 & 8 \\ & 36 & 24 & \\ \hline 32 & 7 & 6 & 8 \end{array}$$

(15) $$9^3 = (10-1)^3$$
$$\left.\begin{array}{r} = 1000 - 100 + 10 - 1 \\ -200 + 20 \end{array}\right\} = 1000 - 300 + 30 - 1 = 729$$

(16) $$97^3 = \begin{array}{rrrr} 729 & 567 & 441 & 343 \\ & 1134 & 882 & \\ \hline 912 & 6 & 7 & 3 \end{array}$$

or, better still, $97^3 = (100-3)^3$

$$\begin{array}{rrrr} = 100000 & -30000 + & 900 & -27 \\ & -60000 + & 1800 & \\ \hline 100000 & -90000 + & 2700 & -27 \\ = 912 & 6 & 7 & 3 \end{array}$$

*N.B.*: If you start with the cube of the first digit and take the next three numbers in the top row in a geometrical proportion in the ratio of the original digits themselves you will find that the 4th figure on the right end is just the cube of the second digit!

The Algebraical explanation hereof is very simple:
If $a$ and $b$ are the two digits, then our chart reads:

$$\begin{array}{l} a^3 + \ a^2b + \ ab^2 + b^3 \\ \quad\ \ 2a^2b + 2ab^2 \\ \hline a^3 + 3a^2b + 3ab^2 + b^3 \\ \hline \end{array}$$

and this is exactly $(a+b)^3$!

Almost every mathematical student *knows* this; but very few people *apply* it! This is the whole tragedy and the pathos of the situation!

### The Yāvadūnam Sūtra (for Cubing)

The same *Yāvadūnam Sūtra* can, in view of the above, be applied for cubing too. The only difference is that we take here not the deficit or the surplus but exactly twice the deficit or the surplus as the case may be and make a few corresponding alterations in the other portions also, as follows:

Suppose we wish to ascertain the cube of 104. Our base being 100, the excess is 4. So we add not 4 as we did in the squaring operation but double that, i.e. 8 and thus have $104+8\ (=112)$ as the left-hand-most portion of the cube. Thus we obtain 112.

Then we put down the new excess multiplied by the original excess (i.e. $12 \times 4 = 48$) and put that down as the middle portion of the product. 112/48

And then we affix the cube of the original excess (i.e. 64) as the last portion thereof. And the answer is complete. 112/48/64

Some more illustrative instances are given below for familiarising the student with the new process which is not *really* new but only a very useful practical application of the $(a+b)^3$ formula described above:

(1) $103^3 = 109/27/27$ (because $9 \times 3 = 27$; and $3^3 = 27$)

(2) $113^3 = 139/\overset{}{\underset{5}{07}}/\underset{21}{97}$ (because $39 \times 13 = 507$ and $13^3 = 2197$)

$= 1442897$

(3) $1004^3 = 1012/048/064$ (because $12 \times 4 = 48$ and $4^3 = 64$)

(4) $10005^3 = 10015/0075/0125$ (because $15 \times 5 = 75$ and $5^3 = 125$)

(5) $996^3 = 988/048/\overline{064} = 998/047/936$ ($\because -12 \times -4 = 48$ and

$-4^3 = -64)$

(6) $93^3 = 79/\underset{1}{47}\ \underset{-3}{}/\overline{43}$ (because $-21 \times -7 = 147$ and $-7^3 = -343$)

$= 804357$

(7) $9991^3 = 9973/0243/\overline{0729}$ (because $-27 \times -9 = 243$ and $-9^3 = -729) = 9973/0242/9271$

(8) $10007^3 = 10021/0147/0343$

(9) $99999^3 = 99997 / 00003 / 0000\overline{1} = 99997 / 00002 / 99999$

(10) $100012^3 = 100036 / 00432 / 01728$

(11) $99998^3 = 99994/00012/00008 = 99994/00011/99992$

(12) $1000007^3 = 1000021/000147/000343$

(13) $999992^3 = 999976/000191/999488$ (because $24 \times 8 = 192$ and $-8^3 = -512$)

### Fourth Power

We know that $(a+b)^4 = a^4 + 4a^3b + 6a^2b^2 + 4ab^3 + b^4$. This gives us the requisite clue for raising any given number to its fourth power. Thus,

(1) $11^4 = 1\ 1\ 1\ 1\ 1$
$\quad\ \ 3\ 5\ 3$
$= 1\ 4\ 6\ 4\ 1$

(2) $12^4 = 1\ 2\ 4\ 8\ 16$
$\quad\ \ 6\ 20\ 24$
$= 2\ 0\ 7\ 3\ 6$

### The Binomial Theorem

The "binomial theorem" is thus capable of practical application and—in its more comprehensive Vedic form—has thus been utilised, to splendid purpose, in the Vedic *Sūtras*. And a huge lot of Calculus work both differential and integral has been and can be facilitated thereby. We shall hold over these details, for a later stage.

# 33

# Straight Squaring

Reverting to the subject of the squaring of numbers, the student need hardly be reminded that the methods expounded and explained in an early chapter and even in the previous chapter are applicable only to special cases and that a general formula capable of universal application is still due.

And, as this is intimately connected with a procedure known as the *Dvandva Yoga* or the Duplex Combination process and as this is of still greater importance and utility at the next step on the ladder, namely, the easy and facile extraction of square roots, we now go on to a brief study of this procedure.

### The Dvandva-Yoga (or the Duplex Combination Process)

The term "*Dvandva Yoga*" or Duplex is used in two different senses. The first one is by squaring; and the second one is by cross-multiplication. And, in the present context, it is used in both the senses ( $a^2$ and 2ab).

In the case of a single central digit, the square ( $a^2$ etc.,) is meant; and in the case of an even number of digits say, a and b equidistant from the two ends, double the cross-product (2ab) is meant.

A few examples will elucidate the procedure.

Denoting the Duplex with the symbol D, we have:

(1) For 2, $D = 2^2 = 4$

(2) For 7, $D = 49$

(3) For 34, $D = 2(12) = 24$

(4) For 74, $D = 2(28) = 56$

(5) For 409, $D = 2(36) + 0 = 72$

(6) For 071, $D = 0 + 49 = 49$

(7) For 713, $D = 2(21) + 1^2 = 43$

(8) For 734, $D = 2(28) + 3^2 = 65$
(9) D for $7346 = 2 \times 42 + 2 \times 12 = 108$
(10) D for $26734 = 16 + 36 + 49 = 101$
(11) D for $60172 = 24 + 0 + 1^2 = 25$
(12) D for $73215 = 70 + 6 + 4 = 80$
(13) D for $80607 = 112 + 0 + 36 = 148$
(14) D for $77 = 2 \times 49 = 98$
(15) D for $521398 = 80 + 36 + 6 = 122$
(16) D for $746213 = 42 + 8 + 24 = 74$
(17) D for $12345679 = 18 + 28 + 36 + 40 = 122$
(18) D for $370415291 = 6 + 126 + 0 + 40 + 1 = 173$
(19) D for $432655897 = 56 + 54 + 32 + 60 + 25 = 227$

This is merely a recapitulation of the *Ūrdhva Tiryak* process of multiplication as applied to squaring and needs no exposition.

*Note*: If a number consists of n digits, its square must have 2n or 2n–1 digits. So, in the following process, take extra dots to the left one less than the number of digits in the given numbers.

*Examples*

(1) $207^2 = \left.\begin{matrix} 40809 \\ 2\ 4 \end{matrix}\right\} = 42849$

$$\begin{array}{r} . \ . \ \ 207 \\ \hline 4\ 0\ \ 809 \\ 2\ \ \ \ 4 \\ \hline 4\ 2\ \ 849 \\ \hline \end{array}$$

(2) $213^2 = 44_1369 = 45369$

(3) $221^2 = 48841$

(4) $334^2 = \begin{matrix} 9\ 8\ 3\ 4\ 6 \\ 1\ 3\ 2\ 1 \end{matrix} = 111556$

(5) $425^2 = \begin{matrix} ..425 \\ 16/16/44/20/25 \end{matrix} = 180625$

(6) $543^2 = \begin{matrix} 25\ 0\ 6\ 49 \\ 4\ 4\ 2 \end{matrix} = 294849$

(7) $897^2 = \begin{matrix} ...8\ 9\ 7 \\ 64/144/193/126/49 \end{matrix} = 80609$

*Or* $1\ \bar{1}\ 0\ \bar{3}^2 = 1\bar{2}1\bar{6}609 = 80609$

Or by *Yāvadūnam Sūtra* $= 784/103^2 = 80609$

(8) $889^2 = \dfrac{\cdot\ \ \cdot\ \ \ 8\ \ 8\ \ 9}{64/128/208/144/81} = 790321$

*Or* $(1\ \bar{1}\ \bar{1}\ \bar{1})^2 = \dfrac{\cdot\ \cdot\ \cdot\ 1\ \bar{1}\ \bar{1}\ \bar{1}}{1\ \bar{2}\ \bar{1}\ 0\ 3\ 2\ 1} = 790321$

*Or by Yāvadūnam Sūtra*

$889^2 = 778/111^2 = 789/\ \ 321 = 790/321.$
$\qquad\qquad\qquad\quad /12$

(9) $1113^2 = \dfrac{\cdot\ \cdot\ \cdot\ 1\ 1\ 1\ 3}{1\ 2\ 3\ 8\ 7\ 6\ 9}$

(10) $2134^2 = 4/4/13/22/17/24/16 = 4553956$

(11) $3214^2 = 9/12/10/28/17/8/16 = 10329796$

(12) $3247^2 = 9/12/28/58/44/56/49 = 10543009$

(13) $6703^2 = 36/84/49/36/42/0/9 = 44930209$

(14) $31.42^2 = 9/6/25/20/20/16/4 = 987.2164$

(15) $.0731^2 = .0049/42/23/6/1 = .00534361$

(16) $8978^2 = 64/144/193/254/193/112/64 = 80604484$

*Or* $(1\ \bar{1}\ 0\ \bar{2}\ \bar{2})^2 = 1/\bar{2}/1/\bar{4}/0/4/4/8/4 = 80604484$

*Or* by *Yāvadūnam Sūtra* $7956/1022^2 = 80604484$

(17) $8887^2 = 64/128/192/240/176/112/49 = 78978769$

*Or* $\left(\dfrac{2}{11113}\right)^2 = 1-2-10-3+8769 = 78978769$

Or by *Yāvadūnam* $7774/1113^2 = 7774/\ \ 8769 = 78978769$
$\qquad\qquad\qquad\qquad\qquad\qquad /123$

(18) $141.32^2 = 1/8/18/14/29/22/13/12/4 = 19971.3424$

(19) $21345^2 = 4/4/13/22/37/34/46/40/25 = 455609025$

(20) $43031^2 = 16/24/9/24/26/6/9/6/1 = 1851666961$

(21) $46325^2 = 16/48/60/52/73/72/34/20/25 = 2146005625$

(22) $73214^2 = 49/42/37/26/66/28/17/8/16 = 5360289796$

# 34

# Vargamūla (Square Root)

Armed with the recapitulation in the last chapter of the "Straight Squaring method" and the practical application of the *Dvandvayoga* (Duplex Process) thereto we now proceed to deal with the *Vargamūla*, i.e. the square root on the same kind of simple, easy and straight procedure as in the case of "Straight Division".

### Well-known First Principles

The basic or fundamental rules governing the extraction of the square root, are as follows:

(1) The given number is first arranged in two-digit groups from right to left; and a single digit if any left over at the left-hand-end is counted as a simple group by itself.

(2) The number of digits in the square root will be the same as the number of digit-groups in the given number itself including a single digit if any such there be. Thus 16 will count as one group, 144 as two groups and 1024 as two.

(3) So, if the square root contains n digits, the square must consist of 2n or 2n-1 digits.

(4) And, conversely, if the given number has n digits, the square root will contain $\frac{n}{2}$ or $\frac{n+1}{2}$ digits.

(5) But, in cases of pure decimals, the number of digits in the square is always double that in the square root.

(6) The squares of the first nine natural numbers are 1, 4, 9, 16, 25, 36, 49, 64 and 81. This means:

(*i*) that an exact square cannot end in 2, 3, 7 or 8;

(*ii*) (*a*) that a complete square ending in 1 must have either 1 or 9 mutual complements from 10 as the last digit of its square root;

(*b*) that a square can end in 4, only if the square root ends in 2 or 8 complements;
(*c*) that the ending of a square in 5 or 0 means that its square root too ends in 5 or 0 respectively;
(*d*) that a square ending in 6 must have 4 or 6 complements as the last digit in its square root; and
(*e*) that the termination of an exact square in 9 is possible, only if the square root ends in 3 or 7 complements.

In other words, this may be more briefly formulated thus,

(*a*) that 1, 5, 6 and 0 at the end of a number reproduce themselves as the last digits in its square;

(*b*) that squares of complements from ten have the same last digit. Thus, $1^2$ and $9^2$; $2^2$ and $8^2$; $3^2$ and $7^2$; $4^2$ and $6^2$; $5^2$ and $5^2$; and $0^2$ and $10^2$ have the same ending, namely, 1, 4, 9, 6, 5 and 0 respectively; and

(*c*) that 2, 3, 7 and 8 are out of court altogether, as the final digit of a perfect square.

## READILY AVAILABLE FIRST DATA

Thus, before we begin the straight extracting of a square root by "straight division" method, we start with previous knowledge of (1) the number of digits in the square root and (2) the first digit thereof. Thus,

(1) 74562814 $N = 8 \therefore$ N in square root $= N/2 = 4$; and the first digit thereof is 8.

(2) 963106713. $N = 9 \therefore$ N in the square root $= \frac{N+1}{2} = 5$; and the first digit thereof is 3.

But(3) $(.7104)^2$ *must* contain 8 decimal digits.

(4) $\sqrt{.16} = .4$

(5) $\sqrt{.0064} = .08$

(6) $\sqrt{.000049} = .007$

(7) $\sqrt{.00007(0)} = .008$ etc.

(8) $\sqrt{.00000007} = .0002$ etc.

(9) $\sqrt{.09} = .3$

But (10) $\sqrt{.9} = \sqrt{.90} = .9$ etc.

## Modus Operandi (of Straight Squaring)

The procedure of straight squaring as inculcated in the Vedic *Sūtras* is precisely the same as in straight division but with this difference, namely, that in the former the divisor should be exactly double the first digit of the square root.

*N.B.*: As a single digit can never be more than 9, it follows therefore that, in our method of straight squaring, no divisor above 18 is necessary. We may, of course, *voluntarily choose* to deal with larger numbers; but there is no *need* to do so.

## Initial Chart

We thus start our operation with an initial chart, like the samples given hereunder:

```
(1)     5 : 29 :            (2)      7 : 31 :
     4:     :1   :                4:     :3   :
       ------------                 ------------
       : 2 :      :                 : 2 :      :

(3)    32 : 49 :            (4)     40 : 96 :
   10 :     :7   :               12:     :4   :
       ------------                 ------------
       : 5 :      :                 : 6 :      :

(5)     1 : 63 84 :         (6)      8 : 31 76 :
     2:    :0                     4:    :4     :
       --------------               -------------
       :1 :          :              2 :         :

(7)    44 : 44 44:          (8)     61 :  13 6:
   12 :    :8       :            14 :    :12     :
       --------------               -------------
       :6 :          :              :7 :          :

(9)    73 : 60 : 84 :       (10)     6 : 00 00 01 :
   16:     :9   :    :            4:    :2         :
       ---------------              ---------------
       :8  :    :    :              :2 :           :

(11)   10 : 73 69 42 :      (12)    90: 61 71 74 :
    6:     :1          :         18  :  :9          :
       -----------------             ----------------
       :3  :           :             :9 :           :
```

## Further Procedure

Let us now take a concrete case the extraction of the square root of, say, 119716 and deal with it:

(*i*) In the above given general chart, we have not only put down the single first digit of the square root wanted but also prefixed to the next dividend-digit, the remainder after our subtraction of the square of that first digit of the left-hand-most digit or digit-group of the given number. And we have also set down as our divisor, the exact double of the first digit of the quotient.

```
     11 : 9 7 1 6 :
 6 :    : 2       :
   : 3  :         :
```

(*ii*) Our next gross dividend-unit is thus 29. Without subtracting anything from it, we simply divide the 29 by the divisor 6 and put down the second quotient-digit 4 and the second remainder 5 in their proper places as usual.

```
     11 : 9 7 1 6 :
 6 :    : 2 5     :
   : 3  : 4       :
```

(*iii*) Thus our third gross dividend is 57. From this we subtract 16 the square of the second quotient-digit, get 41 as the actual dividend, divide it by 6 and set down the Q (6) and R (5) in their proper places as usual.

```
     11 : 9 7 1 6 :
 6 :    : 2 5 5   :
   : 3  : 4 6     :
```

(*iv*) Our third gross dividend-unit is 51. From this we subtract the *Dvandva Yoga* (Duplex) (= 48), obtain 3 as the remainder, divide it by 6 and put down the Q (0) and the R (3) in their proper places.

```
      11 : 9716
 6       : 2553
   : 3   : 46.00
```

(*v*) This gives us 36 as our last gross dividend-unit. From this we subtract 36 the *Dvandva Yoga* of the third quotient-digit 6; get 0 as Q and as R. This means that the work has been completed, that the given expression is a perfect square and that 346 is its square root. And that is all.

```
      11 : 9 7 1 6
 6       : 2 5 5 3
   : 3   : 4 6.0 0
     Complete
```

## PROOF OF COMPLETENESS AND CORRECTNESS

(1) A manifest proof of the complete-squareness of the given expression and of the correctness of the square root ascertained is by squaring the latter and finding the square to be exactly the same as the given complete square. Thus,

$$346^2 = 9/24/52/48/36 = 119716 \quad \therefore \text{ Yes.}$$

(2) But this is too mechanical. We obtain a neat and valid proof from the very fact that, if and when the process is continued into the decimal part, all the quotient-digits in the decimal part are found to be zeroes and the remainders too are all zeroes!

## Proof to the Contrary

A number can *not* be an exact square in the following circumstances :

(1) if it ends in 2, 3, 7 or 8;

(2) if it terminates in an odd number of zeroes;

(3) if its last digit is 6 but its penultimate digit is even;

(4) if its last digit is not 6 but its penultimate digit is odd; and

(5) if, even though the number be even, its last two digits (taken together) are not divisible by 4;

And a square root *cannot* be correct if it fails to fulfil any of the requirements hereinabove indicated:

## Examples

Some instructive illustrative examples are given below:

```
(1)    5 : 2 9              :
    4:     : 1 0            :
     : 2  : 3 (complete) :

(2)    32 : 4 9             :
   10:      : 7 4           :
     :  5 : 7 (complete) :

(3)    40 :  9 6            :
   12:      : 4 1           :
     :  6 : 4 (complete) :

(4)    1  :  6 3 8 4        :
    2:      : 0 2 3 6       :
     : 1  :  2 8 (complete) :

(5)   : 55 :  2 0 4 9
   14:      : 6 6 2
     :  7 :  4 3.00 (complete)

(6)   : 69 :  9 0 6 3 2 1
   16:      : 5 11 5 4 1
     :  8     3 6 1. 0 0 0
   ∴ A Complete square

(7)   : 53 :  1  6 3  2 1 4
   14:      : 4 13 6 13 5 7
     :  7 : 2 9  1 .3 1 ...(incomplete)

(8)   : 14 :  0 4 7 5 0 4
    6:      : 5 8 11 13
     :  3 :  7 4 8 ...(complete)

(9)   : 41 :  2 5 4 9 2 9
   12:      : 5 4 5 2 1 0
     :  6 :  4 2 3 .0 0 0  ∴ A complete square

(10)  : 7 :  3 8 9 1 5 4 8 9
    4:     : 3 5 5 13 6 7 4 0
     : 2 :  7 1 8 3 0 0 0 0 0  ∴ An exact square
```

```
(11)   : 25 :  7 4 5 4 7 6
   10 :       : 0 7 4 5 5 1
      :  5 : 0 7 4 .0 0 0  ∴ A perfect square

(12)   : 45 :  3 1 9 8 2 4
   12 :       : 9 9 6 3 1
      :  6 : 7 3 2.0 0 0  ∴ A complete square

(13)   : 74 :   5 7 5  3  1 4 49
   16 :       : 10 9 13 19 12 7 7 4
      :  8  :  6 3  5 7.  0  0 0    ∴ An exact square

(14)   : 52 :  4 4 3 9 0 7 (to 2 places of decimals)
   14 :       : 3 6 4 13 7
      :  7 :  2 4 1 .. 82 ...

(15)   : 73 :  2  1  0  8   (to 3 decimal - places)
   16 :       : 9 12 16 14 15
      :  8 :  5 5. 6  3  3

(16)   : 18 :  1 3 4  5  1   2  (to 3 decimal - places)
    8 :       : 2 5 9 10 16 17
      :  4 : 2 5 8 .4 6  4

(17)   : 13 :  8 7. 0 0 0    (to 3 decimal - places)
    6 :       : 4 6 6 8 8 12
      :  3 :  7  .24  24  ...

(18)   : 75 :  0  1 .7 1  0   0 0 0  (to 4 places)
   16 :       : 11 14 9 9 11 10
      :  8 :  6 .6 1 24     ...

(19)   : .00 :  09 :  2 4 0 1 6 0  (to 6 places)
    6 :        :      : 0 2 6 6
      : .0  :  3  :  0 3 9 7 ...

(20)   : 16. :  7 9 0 0  0 0 0 0 0  (to 6 places)
    8 :        : 0 7 7 14 19
      :  4. : 0 9 7 5  6        ...

(21)   : 27. :  1 3 0 0  0 0 0 0 0 0 0  (to 5 places)
   10 :        : 2 1 9 10 8 16 7
      :  5.  :  2 0 8 64                ...
```

```
(22)    : .74 :   1  0   7   0   0    000 (to 5 places)
    16 :      : 10 5 14 19 14 16
       : .8  :  6 0 8  54       ...
```

```
(23)    : 19. :  7  0   6   4   1     2  8  14 (to 6 places)
     8 :      : 3  5 10 10 15   17 13
       : 4.  :  4 3 9 1   9 0 ...
```

```
(24)    : 27. :  0   0  0   0   0    0 0 0 (to 6 places)
    10 :      : 2  10 9  12 17 10
       : 5.  : 1  9   6  1  5   2  4 ...
```

```
(25)    : .09 :  0   0  4  5  1  3   (to 5 decimal - places)
     6 :      : 0  0  0  4  3  1
       : .3  : 0 0  07    5    ...
```

```
(26)   : .0009 :  1   3  4  0  0  0  0  0    (to 6 places)
     6 :        : 0  1  1  2
       : .03    : 0  2  2  2        ...
```

```
(27)   : .0039 :  3  0  0  0  0  0  0  0    (to eight places)
    12 :        : 3 9 14 20 24 24 24
       : .06    : 2 6 8  9  7  7  ...
```

```
(28)   : .00000083 :  1  0  0  0  0  0 0 0  (to 8 places)
    18 :            : 2  3 11 18 7  0
       : .00   0 9  :  11 5  9        ...
```

```
(29)   : .000092    :   4  0  1  0  0  0  0  0 (to ten places)
    18 :            : 11  6  6 13 15 14 24 32
       : .0  09     :  6  1  2  5  4  38      ...
```

```
(30)   : 2 : .0  7  3  6
     2 :    : 1  2  3  1
       : 1 : 4  4  0  0    ∴ A complete square.
```

Or, taking the first three digits together at the first step, we have:

```
    : 207 :    3  6
 28 :      : 11  1
    : 14   :   4.0   ∴ An exact square.
```

# 35

# Cube Roots of Exact Cubes

### Mainly by Inspection and Argumentation

*(Well-known) First-Principles*

(1) The lowest cubes, i.e. the cubes of the first nine natural numbers are 1, 8, 27, 64, 125, 216, 343, 512 and 729.

(2) Thus, they all have their own distinct endings; and there is no possibility of overlapping or doubt as in the case of squares.

(3) Therefore, the last digit of the cube root of an exact cube is obvious:

(*i*) Cube ends in 1; ∴ cube root ends in 1;
(*ii*) C ends in 2; ∴ C R ends in 8;
(*iii*) C ends in 3; ∴ C R ends in 7;
(*iv*) C ends in 4; ∴ C R ends in 4;
(*v*) C ends in 5; ∴ C R ends in 5;
(*vi*) C ends in 6; ∴ C R ends in 6;
(*vii*) C ends in 7; ∴ C R ends in 3;
(*viii*) C ends in 8; ∴ C R ends in 2; and
(*ix*) C ends in 9; ∴ C R ends in 9;

(4) In other words,

(*i*) 1, 4, 5, 6, 9 and 0 repeat themselves in the cube-ending; and
(*ii*) 2, 3, 7 and 8 have an inter-play of complements from 10.

(5) The number of digits in a cube root is the same as the number of 3-digit groups in the original cube including a single digit or a double-digit group if there is any :

(6) The first digit of the cube-root will always be obvious from the first group in the cube.

(7) Thus, the number of digits, the first digit and the last digit of the cube root of an exact cube are the data with which we start, when we begin the work of extracting the cube root of an exact cube.

### Examples

Let F, L and n be the symbols for the first digit, the last digit and n the number of digits in the cube root of an exact cube.

(1) For 226, 981, F = 6, L = 1 and n = 2
(2) For 4, 269, 813 F = 1, L = 7 and n = 3
(3) For 1, 728, F = 1, L = 2 and n = 2
(4) For 33, 076, 161, F = 3, L = 1 and n = 3
(5) For 83, 453, 453, F = 4, L = 7 and n = 3
(6) For 105, 823, 817, F = 4, L = 3 and n = 3
(7) For 248, 858, 189, F = 6, L = 9 and n = 3
(8) For 1, 548, 816, 893, F = 1, L = 7 and n = 4
(9) For 73, 451, 930, 798, F = 4, L = 2 and n = 4
(10) For 76, 928, 302, 277, F = 4, L = 3 and n = 4
(11) For 6, 700, 108, 456, 013, F = 1, L = 7 and n = 5
(12) For 62, 741, 116, 007, 421, F = 3, L = 1 and n = 5
(13) For 91, 010, 000, 000, 468, F = 4, L = 2 and n = 5 and so on.

### The Chart-Preliminary and Procedure

The procedure is similar to the one adopted by us in "Straight Division" and particularly in the extraction of square roots. The only difference is that our divisor in this context will not be double the first digit of the root but three times the square thereof. As we know the first digit at the very outset, our chart begins functioning as usual as follows :

```
(1)    : 1 :  7 2 8          (2)     : 13 :  8 2 4
     3 :    : 0 1                 12 :      : 5
     ----------------             -----------------
       : 1 :                         : 2  :

(3)    : 73 :  089149        (4)     : 27 :  8 4 1
    48 :      : 901               27 :      : 0
     ----------------             -----------------
       : 4  :                        : 3  :

(5)    : 21 :  400 713       (6)     : 600 :  132 419
    12 :      : 13               192 :       : 88
     ----------------             -----------------
       : 2  :                        : 8   :

(7)    : 79 :  314 502
    48 :      : 15
     ----------------
       : 4  :
```

### ALGEBRAICAL PRINCIPLE UTILISED

Any arithmetical number can be put into its proper algebraical shape as:

$$a + 10b + 100c + 1000d \text{ etc.}$$

Suppose we have to find the cube of a three-digit arithmetical number. Algebraically, we have to expand $(a+10b+100c)^3$. Expanding it accordingly, we have :

$$(a + 10b + 100c)^3 = a^3 + 100b^3 + 1000000c^3 + 30a^2b + 300ab^2 + 300a^2c + 30000ac^2 + 30000b^2c + 300000bc^2 + 6000abc.$$

Removing the powers of ten and putting the result in algebraical form, we note the following:

(1) The units' place is determined by $a^3$.
(2) The tens' place is determined by $3a^2b$.
(3) The hundreds' place is contributed to by $3ab^2 + 3a^2c$
(4) The thousands' place is formed by $b^3 + 6abc$
(5) The ten thousands' place is given by $3ac^2 + 3b^2c$
(6) The lakhs' place is constituted of $3bc^2$; and
(7) The millions' place is formed by $c^3$.

*N.B.*: The number of zeroes in the various coefficients in the Algebraical expansion will prove the correctness of this analysis.

*Note*: If one wishes to proceed in the reverse direction, one may do so; and, for the sake of facility's the letters substituted (for a, b, c, d etc.) may be conveniently put down as L, K, J, H etc.

### THE IMPLICATIONS OF THE PRINCIPLE

This analytical sorting of the various parts of the algebraical expansion into their respective places, gives us the necessary clue for eliminating letter after letter and determining the previous digit. And the whole procedure is really of an argumentational character. Thus,

(*i*) From the units' place, we subtract $a^3$ (or $L^3$); and that eliminates the last digit.

(*ii*) From the ten's place, we subtract $3a^2b$ (or $3L^2K$) and thus eliminate the penultimate digit.

(*iii*) From the hundreds' place, we subtract $3a^2c + 3ab^2$ (or $3L^2J + 3LK^2$) and thereby eliminate the prepenultimate digit.

(*iv*) From the thousands' place, we deduct $b^2 + 6abc$; and so on

*N.B.*: In the case of perfect cubes we have the additional advantage of knowing the last digit too, beforehand.

Some instructive examples are given below:

(1) Extract the cube root of the exact cube 33, 076, 161.

Here F = 3; L = 1; and n = 3

(L) L=1 ∴ $L^3$=1.

∴ Subtracting 1, we have

```
  1   1
33 076 161
         1
----------
33 076 16
```

(K) $3L^2K = 3K$ (ending in 6) ∴ K = 2

Deducting 3K, we have

```
        6
---------
33 0761
```

(J) $3L^2J + 3LK^2 = 3J + 12$ (ending in 1) } ∴ CR = 321

∴ 3J ends in 9 ∴ J = 3

*N.B.*: (1) The last step is really unnecessary as the first digit is known to us from the outset.

(2) Extract the cube root of the exact cube 1728.

Here, F = 1; L = 2 and n = 2 ∴ CR = 12

(3) Extract the cube root of the exact cube 13,824

Here, F = 2; L = 4; and n = 2 ∴ CR = 24

(4) Determine the cube root of the exact cube 83, 453, 453.

Here F = 4; L = 7; and N = 3

(L) L = 7 ∴ $L^3$ = 343.

Subtracting this, we have

(K) $3L^2K = 147K$ (ending in 1) ∴ K = 3

∴ subtracting 441

(J) $3L^2J + 3LK^2 = 147J + 189$ (ending in 7)

∴ 147J ends in 8 ∴ J = 4

```
83 453 453
      343
----------
83 453 11
      441
----------
83 448 7
```

} ∴ CR = 437

*N.B.*: Exactly as in the previous example.

(5) Find out the cube root of the exact cube 84, 604, 519

Here F = 4; L = 9; and n = 3

(L) L = 9 ∴ $L^3$ = 729 ∴ Subtracting this

(K) $3L^2K = 243K$ (ending in 9) ∴ K = 3

Subtracting 729

```
84 604 519
       729
----------
8460379
    729
----------
845965
```

(J) $3L^2J + 3LK^2 = 243J + 243$ (ending in 5) } ∴ CR = 439

∴ 243J ends in 2 ∴ J = 4

*N.B.*: As before.

(6) Extract the cube root of the exact cube 2488 58189

Here F = 6; L = 9; and $n = 3$.

```
 2488 58189
        729
 ----------
 24885  746
        486
 ----------
 248852   6
```

(L) $L = 9 \therefore L^3 = 729 \therefore$ Subtracting this,

(K) $\therefore 3L^2K = 243K$ (ending in 6)

$\therefore K = 2 \therefore$ Deducting 486

(J) $3L^2J + 3LK^2 = 243J + 108$ (ending in 6)

$\therefore$ 243J ends in 6 $\therefore J = 6$

CR = 629.

*N.B.*: Same as before.

(7) Determine the cube root of the exact cube 105823817

Here F = 4; L = 3; and n = 3

```
 105823817
        27
 ---------
 10582379
      189
 ---------
 1058219
```

(L) $L = 3 \therefore L^3 = 27$. Subtracting this

(K) $3L^2K = 27K$ (ending in 9) $\therefore K = 7$

Subtracting 189, we have

(J) $3L^2J + 3LK^2 = 27J + 441$ (ending in 8)

$\therefore J = 4$

$\therefore$ CR = 473

*N.B.*: As before.

(8) Extract the cube root of the exact cube 143 055 667

Here F = 5; L = 3; and $n = 3$

```
 143 055 667
          27
 -----------
 143 055  64
          54
 -----------
 143 055   1
```

(L) $L = 3 \therefore L^3 = 27$. deducting this

(K) $3L^2K = 27K$ (ending in 4) $\therefore K = 2$

Subtracting 54, we have

(J) $3L^2J + 3LK^2 = 27J + 36$ (ending in 1) $\therefore J = 5$ $\therefore$ CR = 523

*N.B.*: Exactly as before.

(9) Find the cube root of the cube 76, 928, 302, 277.

Here F = 4, L = 3; and n = 4. The last 4 digits are 2277

```
 2277
   27
 ----
  225
```

(L) $L = 3 \therefore L^3 = 27$. Subtracting this,

(K) $3L^2K = 27K$ (ending in 5)

$\therefore K = 5 \therefore$ Subtracting 135

```
 135
 ---
  09
```

(J) $3L^2J + 3LK^2 = 27J + 225$ (ending in 9) $\therefore J = 2$ $\therefore$ CR = 4253

*N.B.*: But, if, on principle, we wish to determine the first digit by the same method of successive elimination of the digits, we shall have to make use of another algebraical expansion, namely, of $(L + K + J + H)^3$. And, on analysing its parts as before into the units, the tens, the hundreds etc., we shall find that the 4th step will reveal $3L^2H + 6LKJ + K^3$ as the portion to be deducted. So,

(H) $3L^2H + 6LKJ + K^3 = 27H + 180$

$+125 = 27H + 305$ (ending in 3)

$\therefore H = 4$: and CR = 4253

$$\begin{array}{r} 30:09 \\ 2:79 \\ \hline 27:3 \end{array}$$

(10) Determine the cube root of the cube 11, 345, 123, 223

Here F = 2; L = 7; and n = 4

$$\begin{array}{r} 11,\ 345,\ 123,\ 223 \\ 343 \\ \hline 11\ 345\ 12288 \end{array}$$

(L) $L = 7 \therefore L^3 = 343$. Subtracting it,

(K) $3L^2K + 147K$ (ending in 8)

$\therefore K = 4$. Deducting 588

$$\begin{array}{r} 588 \\ \hline 11\ 345\ 11\ 70 \end{array}$$

(J) $3L^2J + 3LK^2 = 147J + 336$ (ending in 0) $\therefore J = 2$

$\therefore$ Subtracting 630, we have

$$\begin{array}{r} 330 \\ \hline 1134508\ 4 \end{array}$$

(H) $3L^2H + 6LKJ + K^3 = 147H + 336$
$+ 64 = 147H + 400$ (ending in 4)
$\therefore H = 2$

$\therefore$ CR = 2247

*N.B.*: The last part for ascertaining the first digit is really superfluous.

(11) Extract the cube root of the cube 12, 278, 428, 443

Here F = 2; L = 7; and $n = 4$.

$$\begin{array}{r} 12\ 278\ 428\ 443 \\ 343 \\ 12\ 278\ 428\ 10 \end{array}$$

(L) $L = 7 \therefore L^3 = 343 \therefore$ Deducting it

(K) $3L^2K = 147K$ (ending in 0) $\therefore K = 0$.

$$\begin{array}{r} 0 \\ \hline 12\ 278\ 428\ 1 \end{array}$$

(J) $3L^2J + 3LK^2 = 147J + 0$ (ending in 1)
$\therefore J = 3$

$\therefore$ Subtracting 441, we have

$$\begin{array}{r} 441 \\ \hline 12\ 278\ 384 \end{array}$$

(H) $3L^2H + 6LKJ + K^3 = 147H + 0 + 0$

and ends in 4 $\therefore H = 2$

$\therefore$ CR = 2307

*N.B.*: As in the last example.

(12) Find the cube root of the cube 355 045 312 441

Here F = 7; L = 1; n = 4

```
355045 312  441
              1
---------------
355045312  44
           24
---------------
3550453122
       192
---------------
355045293
```

(L) $L = 1 \therefore L^3 = 1$. Deducting it,

(K) $3L^2K = 3K$ (ending in 4)

$\therefore K = 8 \therefore$ Deducting 24.

(J) $3L^2J + 3LK^2 = 3J + 192$ (ending in 2) $\therefore J = 0$

(H) $3L^2H + 6LKJ + K^3 = 3H + 0 + 512$

and ends in 3 $\therefore H = 7$ $\therefore$ <u>CR = 7081</u>

*N.B.*: Exactly as above.

*Note*: The above method is adapted mainly for odd cubes. If the cube be even, ambiguous values may arise at each step and tend to confuse the student's mind.

(13) Determine the cube root of the cube 792 994219216

Here F = 9; L = 6; and n = 4

```
792 994 219  216
             216
----------------
792 994  219 00
             540
----------------
772 994  2136
```

(L) $L = 6 \therefore L^3 = 216$. Deducting this,

(K) $3L^2K = 108K$ (ending in zero)

$\therefore K = 0$ or 5. Which is it to be?

Let us take 5 (a pure gamble)!

(J) $3L^2J + 3LK^2 = 108J + 450$ (ending in 6) $\therefore J = 2$ or 7!

Which should we prefer? Let us accept 2 (another perfect gamble!)

```
          666
-------------
7729961  47
```

(H) $3L^2H + 6LKJ + K^3 = 108H + 360 + 125 = 108H + 485$ (ending in 7)

$\therefore H = 4$ or 9!

Which should we choose? $\therefore CR = 9256$

Let us gamble again and pitch for 9!

Here, however, our previous knowledge of the first digit may come to our rescue and assure us of its being 9. But the other two were pure gambles and would mean $2 \times 2$, i.e. four different possibilities!

### A Better Method

At every step, however, the ambiguity can be removed by proper and cogent argumentation; and this may also prove interesting. And anything intellectual may be welcomed; but it should not become too stiff and abstract; and an ambiguity in such a matter is wholly *undesirable* to put it mildly. A better method is therefore necessary, is available and is given below.

All that has to be done is to go on dividing by 8 until an odd cube emanates, work the sum out and multiply by the proper multiplier thereafter. Thus,

```
8) 792 994 249 216
  8) 99124 281152
  8) 12390535144
     1 548 816 893
```

Here $F = 1$; $L = 7$; and $n = 4$

(L) $L = 7 \therefore L^3 = 343$. Subtracting this,

(K) $3L^2K = 147K$ (ending in 5) $\therefore K = 5$. Deducting 735.

```
1548816893
       343
 154881655
       735
  15488092
       672
```

(J) $3L^2J + 3LK^2 = 147J + 525$ (ending in 2) $\therefore J = 1$ Deducting 672, we have : 1548 742

(H) $3L^2H + 6LKJ + K^3 = 147H + 210 + 125 = 147H + 335$ (ending in 2) $\therefore H = 1$ $\quad\therefore$ The cube root is 1157

and $\therefore$ CR of the original cube $= 8 \times 1157 = 9256$

*N.B.*: Here too, the last step is unnecessary as the first digit is already known to us.

(14) Determine the cube root of the cube 2, 840, 362, 499, 528

Here $F = 1$; $L = 2$; and $n = 5$

(L) $L = 2 \therefore L^3 = 8$

$\therefore$ Deducting this, we have

```
2840 362 499 528
               8
2840 362 499 52
```

(K) $3L^2K = 12K$ and ends in 2

$\therefore$ K = 1 or 6! Let us take 6!

Deducting 12K

$$\begin{array}{r} 72 \\ \hline 2840\ 362\ 4988 \\ 228 \\ \hline 2840\ 362\ 476 \end{array}$$

(J) $3L^2J + 3LK^2 = 12J + 216$ (ending in 8) $\therefore J = 1$ or 6!

Let us take 1!

(H) $3L^2H + 6LKJ + K^3 = 12H + 72 + 216 = 2H + 238$

(ending in 6) $\therefore$ H = 4 or 9: Let us take 4!

(G) We need not bother ourselves about G and the expansion of $(a+b+c+d+e)^3$ and so on.

Obviously G = 1 $\therefore$ <u>CR = 14162</u>

But the middle three digits have been the subject of uncertainty with $2 \times 2 \times 2 = 8$ different possibilities. We must therefore work this case out by the other—the unambiguous—method.

or 8 : 2 8 4 0 3 6 2 4 9 9 5 2 8

:

3 5 5, 0 4 5, 3 1 2, 4 4 1

Here F = 7; L = 1; and n = 4

(L) $L = 1 \therefore L^3 = 1$

$\therefore$ Subtracting this, we have

$$\begin{array}{r} 355\ 045\ 312\ 441 \\ 1 \\ \hline 355\ 045\ 312\ 44 \\ 24 \\ \hline 355\ 045\ 312\ 2 \\ 19\ 2 \\ \hline 355\ 045\ 13 \end{array}$$

(K) $3L^2K = 3K$ (ending in 4)

$\therefore K = 8 \therefore$ Subtracting 24,

(J) $3L^2J + 3LK^2 = 3J + 192$ and ends in 2 $\therefore J = 0$

$\therefore$ Subtracting 192,

(H) $3L^2H + 6LKJ + K^3 = 3H + 0 + 512$

and ends in 3 $\therefore H = 7$ $\therefore$ Cube Root = 7081

*CR of the original expression* = 14162

(15) Find out the 12-digit exact cube whose last four digits are 6741.

Here F = ?; L = 1; and n = 4

(L) L = 1 ∴ $L^3 = 1$ ∴ Subtracting it,

(K) $3L^2K = 3K$ and ends in 4

K = 8 ∴ Deducting 24.

(J) $3L^2J + 3LK^2 = 3J + 19$ and ends in 5 ∴ J = 1

∴Subtracting 195, we have

(H) $3L^2H + 6LKJ + K^3 = 3H + 48 + 512 = 3H + 560$ and ends in 0.

∴ H = 9

...6741
1
...674
24
...65

...195
... 0

∴ *The original Cube Root is* 0181

*N.B.*: As we did not know the first digit beforehand, all the steps were really necessary.

(16) A 13-digit perfect cube begins with 5 and ends with 0541. Find it and its cube root.

Here F = 1; L = 1; and n = 5.

(L) L = 1 ∴ $L^3 = 1$. Deducting it.

(K) $3L^2K = 3K$ ∴ K = 8

∴ Subtracting 24, we have

(J) $3L^2J + 3LK^2 = 3J + 192$

and ends in 3 ∴ J = 7

∴ Deducting 213, we have

(H) $3L^2H + 6LKJ + K^3 = 3H + 336 + 512 = 3H + 848$ and ends in 9

∴ H = 7

(G) And G = 1

...0541
1
054
24
...03
213
...9

∴ $\frac{\text{CR} = 17781}{\text{And the cube}} = 17781^3$

# 36

# Cube Roots (General)

Having explained an interesting method by which the cube roots of exact cubes can be extracted, we now proceed to deal with cubes in general, i.e. whether exact cubes or not. As all numbers cannot be perfect cubes, it stands to reason that there should be a general provision made for all cases. This, of course, there is; and we now take this up.

### First Principles

It goes without saying that all the basic principles explained and utilised in the previous chapter should hold good here too. We need not, therefore, reiterate all that portion of the last chapter but may just, by way of recapitulation, remind ourselves of the conclusions arrived at and the *modus operandi* in question.

### The Sequence of the Various Digits

(1) The first place by $a^3$
(2) The second place by $3a^2b$
(3) The third place by $3ab^2 + 3a^2c$
(4) The fourth place by $6abc + b^3$
(5) The fifth place by $3ac^2 + 3b^2c$
(6) The sixth place by $3bc^2$
(7) The seventh place by $c^3$; the so on.

### The Dividends, Quotients, and Remainders

(1) The first D, Q and R are available at sight.
(2) From the second dividend, no deduction is to be made.
(3) From the third, subtract $3ab^2$
(4) From the fourth, deduct $6abc + b^3$
(5) From the fifth, subtract $3ac^2 + 3b^2$
(6) From the sixth, deduct $3bc^2$

(7) From the seventh, subtract $c^3$; and so on.

**MODUS OPERANDI**

Let us take a concrete example, namely, 258 474 853 and see the *modus operandi* actually at work step by step

```
        258 :  4   74853
  108 :       : 42 100
      : 6  :  3
```

(*a*) Put down 6 and 42 as first Q and first R by mere *Vilokanaṁ* (inspection).

(*b*) The second Gross Dividend is thus 424. Don't subtract anything thereform. Merely divide it by 108 and put down 3 and 100 as $Q_2$ and $R_2$

```
      : 258 :  4      74853
  108 :       : 42 100
      :  6  :  3  7
```

(*c*) So, the third Gross Dividend is 1007. Subtract therefrom $3ab^2$ (i.e. $3 \times 6 \times 3^2$, i.e. 162). The third actual working dividend therefore is $1007 - 162 = 845$. Divide this by 108 and set down 7 and 89 as $Q_3$ and $R_3$

```
      : 258 :  4     7 4853
  108 :       : 42 100 89
      :  6  : 37
```

(*d*) Thus, the fourth gross dividend is 894. Subtract therefrom $6abc+b^3=756$ $+27=783$). So, the fourth actual working dividend is $894 - 783 = 111$. divide this again by 108 and put down 0 and 111 as $Q_4$ and $R_4$.

```
        258 :  4     7  4  853
  108 :       : 42 100 89 111
      :  6  : 37  0
```

(*e*) Our next gross dividend is now 1118. Subtract therefrom$3ac^2+3b^2c=882+189$ Therefore our fifth actual working dividend is 47. Divide it by 108 and put down 0 and 47 as $Q_5$ and $R_5$

```
        258 : 4    7  4   8  5  3
  108 :      : 42 100 89 111 47
      :  6 : 3   7    0  0   0
```

(*f*) Our sixth gross dividend is 475. Subtract therefrom $3bc^2$(= 441). So, our $Q_6$ and $R_6$ now are 0 and 34.

```
        258 : 4     7   4    8    5   3
  108 :      : 42/100/89/111/47/34
      :  6 : 3   7      0  0      0
               (complete cube)
```

(*g*) Our last gross dividend is thus 343. Subtract $C^3$ (= 343) therefrom and set down 0 and 0 as our $Q_7$ and $R_7$

This means that the given number is a perfect cube, that the work of extracting its cube root is over and that the cube root is 637.

*N.B.*: Proof of the correctness of our answer is, of course, readily available in the shape of the fact that $637^3$ is the given number. But this will be too crudely and cruelly laborious. Sufficient proof, however, is afforded by the very fact that, on going into the decimal part of the answers, we find that all the quotients and all the remainders are zeroes.

An *Incomplete* cube is now dealt with as a sample :

Extract the cube root of 417 to 3 places of decimals.

Here the divisor is 147.

| | | | | | | |
|---|---|---|---|---|---|---|
| | 417. : | | 0 | 0 | 0 | 0 |
| 147 : | | : | 74 | 152 | 155 | 163 |
| | : 7. | : | 4 | 7 | 1 | |

(*a*) Here $Q_1$ and $R_1$ are 7 and 74

(*b*) ∴ The second gross dividend is 740. No subtraction is required. ∴ Dividing 740 by 147, we get 4 and 152 as $Q_2$ and $R_2$.

(*c*) ∴ The third gross dividend is 1520. Subtracting $3ab^2 (= 336)$ therefrom, we have 1184 as our third actual working dividend. We divide it by 147 and put down 7 and 155 as our $Q_3$ and $R_3$.

(*d*) Our fourth gross dividend is 1550. We subtract $6abc + b^3$ $(= 1176 + 64 = 1240)$ therefrom, obtain 310 as our fourth actual working dividend, divide it by 147 and set down 1 and 163 as our $Q_4$ and $R_4$.

(*e*) Our next gross dividend is 1630. We subtract $3ac^2 + 3b^2c$ $(= 1029 + 336 = 1365)$ therefrom, get 265 as our fifth actual working dividend, divide it by 147; and so on.

*Note*: The divisor should *not* be too small. Its ultra-smallness will give rise to big quotients sometimes of several digits, the insufficiency of the remainders for the subtractions to be made and other such complications which will confuse the student's mind.

In case the divisor actually happens to be too small, two simple devices are available for surmounting this difficulty.

(*i*) Take the first four or 5 or 6 digits as *one* group and extract the cube root. For example, suppose we have to find out the cube root of 1346, 085. Our chart will then have to be framed thus:

```
363 : 1, 346 :    085
    : 1 331 : 15
    ---------------
    :    11
    ---------------
```

Let us now take an actual concrete example and apply this method for extracting the cube root of 6334625:

```
    : 6334 :   6    2    5
972 : 5832 : 502  166  312
    -----------------------
    :  18  :   5    0    2...
    -----------------------
```

(*a*) $Q_1 = 18$; $R = 502$; and Divisor (D) $= 972$.

(*b*) No subtraction being needed at this point, divide 5026 by 972 and put down 5 and 166 as $Q_2$ and $R_2$.

(*c*) Our third gross dividend (GD) is 1662; subtract $3ab^2 (= 1350)$ from 1662, divide the resultant. Actual dividend (AD), i.e. 312 by 972 and set down 0 and 312 as $Q_3$ and $R_3$.

(*d*) Our next GD is 3125. Subtract $6abc + b^3 (= 0 + 125 = 125)$ from 3125, divide the AD (3000) by 972 and put down 2 and 1056 as $Q_4$ and $R_4$ ; and so on.

Or, Secondly, multiply E (the given expression) by $2^3, 3^3, 4^3$ or $5^3$ etc., as found necessary and most convenient find the cube root and then divide the CR by 2, 3, 4, 5 etc. For instance, instead of taking 3 as the divisor, take $3 \times 4^3 (= 3 \times 64 = 192)$, find the cube root and divide it by 4.

Here again, a concrete example may be worked out by both the methods :

### First Method

(*a*) $Q_1 = 1$; $R_1 = 1$; and $D = 3$

```
  2 :   0 0 0  0
3 :   : 1 4 10 2
  --------------
  :1:   2 6
  --------------
```

(*b*) Now, GD = AD = 10. Divided by 3, it gives 2 and 4 as $Q_2$ and $R_2$.

(*c*) The third GD is 40. From this subtract $3ab^2 (= 12)$. After this subtraction, the AD is 28. Divide this by 3 and put down 6 and 10 as $Q_3$ and $R_3$.

(*d*) The fourth GD is 100. From this deduct $6abc + b^3$ $(= 72 + 8 = 80)$. The AD is 20.

Now, as for dividing this 20 by 3, the directly *apparent* $Q_4$ and $R_4$ are 6 and 2. But the *actual* quotient and remainder are difficult to determine because of the smallness of the divisor and the insufficiency of the remainders for the next subtractions and a good number of trial digits may fail before one can arrive at the correct figures! This is why the other method is to be preferred in such cases. And then the working will be as follows :

Multiplying 2 by $5^3$, we get 250.

| | | | | | |
|---|---|---|---|---|---|
| | 250 : | 0 | 0 | 0 | |
| 108 : | : | 34 | 124 | 196 | 332 |
| | : 6 : | 2 | 9 | 9 ... | |

(*a*) $Q_1$ and $R_1 = 6$ and 34

(*b*) $E_2 = 340$. Dividing this by 108, we have $Q_2 = 2$ and $R_2 = 124$.

(*c*) $3ab^2 = 72$. Deducting this from 1240, we get 1168. Dividing this by 108, $Q_3 = 9$ and $R_3 = 196$

(*d*) $6abc + b^3 = 648 + 8 = 656$ $\therefore$ The Working Dividend $= 1960 - 656 = 1304$. Dividing this by 108, we have $Q_4 = 9$ and $R_4 = 332$ $\therefore$ The CR $= 6.299$—

$\therefore$ Dividing by 5, the actual cube root $= 1.259$

(2) Let us take another concrete example, i.e. $\sqrt[3]{12}$. We multiply 12 by $5^3$ and put 1500 down as the total dividend. And we take the first four digits as one group.

| | | | | | |
|---|---|---|---|---|---|
| | 1500 : | 0 | 0 | 0 | |
| 363 : | : | 169 | 238 | 400 | 339 |
| | : 11 : | 4 | 4 | 7 | ... |

(*a*) Thus $Q_1 = 11$ and $R_1 = 169$

(*b*) Dividing 1690 by 363, we have $Q_2 = 4$ and $R_2 = 238$

(*c*) $3ab^2 = 528$ $\therefore$ Working Dividend $= 2380 - 528 = 1852$

$\therefore$ Dividing it by 363, we have $Q_3 = 4$ and $R_3 = 400$

(*d*) $6abc + b^3 = 1056 + 64 = 1120$. Deducting this from 4000, we get 2880. Dividing this by 363, $Q_4 = 7$ and $R = 339$

$\therefore$ The CR $= 11.447$ etc.

$\therefore$ The cube root of the original $E = 2.289...$

Some more examples may be taken :

(1) (*a*) E = 1728; $Q_1 = 1$;

D = 3; and $R_1 = 0$

```
    1 :  7 2 8
3 :   : 0 1 0
  : 1 :  2 .0 0
  (exact cube)
```

(*b*) 7 divided by 3 gives 2 and 1 as $Q_2$ and $R_2$

(*c*) Third Gross Dividend = 12; $3ab^2 = 12$; ∴ Actual dividend = 0 ∴ $Q_3 = 0$ and $R_3 = 0$

(*d*) Fourth gross dividend = 8; $6abc + b^3 = 0 + 8 = 8$ ∴ Subtracting the latter from the former, $Q_4 = 0$ and $R_4 = 0$

∴ The CR = 12

*N.B.*: The obvious second proof speaks for itself.

(2) (*a*) Here E = 13824;

$Q_1 = 2$; and $R_1 = 5$

D = 12

```
     13 :  8  2 4
12 :    : 5 10 6
   :  2 :  4. 0 0
   (Perfect cube)
```

(*b*) $\frac{58}{12}$ gives $Q_2 = 4$; and $R_2 = 10$

(*c*) Gross Dividend = 102; $3ab^2 = 96$.

∴ Actual dividend = 6. Divided by 12, this gives $Q_3 = 0$ and $R_3 = 6$

(*d*) GD = 64; $6abc + b^3 = 0 + 64$ ∴ A·D = 0 ∴ $Q_4 = 0$ and $R_4 = 0$

∴ <u>The CR is 24.</u>

(3) Here E = 33, 076, 161; $Q_1 = 3$; D = 27; and $R_1 = 6$

(*a*) $Q_1 = 3$; and $R_1 = 6$

```
     33 :  0 7 6 1 6 1
27 :    : 6 6 4 2 0 0
   : 3  :  2 1 0 0 0
   (complete cube).
```

(*b*) GD = AD = 60; Divided by 27, this gives 2 & 6 as $Q_2$ and $R_2$

(*c*) GD is now 67; $3ab^2 = 36$ ∴ A.D = 31. Divided by 27, this gives us 1 and 4 as $Q_3$ and $R_3$.

(*d*) GD is 46; $6abc + b^3 = 36 + 8 = 44$ ∴ AD = 2. And, divided by 27, this gives 0 and 2 as $Q_4$ and $R_4$.

(*e*) GD is 21; $3ac^2 + 3b^2c = 9 + 12 = 21$ ∴ AD = 0; and, divided by 27, this gives 0 and 0 as $Q_5$ and $R_3$.

(*f*) GD is 6; $3bc^2 = 6$ ∴ AD = 0. And, divided by 27, this gives us 0 and 0 as $Q_6$ and $R_6$.

(*g*) GD = 1; $c^3 = 1$; AD = 0. Divided by 27, this gives us 0 and 0 as $Q_7$ and $R_7$ ∴ The CR is 321.

*N.B.*: The second proof is clearly there before us. :

(4) E = 101

```
         101 :     0  0   0   0
    48 :     : 37 82 148 112 40
    ---------------------------
       : 4. :  6  5   9   5
```

(*a*) $Q_1 = 4$; $R_1 = 37$; and D = 48

(*b*) GD = AD = 370; and, divided by 48, this gives us 6 and 82 as $Q_2$ and $R_2$

(*c*) GD = 820; $3ab^2 = 432$ ∴ AD = 388. and, divided by 48, this gives us 5 and 148 as $Q_3$ and $R_3$

(*d*) GD = 1480; and $6abc + b^3 = 720 + 216 = 936$;

∴ AD = 544. And this, divided by 48, this gives us 9 and 112 as $Q_4$ and $R_4$

(*e*) GD = 1120; and $3ac^2 + 3b^2c = 300 + 540 = 840$

∴ AD = 280. And, divided by 48, this gives us 5 and 40 as $Q_5$ and $R_5$; and so on.

(5) E = 29791

(*a*) Here $Q_1 = 3$; and $R_1 = 2$; and D = 27

(*b*) GD = AD = 27; and, divided by 27, this gives us 1 and 0 as $Q_2$ and $R_2$.

```
        29 :  7 9 1
   27 :    :  2 0
   ---------------
      :  3 :  1 .0
   (complete cube)
```

(*c*) GD = 9; and $3ab^2 = 9$; ∴ AD = 0, and divided by 27, this gives us 0 and 0 as $Q_3$ and $R_3$ ∴ The CR is 31.

*N.B.*: The proof is there before us as usual.

(6)The given expression (E) = 83, 453, 453

(*a*) $Q_1 = 4$; $R_1 = 19$; and D = 48.

(*b*) GD = AD = 194. And, divided by 48, this gives us 3 and 50 as $Q_2$ and $R_2$

| | :83 : | 4 | 5 | 3 | 4 | 5 | 3 |
|---|---|---|---|---|---|---|---|
| 48 : | : | 19 | 50 | 61 | 82 | 47 | 34 |
| | : 4 : | 3 | 7. | 0 | 0 | 0 (exact cube) | |

(*c*) GD = 505; and $3ab^2 = 108$ $\therefore$ AD = 397. And, divided by 48, this gives us 7 and 61 as $Q_3$ and $R_3$

(*d*) GD = 613; and $6abc + b^3 = 504 + 27 = 531$ $\therefore$ AD = 82. And, divided by 48, this gives us 0 and 82 as $Q_4$ & $R_4$

(*e*) GD=824; and $3ac^2 + 3b^2c = 588 + 189 = 777$ $\therefore$ AD=47 And this, divided by 48, gives us 0 and 47 as $Q_5$ & $R_5$

(*f*) $GD - 3bc^2 = 475 - 441 = 34$. $\therefore Q_6 = 0$ and $R_6 = 34$

(*g*) GD = 343; and $c^3 = 343$ $\therefore$ AD = 0 $\therefore Q_7 = R_7 = 0$

$\therefore$ The CR is 437.

*N.B.*: The proof is there as usual.

(7) E = 84, 604, 519

(*a*) $Q_1 = 4$;
D = 48;
and $R_1 = 20$

| | 84 : | 6 | 0 | 4 | 5 | 1 | 9 |
|---|---|---|---|---|---|---|---|
| 48 : | : | 20 | 62 | 80 | 129 | 80 | 72 |
| | : 4 : | 3 | 9 . | 0 | 0 | 0 (perfect cube) | |

(*b*) GD=AD=206. And, divided by 48, this gives us 3 and 62 as $Q_2$ and $R_2$

(*c*) GD = 620; and $3ab^2 = 108$ $\therefore$ AD = 512. And, divided by 48, this gives us 9 and 80 as $Q_3$ and $R_3$

(*d*) GD = 804; and $6abc + b^3 = 648 + 27 = 675$ $\therefore$ AD = 129. And, divided by 48, this gives us 0 and 129 as $Q_4$ and $R_4$

(*e*) GD = 1295; and $3ac^2 + 3b^2c = 972 + 243 = 1215$ $\therefore$ AD = 80. And, divided by 48, this gives us 0 and 80 as $Q_5$ and $R_5$

(*f*) GD = 801; and $3bc^2 = 729$ ∴ AD = 72. And, divided by 48, this gives us 0 and 72 as $Q_6$ and $R_6$

(*g*) GD = 729; and $C^3 = 729$ ∴ AD = 0 ∴ $Q_7 = 0$ and $R_7 = 0$

∴ The CR is 439

*N.B.*: The proof is there as usual.

(8) E = 105, 823, 817

(*a*) $Q_1 = 4$;
$R_1 = 41$;
and D = 48

| | : 105 : | 8 2 3 8 1 7 |
|---|---|---|
| 48 : | : | 41 82 90 56 19 2 |
| | : 4 : | 7 3. 0 0 0 (complete cube) |

(*b*) GD = AD = 418. And, divided by 48, this gives us 7 and 82 as $Q_2$ and $R_2$

(*c*) GD = 822; and $3ab^2 = 588$ ∴ AD = 234; and, divided by 48, this gives us 3 and 90 as $Q_3$ and $R_3$

(*d*) GD = 903; and $6abc + b^3 = 504 + 343 = 847$ ∴ AD = 56. And, divided by 48, this gives us zero and 56 as $Q_4$ & $R_4$

(*e*) GD = 568; and $3ac^2 + 3b^2c = 108 + 441 = 549$ ∴ AD = 19. And, divided by 48, this gives us zero and 19 as $Q_5$ & $R_5$

(*f*) GD = 191; and $3bc^2 = 189$ ∴ AD = 2; and, divided by 48, this gives us zero and 2 as $Q_6$ and $R_6$

(*g*) GD = 27; and $C^3 = 27$ ∴ AD = 0 ∴ $Q_7 = 0$ and $R_7 = 0$

∴ The CR = 473

*N. B.*: The proof is there as usual.

(9) E = 143, 055, 667

| | 143 : | 0 5 5 6 6 7 |
|---|---|---|
| 75 : | : | 18 30 20 17 5 2 |
| | : 5 : | 2 3 . 0 0 0 (exact cube) |

(*a*) $Q_1 = 5$; $R_1 = 18$; and D = 75

(*b*) GD = AD = 180; and, divided by 75, this gives us 2 and 30 as $Q_2$ and $R_2$

(*c*) GD = 305; and $3ab^2 = 60$ ∴ AD = 245; and, divided by 75, this gives us 3 and 20 as $Q_3$ and $R_3$

(*d*) GD = 205; and $6abc + b^3 = 180 + 8 = 188$ ∴ AD = 17.

And, divided by 75, this gives us 0 and 17 as $Q_4$ and $R_4$

(*e*) GD = 176; and $3ac^2 + 3b^2c = 135 + 36 = 171$ ∴ AD = 5. And, divided by 75, this gives 0 and 5 as $Q_5$ and $R_5$

(*f*) GD = 56; and $3bc^2 = 54$ ∴ AD = 2; and, divided by 75, this gives 0 and 2 as $Q_6$ and $R_6$.

(*g*) GD = 27; and $c^3 = 27$ ∴ AD = 0 ∴ $Q_7 = 0$ and $R_7 = 0$

∴ The CR is 523

*N.B.*: The proof is there as usual.

(10) E = 248, 858, 189.

| | : 248 : | 8 | 5 | 8 | 1 | 8 | 9 |
|---|---|---|---|---|---|---|---|
| 108 : | : | 32 | 112 | 81 | 162 | 55 | 72 |
| | : 6 : | 2 | 9. | 0 | 0 | 0 (perfect cube) | |

(*a*) $Q_1 = 6$; $R_1 = 32$; and D = 108.

(*b*) GD = AD = 328. And, divided by 108, this gives us 2 and 112 as $Q_2$ and $R_2$

(*c*) GD = 1125; and $3ab^2 = 72$ ∴ AD = 1053; and, divided by 108, this gives us 9 and 81 as $Q_3$ and $R_3$.

(*d*) GD = 818; and $6abc + b^3 = 648 + 8 = 656$ ∴ AD = 162. And, divided by 108, this gives 0 and 162 as $Q_4$ and $R_4$

(*e*) GD = 1621; and $3ac^2 + 3b^2c = 1458 + 108 = 1566$

∴ AD = 55. And, divided by 108, this gives us 0 and 55 as $Q_5$ and $R_5$

(*f*) GD = 558; and $3bc^2 = 486$ ∴ AD = 72; and, divided by 108, this gives us 0 and 72 and $Q_6$ and $R_6$

(*g*) GD = 729; and $c^3 = 729$ ∴ AD = 0; $Q_7 = 0$ and $R_7 = 0$

∴ The CR is 629

*N.B.*: The proof is there as usual.

(11) 11, 345, 123, 223

*Note*: The cube root in this case being of four digits, the method obtained from the expansion of $(a + b + c)$ will naturally not suffice for this purpose; and we shall have to expand $(a+b+c+d)^3$ and vary the above procedure in accordance therewith. This is, of course, perfectly reasonable.

### THE SCHEDULE OF DIGITS

The analytical digit-schedule for $(a+b+c+d)^3$ now stands as follows:

(*a*) First digit (9 zeros) = $a^3$—
(*b*) Second digit (8 zeros) = $3a^2b$ —
(*c*) Third digit (7 zeros) = $3ab^2 + 3a^2c$
(*d*) Fourth digit (6 zeros) = $6abc + b^3 + 3a^2d$
(*e*) Fifth digit (5 zeros) = $6abd + 3ac^2 + 3b^2c$ —
(*f*) Sixth digit (4 zeros) = $6acd + 3bc^2 + 3b^2d$ —
(*g*) Seventh digit (3 zeros) = $6bcd + 3ad^2 + c^3$ —
(*h*) Eighth digit (2 zeros) = $3bd^2 + 3c^2d$ —
(*i*) Ninth digit (1 zero) = $3cd^2$ —
(*j*) Tenth digit (no zero) = $d^3$

### CONSEQUENT SUBTRACTIONS

(1) $Q_1$ and $R_1$ by mere inspection.
(2) $Q_2$ and $R_2$ by simple division without any subtraction whatsoever.

From all the other gross dividends, subtract :

(3) $3ab^2$
(4) $6abc + b^3$
(5) $6abd + 3ac^2 + 3b^2c$
(6) $6acd + 3bc^2 + 3b^2d$
(7) $6bcd + 3ad^2 + c^3$
(8) $3bd^2 + 3c^2d$
(9) $3cd^2$
(10) $d^3$

respectively, in order to obtain the actual working dividend and thence deduce the required Q and R.

*Note*: It will be noted that, just as the equating of d to zero in $(a+b+c+d)^3$ will automatically give us $(a+b+c)^3$ exactly so will the substitution of zero for d in the above schedule give us the necessary schedule for the three-digit cube root.

As we go higher and higher up with the number of digits in the cube root, the same process will be found at work. In other words, there is a general formula for n terms n being any positive integer; and all these are only special applications of that formula with n equal to 2, 3, 4 and so on. We shall take up and explain this general form of the formula at a later stage in the student's progress.

In the meantime, just now, we explain the application of the $(a+b+c+d)^3$ schedule to the present case:

APPLICATION TO THE PRESENT CASE

| | 11 : | 3 | 4 | 5 | 1 | 2 | 3 | 2 | 2 | 3 |
|---|---|---|---|---|---|---|---|---|---|---|
| 12 : | : | 3 | 9 | 22 | 37 | 59 | 76 | 69 | 62 | 34 |
| | : 2 : | 2 | 4 | 7. 0 | 0 | 0 | 0 (exact cube) | | | |

(*a*) $Q_1 = 2$; $R_1 = 3$; and Divisor is 12

(*b*) $GD = AD = 33$. Dividing this by 12, we get $Q_2 = 2$ and $R_2 = 9$

(*c*) $GD = 94$; and $3ab^2 = 24$.

$\therefore AD = 70 \therefore Q_3 = 4$ and $R_3 = 22$

(*d*) $GD = 225$; and $6abc + b^3 = 96 + 8 = 104$; $\therefore$ $AD = 121$

$\therefore$ $Q_4 = 7$ and $R_4 = 37$

(*e*) $GD = 371$; and $6abd + 3ac^2 + 3b^2c = 168 + 96 + 48 = 312$.

$\therefore AD = 50 \therefore$ $Q_5 = 0$ and $R_5 = 59$.

(*f*) $GD = 592$; and $6acd + 3bc^2 + 3b^2d = 336 \quad 96 + 14 = 516$

$D = 76 \therefore Q_6 = 0$ and $R_6 = 76$.

(*g*) $GD = 763$; and $6bcd + 3ad^2 + c^3 = 336 + 294 + 64 = 694$

$\therefore AD = 69 \therefore Q_7 = 0$ and $R_7 = 69$

(*h*) $GD = 692$; and $3bd^2 + 3c^2d = 294 + 336 = 630$

$\therefore AD = 62. \therefore Q_8 = 0$ and $R_8 = 62$

(*i*) $GD = 622$; and $3cd^2 = 588 \therefore AD = 34 \therefore Q_9 = 0$ and $R_9 = 34$

(*j*) $GD = 343$; and $d^3 = 343 \therefore AD = 0 \therefore Q_{10} = 0$ and $R_{10} = 0$

$\therefore$ The cube root is 2247.

*N.B.*: The ocular proof is there, as usual. This is the usual procedure.

There are certain devices, however, which can help us to overcome all such difficulties; and, if and when a simple device is available and can serve our purpose, it is desirable for us to adopt it and minimise the mere mechanical labour involved and not resort to the other ultra-laborious method.

The devices are therefore explained hereunder:

### The First Device

The first device is one which we have already made use of, namely, the reckoning up of the first 4, 5 or 6 digits as a *group* by itself. Thus, in this particular case :

| | | | | | | | |
|---|---|---|---|---|---|---|---|
| | : 11, 345 : | 1 | 2 | 3 | 2 | 2 | 3 |
| 1452 | : 10 648 : | 697 | 1163 | 412 | 363 | 62 | 34 |
| | : 22 : | 4 | 7. | 0 | 0 | 0 (complete cube) | |

(*a*) $Q_1$ (by the same process) is the double-digit number 22; $R_1 = 697$; and $D = 1452$

(*b*) $GD = AD = 6971$; and, divided by 1452, this gives us 4 and 1163 as $Q_2$ and $R_2$

(*c*) $GD = 11632$; and $3ab^2 = 1056$ $\therefore$ $AD = 10576$; and, divided by 1452, this gives us 7 and 412 as $Q_3$ and $R_3$

(*d*) $GD = 4123$; and $6abc + b^3 = 3696 + 64 + 3760$ $\therefore$ $AD = 363$; and, divided by 1452, this gives us 0 and 363 as $Q_4$ and $R_4$

(*e*) $GD = 3632$; and $3ac^2 + 3b^2c = 3234 + 336 = 3570$ $\therefore$ $AD = 62$; and, divided by the same divisor 1452, this gives us zero and 62 as $Q_5$ and $R_5$

(*f*) $GD = 622$; and $3bc^2 = 588$

$\therefore$ $AD = 34$. So $Q_6 = 0$ and $R_6 = 34$

(*g*) $GD = 343$; and $c^3 = 343$ $\therefore$ $AD = 0$ $\therefore$ $Q_7 = 0$ and $R_7 = 0$

$\therefore$ The CR = 2247

*N.B.*: (1) And the proof is there before us, as usual.

(2) By this device, we avoid the complication caused by shifting from $(a+b+c)^3$ to $(a+b+c+d)^3$. It, however, suffers from the drawback that we have first to find the double-digit Q, cube

it and subtract it from the first five-digit portion of the dividend and that all the four operations are of big numbers.

### SECOND DEVICE

This is one in which we do not magnify the first group of digits but substitute $(c + d)$ for c all through and thus have the same $(a + b + c)^3$ procedure available to us. But, after all, it is only a slight alteration of the first device, whereby, instead of a two-digit quotient-item at the commencement, we will be having exactly the same thing at the end.

The real desideratum is a formula which is applicable not only to two-digit, three-digit, or four-digit cube roots, but one which will be automatically and universally applicable. But we shall go into this at a later stage of the student's progress.

In the meantime, a few more illustrative instances are given hereunder for further elucidation of—or at least, the student's practice in, the methods hereinabove explained :

(1) E=12, 278, 428, 443.

Here too we may follow the full procedure or first ascertain the first two-digit portion of the cube root of 12, 278, treat the whole five-digit group as one packet and extract the cube root of the whole given expression in the usual way. The procedure will then be as follows :

(*i*) *Single-digit method* :

```
        12 :   2 7  8   4   2   8  4 4  3
12 :       : 4 6 13 27 22 33 44  3 34
     : 2 :   3 0  7. 0  0  0   0  (perfect cube)
```

(*a*) $Q_1 = 2$; $R_1 = 4$; and $D = 12$

(*b*) $GD = AD = 42 \therefore Q_2 = 3$ and $R_2 = 6$

(c) $GD = 67$; and $3ab^2 = 54$; $\therefore AD = 13 \therefore Q_3 = 0$ and $R_3 = 13$

(*d*) $GD = 138$; and $6abc + b^3 = 0 + 27 = 27 \therefore AD = 111$

$\therefore Q_4 = 7$ and $R_4 = 27$

(*e*) $GD = 274$; and $6abd + 3ac^2 + 3b^2c = 252 + 0 + 0 = 252$;

$\therefore AD = 22 \therefore Q_5 = 0$ and $R_5 = 22$

(*f*) $GD=222$; and $6acd+3bc^2+3bc^2+3b^2d=0+0+189 \therefore AD = 33$

$\therefore Q_6 = 0$ and $R_6 = 33$

(*g*) GD = 338; and $6bcd + 3ad^2 + c^3 = 0 + 294 + 0 = 294$

$\therefore$ AD = 44 $\therefore Q_7 = 0$ and $R_7 = 44$

(*h*) GD = 444; and $3bd^2 + 3c^2d = 441 + 0 = 441$ $\therefore$ AD = 3

$\therefore Q_8 = 0$ and $R_8 = 3$

(*i*) GD = 34; and $3cd^2 = 0$ $\therefore$ AD = 34 $\therefore Q_9 = 0$ and $R_9 = 34$

(*j*) GD = 343; and $d^3 = 343$ $\therefore$ AD = 0, $Q_{10} = 0$ and $R_{10} = 0$

$\therefore$ The CR = 2307

N.B. : The proof is before us, as usual.

(*ii*) *Two-Digit method* :

*Preliminary Work*

| | | |
|---|---|---|
| | : 12 : | 278 |
| 12 : | : | 4 |
| | : 2 : | 3 ... |

$\therefore Q_1$ (of two digits) is 23;

| | | | | | | | | | |
|---|---|---|---|---|---|---|---|---|---|
| | 12278 : | 4 | 2 | 8 | 4 | 4 | 3 | | |
| 1587 : | 13167 : | 111 | 1114 | 33 | 338 | 3 | 34 | | |
| | : 23 : | 0 | 7 . | 0 | 0 | 0 | 0 | | |

(*a*) $Q_1$ (of two digits) = 23; $R_1 = 111$; and D = 1587

(*b*) GD = AD = 1114 $\therefore Q_2 = 0$; and $R_2 = 1114$

(*c*) GD = 11142; and $3ab^2 = 0$ $\therefore$ AD = 11142

$\therefore Q_3 = 7$ & $R_3 = 33$

(*d*) GD = 338; and $6abc + b^3 = 0$ $\therefore$ AD = 338

$\therefore Q_4 = 0$ & $R_4 = 338$

(*e*) GD = 3384; and $3ac^2 + 3b^2c = 3381 + 0 = 3381$

$\therefore$ AD = 3 $\therefore Q_5 = 0$ and $R_5 = 3$

(*f*) GD = 34; and $3bc^2 = 0$ $\therefore$ AD = 34 $\therefore Q_6 = 0$ & $R_6 = 34$

(*g*) GD = 343; and $d^3 = 343$ $\therefore$ AD = 0, $Q_7 = 0$ and $R_7 = 0$

$\therefore$ The CR is 2307

*N.B.* : The proof is before us, as usual.

(2) E = 76, 928, 302, 277.

(*i*) *Single-digit method :*

```
   : 76 :    9  2  8  3  0  2  2  7 7
48 :      : 12 33 44 56 59 44 29 13 2
   : 4  :    2  5  3.  0  0  0  0 (Exact cube)
```

(*a*) $Q_1 = 4$; $R_1 = 12$; and $D = 48$

(*b*) $GD = AD = 129$ $\therefore$ $Q_2 = 2$ and $R_2 = 33$

(*c*) $GD = 332$; and $3ab^2 = 48$ $\therefore$ $AD = 284$ $\therefore$ $Q_3 = 5$ and $R_3 = 44$

(*d*) $GD = 448$; and $6abc + b^3 = 240 + 8 = 248$ $\therefore$ $AD = 200$

$\therefore$ $Q_4 = 3$ and $R_4 = 56$

(*e*) $GD = 563$; and $6abd + 3ac^2 + 3b^2c = 144 + 300 + 60 = 504$

$\therefore$ $AD = 59$ $\therefore$ $Q_5 = 0$ and $R_5 = 59$

(*f*) $GD = 590$; & $6acd + 3bc^2 + 3b^2d = 360 + 150 + 36 = 546$

$\therefore$ $AD = 44$; $\therefore$ $Q_6 = 0$ and $R_6 = 44$

(g) $GD = 442$; and $6bcd + 3ad^2 + c^3 = 180 + 108 + 125 = 413$

$\therefore$ $AD = 29$; $\therefore$ $Q_7 = 0$ and $R_7 = 29$

(h) $GD = 292$; and $3bd^2 + 3c^2d = 54 + 225 = 279$

$\therefore$ $AD = 13$ $\therefore$ $Q_8 = 0$ and $R_8 = 13$

(i) $GD = 137$; and $3cd^2 = 135$ $\therefore$ $AD = 2$ $\therefore$ $Q_9 = 0$ and $R_9 = 2$

(J) $GD = 27$; and $d^3 = 27$ $\therefore$ $AD = 0$ $\therefore$ $Q_{10} = 0$ and $R_{10} = 0$

$\therefore$ The CR = 4253

*N.B.* : Proof as usual.

(*ii*) *Double-digit method* :

or, secondly,

$\therefore$ $Q_1$ (double-digit)=42

```
   : 76 :    9
48 :      : 12 33
   : 4  :  2
```

```
     : 769 28 :    3    0    2    2   7 7
5292 : 740 88 : 2840 1943  404  137  13 2
     :  42    :    5    3    0.   0   0
```

(*a*) $Q_1 = 42$; $R_1 = 2840$; and $D = 5292$

(*b*) GD = AD = 28403 ∴ $Q_2 = 5$; and $R_2 = 1943$

(*c*) GD = 19430; and $3ab^2 = 3150$ ∴ AD = 16280

∴ $Q_3 = 3$; and $R_3 = 404$

(*d*) GD = 4042; and $6abc + b^3 = 3780 + 125 = 3905$ ∴ AD = 137

∴ $Q_4 = 0$ and $R_4 = 137$

(*e*) GD = 1372; and $3ac^2 + 3b^2c = 1134 + 225 = 1359$ ∴ AD = 13

∴ $Q_5 = 0$ and $R_5 = 13$

(*f*) GD = 137; and $3bc^2 = 135$ ∴ AD = 2 ∴ $Q_6 = 0$ and $R_6 = 2$

(*g*) GD = 27; and $c^3 = 27$ ∴ AD = 0; $Q_7 = 0$ and $R_7 = 0$

∴ The CR is 4253

*N.B.* : Exactly as above.

(3) E = 355, 045, 312, 441

(*i*) *Single-Digit method.*

| | | | | | | | | | | |
|---|---|---|---|---|---|---|---|---|---|---|
| | : 355 : | 0 | 4 | 5 | 3 | 1 | 2 | 4 | 4 | 1 |
| 147 : | | : 12 | 120 | 28 | 138 | 39 | 55 | 19 | 2 | 0 |
| | : 7 | : 0 | 8 | 1 | .0 | | | | | |

(*a*) $Q_1 = 7$; $R_1 = 12$; and D = 147

(*b*) GD = AD = 120; ∴ $Q_2 = 0$ and $R_2 = 120$

(*c*) GD = 1204; and $3ab^2 = 0$ ∴ AD = 1204 ∴ $Q_3 = 8$ & $R_3 = 28$

(*d*) GD = 285; and $6abc + b^3 = 0$

∴ AD = 285; $Q_4 = 1$ and $R_4 = 138$

(*e*) GD = 1383 and $6abd + 3ac^2 + 3b^2c = 0 + 1344 + 0 = 1344$

∴ AD = 39 ∴ $Q_5 = 0$ and $R_5 = 39$

(*f*) GD = 391; and $6acd + 3bc^2 + 3b^2d = 336 + 0 + 0 = 336$

∴ AD = 55 ∴ $Q_6 = 0$ and $R_6 = 55$

(*g*) GD = 552; and $6bcd + 3ad^2 + c^3 = 0 + 21 + 512 = 533$

∴ AD = 19 ∴ $Q_7 = 0$ and $R_7 = 19$

(*h*) GD = 194; and $3bd^2 + 3c^2d = 0 + 192 = 192$ ∴ AD = 2

∴ $Q_8 = 0$ and $R_8 = 2$

($i$) GD = 24; and $3cd^2 = 24$ ∴ AD = 0; Q = 0 and R = 0

($j$) GD = 1; and $d^3 = 1$ ∴ AD = 0, Q = 0 and R = 0

∴ The CR is 7081

*N.B.* : As above.

($ii$) *Double-digit method* :

$$\begin{array}{r|l|llllll} & :355045: & 3 & 1 & 2 & 4 & 4 & 1 \\ 14700 & :343000: & 12045 & 2853 & 391 & 40 & 2 & 0 \\ \hline & :\ 70\ : & 8 & 1 & .0 & 0 & & 0 \end{array}$$

($a$) $Q_1$ (of 2 digits) = 70; $R_1 = 12045$; and D = 14700

($b$) GD = AD = 120453; $Q_2 = 8$ and $R_2 = 2853$

($c$) GD = 28531; and $3ab^2 = 13440$ ∴ AD = 15091 ∴ $Q_3 = 1$ and $R_3 = 391$

($d$) GD = 3912 ; & $6abc + b^3 = 3360 + 512 = 3872$ ∴ AD = 40 ∴ $Q_4 = 0$ and $R_4 = 40$

($e$) GD = 404; & $3ac^2 + 3b^2c = 210 + 192 = 402$ ∴ AD = 2 ∴ $Q_5 = 2$ and $R_5 = 2$

($f$) GD = 24; and $3bc^2 = 24$ ∴ AD = 0 ∴ $Q_6 = 0$ and $R_6 = 0$

($g$) GD = 1; and $c^3 = 1$ ∴ AD = 0 ∴ $Q_7 = 0$ and $R_7 = 0$

∴ The CR is 7081

*N.B.* : As above.

(4) E = 792, 994, 249, 216

($i$) *Single-Digit method* :

$$\begin{array}{r|l|lllllllll} & 792: & 9 & 9 & 4 & 2 & 4 & 9 & 2 & 1 & 6 \\ 243: & : & 63 & 153 & 216 & 158 & 199 & 152 & 72 & 56 & 21 \\ \hline & :\ 9\ : & 2 & & 5 & 6 & .0 & 0 & 00 & & \end{array}$$

($a$) $Q_1 = 9$; $R_1 = 63$; and D = 243

($b$) GD = AD = 639 ∴ $Q_2 = 2$; and $R_2 = 153$

($c$) GD = 1539; and $3ab^2 = 108$ ∴ AD = 1431 ∴ $Q_3 = 5$ and $R_3 = 216$

($d$) GD = 2164; and $6abc + b^3 = 540 + 8 = 548$ ∴ AD = 1616

$\therefore$ $Q_4 = 6$ and $R_4 = 158$

(*e*) GD = 1582; and $6abd + 3ac^2 + 3b^2c = 648 + 675 + 60 = 1383$

$\therefore$ AD = 199 $\therefore$ $Q_5 = 0$ and $R_5 = 199$

(*f*) GD = 1994; and $6acd + 3bc^2 + 3b^2d = 1620 + 150 + 72 = 1842$

$\therefore$ AD = 152 $\therefore$ $Q_6 = 0$ and $R_6 = 152$

(*g*) GD = 1529; and $6bcd + 3ad^2 + c^3 = 360 + 972 + 125 = 1457$

$\therefore$ AD = 72 $\therefore$ $Q_7 = 0$ and $R_7 = 72$

(*h*) GD = 722; & $3bd^2 + 3c^2d = 216 + 450 = 666$ $\therefore$ AD = 56

$\therefore$ $Q_8 = 0$ and $R_8 = 56$

(*i*) GD = 561; and $3cd^2 = 540$ $\therefore$ AD = 21 $\therefore$ $Q_9 = 0$ and $R_9 = 21$

(*j*) GD = 216; and $d^3 = 216$ $\therefore$ AD = 0 $\therefore$ $Q_{10} = 0$ and $R_{10} = 0$

$\therefore$ The CR = 9256

*N.B.* : As above.

## DOUBLE-DIGIT METHOD

| | | | |
|---|---|---|---|
| | 792 : | 9 | : |
| 243 : | | : 63 153 | : |
| | : 9 : | 2 | : |

$\therefore$ $Q_1$ (of two digits) is 92 and

$92^3 = 778688$. And D = 25392

| | | | | | | | |
|---|---|---|---|---|---|---|---|
| | 792994 : | 2 | 4 | 9 | 2 | 1 | 6 |
| 25392 : | 778688 : | 14306 | 16102 | 1772 | 1044 | 56 | 21 |
| | : 92 : | 5 | 6 | 0 | 0 0 | 0 | 0 |

(*a*) $Q_1 = 92$; and $R_1 = 14306$

(*b*) GD = AD = 143062 $\therefore$ $Q_2 = 5$; and $R_2 = 16102$

(*c*) GD = 161024; and $3ab^2 = 6900$ $\therefore$ AD = 154124 $\therefore$ $Q_3 = 6$;

and $R_3 = 1772$

(*d*) GD = 17729; and $6abc + b^3 = 16560 + 125 = 16685$

$\therefore$ AD = 1044 $\therefore$ $Q_4 = 0$ and $R_4 = 1044$

(*e*) GD = 10442; and $3ac^2 + 3b^2c = 9936 + 450 = 10386$

$\therefore$ AD = 56 $\therefore$ $Q_5 = 0$ and $R_5 = 56$

(*f*) GD = 561; and $3bc^2 = 540$ $\therefore$ AD = 21 $\therefore$ $Q_6 = 0$ & $R_6 = 21$

(*g*) GD = 216; and $c^3 = 216$ ∴ AD = 0 ∴ $Q_7 = 0$ and $R_7 = 0$

∴ The CR is 9256

*N.B.*: As above.

*Note*: It must be admitted that, although the double-digit method uses the $(a+b+c)^3$ schedule and avoids the $(a+b+c+d)^3$ one, yet it necessitates the division, multiplication and subtraction of big numbers and is therefore likely to cause more mistakes. It is obviously better and safer to use the $(a+b+c+d)^3$ and deal with smaller numbers.

In this particular case, however, as the given number terminates in an even number and is manifestly divisible by 8 and perhaps 64 or even 512, we can in this case utilise a third method which has already been explained in the immediately preceding chapter, namely, divide out by 8 and its powers and thus diminish the magnitude of the given number. We now briefly remind the student of that method.

### Third Method

```
8 : 7  9 2  994  249  216
    8 : 9 9  124  281  152
       8 : 12  390  535  144
             1,  548,  816,  893
```

```
   : 1 :   5   4   8   8   1   6   8   9   3
3 :    :   0   2   6  16  36  55  74  76  34
   : 1 :   1   5   7   0   0   0   0
```

(*a*) $Q_1 = 1$ and $R_1 = 0$

(*b*) GD = AD = 05 ∴ $Q_2 = 1$ and $R_2 = 2$

(*c*) GD = 24; and $3ab^2 = 3$ ∴ AD = 21 ∴ $Q_3 = 5$ and $R_3 = 6$

(*d*) GD = 68; and $6abc + b^3 = 30 + 1 = 31$ ∴ AD = 37 ∴ $Q_4 = 7$; and $R_4 = 16$

(*e*) GD = 168; and $6abd + 3ac^2 + 3b^2c = 42 + 75 + 15 = 132$

∴ AD = 36 ∴ $Q_5 = 0$ and $R_5 = 36$

(*f*) GD = 361; and $6acd + 3bc^2 + 3b^2d = 210 + 75 + 21 = 306$

∴ AD = 55 ∴ $Q_6 = 0$ and $R_6 = 55$

(*g*) GD = 556; and $6bcd + 3ad^2 + c^3 = 210 + 147 + 125 = 482$

$\therefore$ AD = 74 $\therefore$ $Q_7 = 0$ and $R_7 = 74$

(*h*) GD = 748; and $3bd^2 + 3c^2d = 147 + 525 = 672$ $\therefore$ AD = 76

$\therefore$ $Q_8 = 0$ and $R_8 = 76$

(*i*) GD = 769; and $3cd^2 = 735$ $\therefore$ AD = 34 $\therefore$ $Q_9 = 0$; and $R_9 = 34$

(*j*) GD = 343; and $d^3 = 343$ $\therefore$ AD = 0 $\therefore$ $Q_{10} = 0$; and $R_{10} = 0$

$\therefore$ The C.R. (of the sub-multiple)=1157

$\therefore$ The C.R. of the given number=9256

Or, Fourthly, the derived sub-multiple may be dealt with (by the two-digit method) thus,

```
       1548 :      8    1    6    8    9    3
 363 : 1331 :    217  363  265  221   76   34
     :   11 :      5    7  0  0  0
```

(*a*) $Q_1 = 11$; $R_1 = 217$; and D = 363

(*b*) GD = AD = 2178 $\therefore$ $Q_2 = 5$ and $R_2 = 363$

(*c*) GD = 3631; and $3ab^2 = 825$ $\therefore$ AD = 2806

$\therefore$ $Q_3 = 7$ & $R_3 = 265$

(*d*) GD = 2656; and $6abc + b^3 = 2310 + 125 = 2435$ $\therefore$ AD = 221

$\therefore$ $Q_4 = 0$ and $R_4 = 221$

(*e*) GD = 2218; and $3ac^2 + 3b^2c = 1617 + 525 = 2142$ $\therefore$ AD = 76

$\therefore$ $Q_5 = 0$ and $R_5 = 76$

(*f*) GD = 769; and $3bc^2 = 735$ $\therefore$ AD = 34 $\therefore$ $Q_6 = 0$ & $R_6 = 34$

(*g*) GD = 343; and $c^3 = 343$ $\therefore$ AD = 0 $\therefore$ $Q_7 = 0$ and $R_7 = 0$

$\therefore$ The cube root of the sub-multiple is 1157

$\therefore$ The CR of the original number=9256

*N.B.* : As above.

(5) E =

```
      : 2, 840, 362, 499, 528
    8 :
      :   355, 045, 312, 441,
  355 :     0    4
147 :     : 12  120
    : 7 :     0  8 1
```

This very number having already been dealt with in example 3 of this very series, in this very portion of this subject, we need not work it all out again. Suffice it to say that, because, 7081 is the cube root of this derived sub-multiple.

∴ The C.R. of the original number is 14162

*Note* : All these methods, however, fall in one way or another, short of the Vedic ideal of ease and simplicity. And the general formula which is simultaneously applicable to all cases and free from all flaws is yet ahead. But we shall go into these matters later.

# 37

# Pythagoras' Theorem etc.

Modern Historical Research has revealed—and all the modern historians of mathematics have placed on record the historical fact that the so-called "Pythogoras' Theorem" was known to the ancient Indians long long before the time of Pythagoras and that, just as although the Arabs introduced the Indian system of numerals into the Western world and distinctly spoke of them as the "*Hindu*" numerals, yet, the European importers thereof undiscerningly dubbed them as the Arabic numerals and they are still described everywhere under that designation; similarly exactly it has happened that, although Pythagoras introduced his theorem to the Western mathematical and scientific world long long afterwards, yet that Theorem continues to be known as Pythagoras' Theorem!

This theorem is constantly in requisition in a vast lot of practical mathematical work and is acknowledged by all as practically the real foundational pre-requisite for Higher Geometry including Solid Geometry, Trigonometry both Plane and Spherical, Analytical Conics, Calculus Differential and Integral and various other branches of mathematics, Pure and Applied. Yet, the proof of such a basically important and fundamental theorem as presented, straight from the earliest sources known to the scientific world, by Euclid etc., and as still expounded by the most eminent modern geometricians all the world over is ultranotorious for its tedious length, its clumsy cumbrousness etc., and for the time and toil entailed on it!

There are several Vedic proofs, every one of which is much simpler than Euclids' etc. A few of them are shown below:

### First Proof

Here, the square AE = the square KG + the four congruent right-angled triangles all around it.

Their areas are $c^2$, $(b-a)^2$ and $4 \times \frac{1}{2}ab$ respectively.

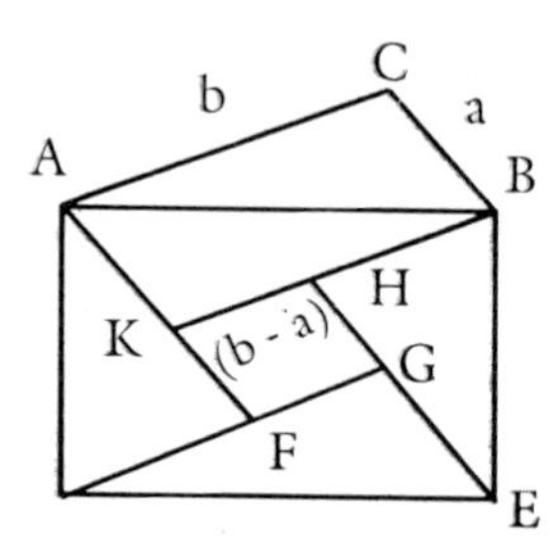


$\therefore c^2 = a^2 - 2ab + b^2 + 4(\frac{1}{2}ab) = a^2 + b^2$

Q.E.D.

### Second Proof

*Construction*

CD = AB = m; and DE = BC = n.

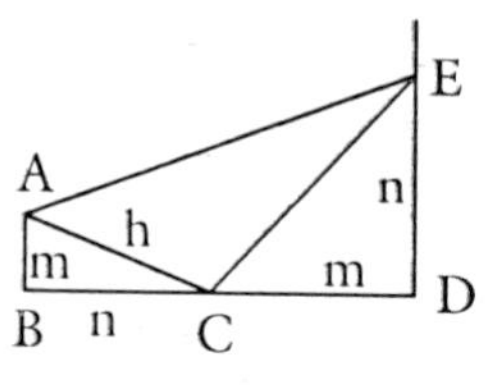


$\therefore$ ABC and CDE are Congruent; and ACE is right-angled Isosceles. Now, the trapezium ABDE = ABC + CDE + ACE

$$\therefore \tfrac{1}{2}mn + \tfrac{1}{2}h^2 + \tfrac{1}{2}mn = \tfrac{1}{2}(m+n)\times(m+n)$$

$$= \tfrac{1}{2}m^2 + mn + \tfrac{1}{2}n^2 \therefore \tfrac{1}{2}h^2$$

$$= \tfrac{1}{2}m^2 + \tfrac{1}{2}n^2 \therefore h^2 = m^2 + n^2$$

Q.E.D.

*N.B.*: Here we have utilised the proposition that the area of trapezium $= \frac{1}{2}$ the altitude $\times$ the sum of the parallel sides.

### Third Proof

Here, AE = BF = CG= DH = m and EB = FC = GD = HA = *n*

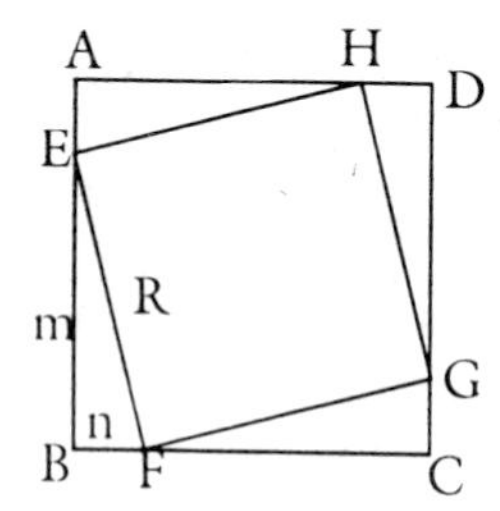


Now, the square AC = the square EG + the 4 congruent right-angled triangles around it

$$\therefore h^2 + 4(\tfrac{1}{2}mn) = (m+n)^2$$

$$= m^2 + 2mn + n^2 \therefore h^2 = m^2 + n^2$$

Q.ED.

### Fourth Proof

The proposition to be used here is that the areas of similar triangles are proportional to the squares on the homologous sides. Here, BD is $\perp$ to AC

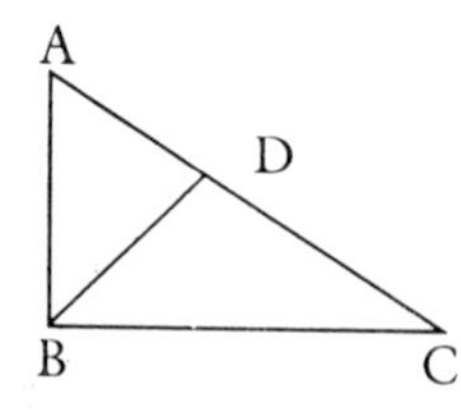


$\therefore$ Thee triangles ABC, ABD and BCA are similar.
$\therefore$ As between (1) the first two triangles and (2) the first and third ones,

$$\frac{AB^2}{AC^2} = \frac{ADB}{ABC}; \text{ and } \frac{BC^2}{AC^2} = \frac{BCD}{ABC}$$

$\therefore$ By addition, $\frac{AB^2 + BC^2}{AC^2} = \frac{ADB + BCD}{ABC} = \frac{ABC}{ABC} = 1$

$\therefore AB^2 + BC^2 = AC^2$ Q.E.D.

### FIFTH PROOF

This proof is from Co-ordinate Geometry. And, as modern Conics and Co-ordinate Geometry and even Trigonometry take their genesis from Pythagoras' Theorem, this process would be objectionable to the modern mathematician. But, as the Vedic *Sūtras* establish their Conics and Co-ordinate Geometry and even their Calculus, at a very early stage, on the basis of first principles and not from Pythagoras' Theorem (*sic*), no such objection can hold good in this case.

The proposition is the one which gives us the distance between two points whose co-ordinates have been given. Let the points be A and B and let their co-ordinates be (a, 0) and (0, b) respectively.

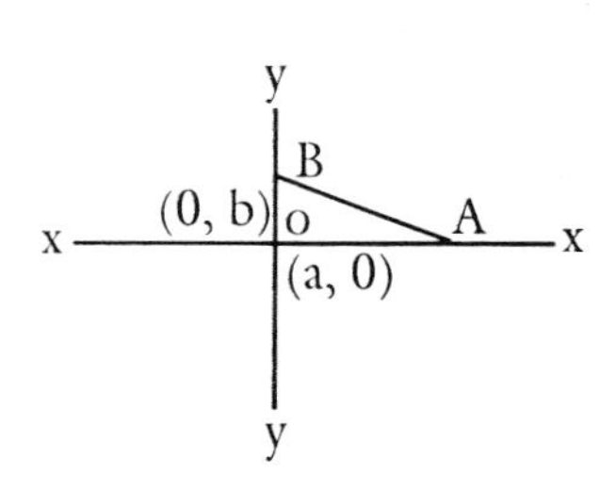


Then, $BA = \sqrt{(a-0)^2 + (0-b)^2} = \sqrt{a^2 + b^2}$ $\therefore BA^2 = a^2 + b^2$ Q.E.D.

*Note:* The Apoilonius' Theorem, Ptolemy's Theorem and a vast lot of other Theorems are similarly easy to solve with the aid of the Vedic *Sūtras*. We shall not, however, go into an elaborate description thereof except of the Apollonius' Theorem just now but shall reserve them for a higher stage in the student's studies.

# 38

# Apollonius' Theorem

Apollonius' Theorem (*sic*) is practically a direct and elementary corollary or offshoot from Pythagoras' Theorem. But, unfortunately, its proof too has been beset with the usual flaw of irksome and needless length and laboriousness.

The usual proof is well-known and need not be reiterated here. We need only point out the Vedic method and leave it to the discerning reader to do all the contrasting for himself. And, after all, that is the best way. Isn't it?

Well, in any triangle ABC, if D be the mid-point of BC, then $AB^2+AC^2 = 2(AD^2+BD^2)$. This is the proposition which goes by the name of Apollonius' Theorem and has now to be proved by us by a far simpler and easier method than the one employed by him.

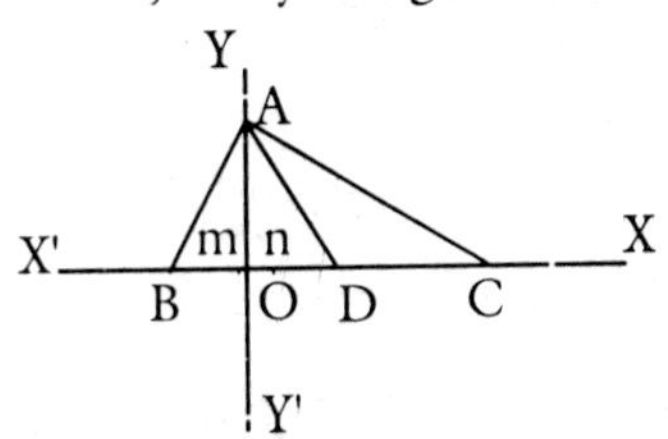


Let AO be the perpendicular from A on BC; let $XOX'$ and $YOY'$ be the axes of co-ordinates; and let BO, OD and OA be m, n and p respectively.

$\therefore DB = DC = m + n$

$\therefore AB^2 + AC^2 = (p^2 + m^2) + (m^2 + 4mn + 4n^2 + p^2)$

$= 2p^2 + 2m^2 + 4mn + 4n^2$

and $2\ (AD^2 + BD^2) = 2[(p^2 + n^2) + (m^2 + 2mn + n^2)]$

$= 2p^2 + 2m^2 + 4mn + 4n^2$

$\therefore AB^2 + AC^2 = 2(AD^2 + DB^2)$ Q.E.D.

*Note* : We faintly remember to have read a proof of Apollonius' Theorem on these lines in some publication of Prof. S. L. Loney; but we are not sure. However that may be, this proof by means of Co-ordinate Geometry was well-known to the ancient Indian mathematicians and specifically finds its place in the Vedic *Sūtras*. And all the Geometrical Theorems about the concurrency of certain straight lines and about the orthocentre, the circumcentre, the in-centre, the ex-centre, the centroid, the Nine-Points-circle etc., can be similarly proved very simply and very easily by means of Co-ordinate Geometry.

We shall go into details of these theorems and their Vedic proofs later on; but just now we would merely point out that, like the "Arabic numerals" and "Pythogoras' Theorem", the "Cartesian" co-ordinates are a historical misnomer, no more, and no less.

# 39

# Analytical Conics

Analytical Conics is a very important branch of mathematical study and has a direct bearing on practical work in various branches of mathematics. It is in the fitness of things, therefore, that Analytical Conics should find an important and predominating position for itself in the Vedic system of mathematics as it actually does.

A few instances relating to certain very necessary and very important points connected with Analytical Conics are therefore given hereunder merely by way, of illustration.

## 1. Equation to the Straight Line

For finding the Equation of the straight line passing through two points whose co-ordinates are given:

say, (9, 17) and (7, –2).

The current method tells us to work as follows:

Take the general equation $y=mx+c$.

Substituting the above values therein,

we have: $9m+c=17$; and $7m+c=-2$.

Solving this simultaneous equation in m and c, we have;

$9m+c=17$

$7m+c=-2$

$\therefore 2m = 19 \therefore m = 9\frac{1}{2}$

Substituting this value of m in either of the above two equations we have, $66\frac{1}{2}+c=-2 \therefore c=-68\frac{1}{2}$. Substituting these values of m and c in the Original General Equation ($y=mx+c$), we get $y = 9\frac{1}{2} \times -68\frac{1}{2} \therefore$ Removing fractions, we have $2y=19x-137$. And then, by transposition, we say, 19x –

$2y = 137$. But this method is decidedly too long and cumbrous and especially for such a petty matter!

And the *Second Current Method* which uses the formula $y - y_1 = \frac{y_2 - y_1}{x_2 - x_1}(x - x_1)$ is equally cumbrous and confusing. It ultimately amounts to the right thing; but it does not make it *clear* and requires several more steps of working!

But the Vedic *at-sight, one-line, mental method by the Parāvartya Sūtra* enables us to write the answer mechanically down by a *mere casual look* at the given co-ordinates. And it is as follows:

The General Equation to the straight line in its final form is ..x– ..y = .. where the co-efficients of x and y on the left hand side and the independent on the right hand side have to be filled in. The *Sūtra* tells us to do this very simply by:

(*i*) putting the difference of the y-co-ordinates as the x-co-efficient and *vice versa*; and

(*ii*) evaluating the independent term on that basis.

For example, in the above example, the co-ordinates are: (9,17) and (7, –2).

(*i*) so our x-coefficient is 17–(–2)=19

(*ii*) and our y-coefficient is 9–7=2.

Thus we have 19 × –2y as our L.H.S. straightaway.

(*iii*) As for the absolute term on the R.H.S., as the straight line in question passes through the two given points, the substitution of the original co-ordinates of each of the points must give us the independent term.

So, the substitution of the values 9 and 17 in the L.H.S. of the equation gives us 19 × 9–2 × 17=171–34=137!

Or substituting the values 7 and –2 therein, we get 19 × 7 – 2x – 2 = 133 + 4 = 137! And that is additional confirmation and verification!

But this is not all. There is also a third method by which we can obtain the independent term on the R.H.S. And this is with the help of the rule about *Ādyam, Antyam* and *Mādhyam*, i.e. bc–ad, i.e. the product of the means *minus* the product of the extremes! So, we have 17 × 7 – 9 × –2=119 + 18= 137! And this is still further additional confirmation and verification!

So, the equation is : –19x – 2y = 137 which is exactly the same as the one obtained by the elaborate current method with its simultaneous equations transpositions and substitutions etc., *galore*! And all the work involved in the Vedic method has been purely *mental, short, simple* and *easy*!

A few more instances are given below:

(1) Points (9,7) and (–7,2)

∴ The equation to the straight line joining them is :

$5x - 16y = -67$

(2) (10,5) and (18,9) ∴ $x=2y$ (by *Vilokanaṁ* too)

(3) (10, 8) and (9, 7) ∴ $x - y = 2$ (by *Vilokanaṁ* too)

(4) (4, 7) and (3, 5) ∴ $2x - y = 1$

(5) (9, 7) and (5, 2) ∴ $5x - 4y = 17$

(6) (9, 7) and (4, – 6) ∴ $13x - 5y = 82$

(7) (17, 9) and (13, – 8) ∴ $17x - 4y = 253$

(8) (15, 16) and (9, –3) ∴ $19x - 6y = 189$

(9) (a, b) and (c, d)

∴ $x(b-d) - y(a-c) = bc - ad$

## II The General Equation and Two Straight Lines

The question frequently arises:— when does the general equation to a straight line represent two straight lines?

Say, $12x^2 + 7xy - 10y^2 + 13x + 45y - 35 = 0$

Expounding the current conventional method, Prof. S.L. Loney the world-reputed present-day authority on the subject devotes about 15 lines not of argument or of explanation but of hard solid working in section 119, example 1 on page 97 of his "Elements of Co-ordinate Geometry", to his model solution of this problem as follows:

$$\text{Here } a = 12, h = \frac{7}{2}, b = -10, g = \frac{131}{2}, f = \frac{45}{2} \text{ and } c = -35$$

$$\therefore abc + 2fgh - af^2 - bg^2 - ch^2$$

$$= 12 \times (-10) \times (-35) + 2 \times \frac{45}{2} \times \frac{13}{2} \times \frac{7}{2} - 12\left(\frac{45}{2}\right)^2$$

$$-(-10)\left(\frac{13}{2}\right)^2 - (-35)\left(\frac{7}{2}\right)^2$$

$$= 4200 + \frac{4095}{4} - 6075 + \frac{1690}{4} + \frac{1715}{4} = -1875 + \frac{7500}{4} = 0$$

∴ The equation represents two straight lines.

Solving it for x, we have:

$$x^2 + x\frac{7y+13}{12} + \left(\frac{7y+13}{24}\right) = \frac{10y^2 - 45y + 35}{12} + \left(\frac{7y+13}{24}\right)^2$$

$$= \left(\frac{23y - 43}{24}\right)^2$$

$$\therefore\ x + \frac{7y + 13}{24} = \mp\frac{23y - 43}{24}$$

$$\therefore\ x = \frac{2y - 7}{3} \text{ or } \frac{-5y + 5}{4}$$

$\therefore$ The two straight lines are $3x = 2y - 7$ and $4x = -5y + 5$

*Note*: The only comment possible for us to make hereon is that the very *magnitude of the numbers* involved in the fractions, their multiplications, subtractions etc., *ad infinitum* is appalling and panic-striking and that it is such asinine burden-bearing labour that is responsible for, not as a justification for, but, at any rate, an extenuation for the inveterate hatred which many youngsters develop for mathematics as such and for their mathematics-teachers as well!

We make no reflection on Prof. Loney. He is perhaps one of the best, the finest and the most painstaking of mathematicians and is very highly esteemed by us as such and for his beautiful publications which are standard authorities on the various subjects which they deal with. It is the *system* that we are blaming, or, at any rate, comparing and contrasting with the Vedic system.

Now, the Vedic method herein is one by which we can immediately, apply the *Ūrdhva Sūtra* the *Ādyam Ādyena Sūtra* and the *Lopana Sthāpana Sūtra* and by merely looking at the frightful but really harmless quadratic before us, readily by *mere mental arithmetic,* write down the answer to this question and say:— "Yes, and the straight lines are $3x - 2y + 7 = 0$ and $4x + 5y - 5 = 0$." How exactly we do this by mental arithmetic, we proceed to explain presently.

## The Vedic Method

(1) by the *Ūrdhva Tiryak*, the *Lopana Sthāpana* and the *Ādyam Ādyena Sūtras* as explained in some of the earliest chapters, we have mentally : $12x^2 + 7xy - 10y^2 = (3x - 2y)(4x+5y)$ and we find 7 and $-5$ to be the absolute terms of the two factors. We thus get $(3x-2y+7)=0$ and $(4x+5y-5)=0$ as the two straight lines represented by the given equation. And that is all there is to it.

$$\begin{array}{r} 3x - 2y + 7 \\ 4x + 5y - 5 \\ \hline 12x^2 - 7xy - 10y^2 + 13x + 45y - 35 \end{array}$$

## The Hyperbolas and the Asymptotes

Dealing with the same principle and adopting the same procedure in connection with the Hyperbola, the Conjugate Hyperbola and the

Asymptotes, in articles 324 and 325 on pages 293 and 294 of his "Elements of Co-ordinate Geometry", Prof. S.L. Loney devotes 27+14(=41) lines in all to the problem and concludes by saying:

"As $3x^2-5xy-2y^2+5x+11y+c=0$ is the equation of the Asymptotes,

$$\therefore 3(-2)\,c+2\frac{5}{2}\cdot\frac{11}{2}\left(\frac{-5}{2}\right)-3\left(\frac{11}{2}\right)^2-(-2)\left(\frac{5}{2}\right)^2-c\left(\frac{-5}{2}\right)^2=0$$

$\therefore c=-12$

$\therefore$ The Equation to the Asymptotes is $3x^2-5xy-2y^2+5x+11y-12=0$

And consequently the Equation to the Conjugate Hyperbola is $3x^2-5xy-2y^2+5x+11y-16=0$".

Well, all this is not so terrific-looking, because of the very simple fact that all the working according to Art. 116 on pages 95 etc. has been taken for granted and done "out of Court" or in private, so to speak. But even then the substitution of the values of a, b, c, f, g, and h in the Discriminant to the General Equation and so on is, from the Vedic standpoint, wholly supererogatory toil and therefore to be avoided.

$$\begin{array}{r} 3x+y-4 \\ x-2y+3 \\ \hline 3x^2-5xy-2y^2+5x+11y-12 \end{array}$$

By the Vedic method, however, we use the same *Lopana Sthāpana*, the *Ūrdhva Tiryak* and the *Ādyam Ādyena Sūtras;* we first get mentally $3x+y$ and $x-2y$ and then $-4$ and 3 as the only possibilities in the case; and as this gives us $-12$ in the product, we get this product = 0 as the Equation to the Asymptotes; and, as the Conjugate Hyperbola is at the same distance in the opposite direction from the Asymptotes, we put down the same equation with only $-16$ instead of $-8$ as the required Equation to the Conjugate Hyperbola and have not got to bother about the complexities of the Discriminants, the inevitable substitutions and all the rest of it! And that is all.

A few more illustrative instances will not be out of place:

(1) $$\left.\begin{array}{l} 8x^2+10xy-3y^2-2x+4y-2=0 \\ \therefore (2x+3y)(4x-y)-2x+4y-2=0 \end{array}\right\}$$

$$\begin{array}{r} 4x-\ \ y+1 \\ 2x+3y-1 \\ \hline 8x^2+10xy-3y^2-2x+4y-1=0 \end{array}$$

$\therefore$ The Equation to the Asymptotes is $8x^2+10xy-3y^2-2x+4y=1$;

and the Equation to the Conjugate Hyperbola is

$8x^2+10xy-3y^2-2x+4y=0$

(2) $y^2 - xy - 2x^2 - 5y + x - 6 = 0 \therefore y + x - 2$
$y - 2x - 3$

$y^2 - xy - 2x^2 + x - 5y + 6 = 0$

$\therefore$ The Asymptotes are $(y + x - 2)(y - 2x - 3) = 0$
And the Conjugate Hyperbola is $y^2 - xy - 2x^2 + x - 5y + 18 = 0$

(3) $55x^2 - 120xy + 20y^2 + 64x - 48y = 0 \therefore 11x - 2y + 4$
$5x - 10y + 4$

$\therefore 55x^2 - 120xy + 20y^2 + 64x - 48y + 16 = 0$

$\therefore$ This is the Equation to the Asymptotes; and the Equation to the Conjugate Hyperbola is $55x^2 - 120xy + 20y^2 + 64x - 48y + 32 = 0$

(4) $12x^2 - 23xy + 10y^2 - 25x + 26y - 14 \therefore 4x - 5y - 3$
$3x - 2y - 4$

The Asymptotes are : $12x^2 - 23xy + 10y^2 - 25x + 26y + 12 = 0$

And the Conjugate Hyperbola is $12x^2 - 23xy + 10y^2 - 25x + 26y + 38 = 0$

(5) $6x^2 - 5xy - 6y^2 + 14x + 5y + 4 \therefore 2x - 3y + 4$
$3x + 2y + 1$

$\therefore$ Independent term $= 4$

$\therefore$ Two straight lines.

# 40

# Miscellaneous Matters

There are also various subjects of a miscellaneous character which are of great practical interest not only to mathematicians and statisticians as such but also to ordinary people in the ordinary course of their various businesses etc., to which the modern system of accounting etc., does scant justice and in which the Vedic *Sūtras* can be very helpful to them. We do not propose to deal with them now, except to *name* a few of *them:*

(1) Subtractions;
(2) Mixed additions and subtractions;
(3) Compound additions and subtractions;
(4) Additions of Vulgar Fractions etc.;
(5) Comparison of Fractions;
(6) Simple and compound practice without taking Aliquot parts etc.;
(7) Decimal operations in all decimal work;
(8) Ratios, Proportions, Percentages, Averages etc.;
(9) Interest, Annuities, Discount etc.;
(10) The Centre of Gravity of Hemispheres etc.;
(11) Transformation of Equations; and
(12) Dynamics, Statistics, Hydrostatics, Pneumatics etc., Applied Mechanics etc.

*N.B.*: There are some other subjects, however, of an important character which need detailed attention but which owing to their being more appropriate at a later stage we do not now propose to deal with but which, at the same time, in view of their practical importance and their absorbingly interesting character, *do* require a brief description. We deal with them, therefore, briefly hereunder.

SOLIDS, TRIGONOMETRY, ASTRONOMY ETC.

In Solid Geometry, Plane Trigonometry, Spherical Trigonometry and

Astronomy too, there are similarly huge masses of Vedic material calculated to lighten the mathematics students' burden. We shall not, however, go here and now into a detailed disquisition on such matters but shall merely *name* a few of the important and most interesting *headings* under which these subjects may be usefully studied:

(1) The Trigonometrical Functions and their inter-relationships, etc.;
(2) Arcs and chords of circles, angles and sines of angles etc.;
(3) The converse, i.e. sines of angles, the angles themselves, chords and arcs of circles etc.;
(4) Determinants and their use in the Theory of Equations, Trigonometry, Conics, Calculus etc.;
(5) Solids and why there can be only five regular Polyhedrons, etc.;
(6) The Earth's daily Rotation on its own axis and her annual relation around the Sun;
(7) Eclipses;
(8) The Theorem in Spherical Triangles relating to the product of the sines of the Alternate Segments, i.e.

$$\text{about: } \frac{\text{Sin BD}}{\text{Sin DC}} \cdot \frac{\text{Sin CE}}{\text{Sin EA}} \frac{\text{Sin AF}}{\text{Sin FB}} = 1 \text{ and}$$

(9) The value of $\pi$, i.e. the ratio of the circumference of a circle to its Diameter.

*N.B.*: The last item, however, is one which we would like to explain in slightly greater details.

Actually, the value of $\frac{\pi}{10}$ is given in the well-known *Anuṣṭub* metre and is couched in the Alphabetical Code-Language described in an earlier chapter :

गोपीभाग्यमधुव्रात-श्रृङ्गि.शोदधिसन्धिग ॥
खलजीवितखाताव गलहालारसंघर ॥

It is so worded that it can bear three different meanings—all of them quite appropriate. The first is a hymn to Lord Śrī Kṛṣṇa; the second is similarly a hymn in praise of Lord Shri Śaṅkara; and the third is an evaluation of $\frac{\pi}{10}$ to 32 places of Decimals! with a "Self-contained master-key" for extending the evaluation to *any* number of decimal places!

As the student and especially the non-Sanskrit knowing student is not likely to be interested in and will find great difficulty in understanding the puns and other literary beauties of the verse in respect of the first two meanings but will naturally feel interested in and can easily follow the third meaning, we give only that third one here:

$$\frac{\pi}{10} = \left.\begin{array}{l} .31415926535897 93 \\ 23846264338327 92\ldots \end{array}\right\}$$

on which, on understanding it, Dr. V.P. Dalal of the Heidelburg University, Germany felt impelled—as a mathematician and physicist and also as a Sanskrit scholar—to put on record his comment as follows:

"It shows how deeply the ancient Indian mathematicians penetrated, in the subtlety of their calculations, even when the Greeks had no numerals above 1000 and their multiplications were so very complex, which they performed with the help of the counting frame by adding so many times the multiplier! 7 × 5 could be done by adding 7 on the counting frame 5 times!" etc.

# Recapitulation and Conclusion

In these pages, we have covered a large number of branches of mathematics and sought, by comparison and contrast, to make the exact position clear to all seekers after knowledge. Arithmetic and Algebra being the basis on which all mathematical operations have to depend, it was and is both appropriate and inevitable that, in an introductory and preliminary volume of this particular character, Arithmetic and Algebra should have received the greatest attention in this treatise. But this is only a kind of preliminary "Prolegomena" and sample type of publication and has been intended solely for the purpose of giving our readers a foretaste of the delicious delicacies in store for them in the volumes ahead.* If this volume achieves this purpose and stimulates the reader's interest and prompts him to go in for a further detailed study of Vedic Mathematics we shall feel more than amply rewarded and gratified thereby.

* No subsequent volume has been left by the author.—Editor.

# A REPRESENTATIVE PRESS OPINION

*Reproduced from the Statesman, India, dated 10th January, 1956.*

## EVERY MAN A MATHEMATICIAN

(MR. DESMOND DOIG)

Now in Calcutta and peddling a miraculous commodity is His Holiness Jagad Guru Sri Shankaracharya of the Govardhan Peeth, Puri.

Yet Sri Shankaracharya denies any spiritual or miraculous powers, giving the credit for his revolutionary knowledge to anonymous ancients who in 16 *Sutras* and 120 words laid down simple formulae for all the world's mathematical problems.

The staggering gist of Sri Shankaracharya's peculiar knowledge is that he possesses the knowhow to make a mathematical vacuum like myself receptive to the high voltage of higher mathematics. And that within the short period of one year. To a person who struggled helplessly with simple equations and simpler problems, year after school-going year and without the bolstering comfort of a single credit in the subject, the claim that I can face M.A. Mathematics fearlessly after only six months of arithmetical acrobatics, makes me an immediate devotee of His Holiness Jagad Guru Sri Shankaracharya of the Govardhan Peeth, Puri.

I was introduced to him in a small room in Hastings, a frail but young 75 year-old, wrapped in pale coral robes and wearing light spectacles. Behind him a bronze Buddha caught the rays of a trespassing sun, splintering them into a form of aura; and had 'His Holiness' claimed divine inspiration, I would have believed him. He is that type of person, dedicated as, much as I hate using the word; a sort of saint in saint's clothing, and no inkling of anything so mundane as a mathematical mind.

### ASTOUNDING WONDERS

My host, Mr. Sitaram, with whom His Holiness Sri Shankaracharya is staying, had briefly prepared me for the interview. I could pose any question

I wished, I could take photographs, I could read a short descriptive note he had prepared on "The Astounding Wonders of Ancient Indian Vedic Mathematics". His Holiness, it appears, had spent years in contemplation, and while going through the Vedas had suddenly happened upon the key to what many historians, devotees and translators had dismissed as meaningless jargon. There, contained in certain Sūtras, were the processes of mathematics, psychology, ethics, and metaphysics.

"During the reign of King Kaṃsa" read a Sutra, "rebellions, arson, famines and insanitary conditions prevailed". Decoded, this little piece of libellous history gave decimal answer to the fraction 1/17; sixteen processes of simple mathematics reduced to one.

The discovery of one key led to another, and His Holiness found himself turning more and more to the astounding knowledge contained in words whose real meaning had been lost to humanity for generations. This loss is obviously one of the greatest mankind has suffered; and, I suspect, resulted from the secret being entrusted to people like myself, to whom a square root is one of life's perpetual mysteries. Had it survived, every-*educated "soul"* would be a, *mathematical-"wizard";* and, *maths "masters" would "starve"*. For my note reads "Little children merely look at the sums written on the black-board and immediately shout out the answers...they... have merely to go on reeling off the digits, one after another forwards or backwards, by mere mental arithmetic without needing pen, pencil, paper or slate." This is the sort of thing one usually refuses to believe. I did. Until I actually met His Holiness.

On a child's blackboard, attended with devotion by my host's wife; His Holiness began demonstrating his peculiar skill; multiplication, division, fractions, algebra, and intricate excursions into higher mathematics for which I cannot find a name, all were reduced to a disarming simplicity. Yes, I even shouted out an answer. ('Algebra for High Schools', page 363, exercise 70, example ten). More, I was soon tossing off answers to problems, which; *official-Maths-books "described"*, as *"advanced"*, difficult, and very difficult. Cross my heart!

His Holiness' ambition is to restore this lost art to the world, certainly to India. That India should today be credited with having given the world, via Arabia, the present numerals we use, especially the epochmaking "zero", is not enough. India apparently once had the knowledge which we are today rediscovering. Somewhere along the forgotten road of history, calamity, or deliberate destruction, lost to man the secrets he had amassed. It might happen again.

In the meantime, people like His Holiness Jagad Guru Sri Shankaracharya of the Govardhan Peeth, Puri, are by a devotion to true knowledge, endeavouring to restore to humanity an interest in great wisdom by making that wisdom more easily acceptable. Opposition there is, and will be. But eminent mathematicians both here and abroad are taking more than a passing interest in this gentle ascetic's discoveries. I for one, as a representative of all the mathematically despairing, hope, sincerely hope, that his gentle persuasion will prevail.

## Vedic Mathematics for Schools (Book 1)

*J. T. Glover*

*Vedic Mathematics for Schools* offers a fresh and easy approach to learning mathematics. It is intended for primary schools in which many of the fundamental concepts of mathematics are introduced. It has been written from the classroom experience of teaching Vedic mathematics to eight and nine year-olds. At this age a few of the Vedic methods are used, the rest being introduce at a later stage.

## Vedic Mathematics for Schools (Book 2)

*J. T. Glover*

*Vedic Mathematics for Schools* Book 2 is intended as first year textbook for senior schools or for children aiming for examination at 11+. Topics covered include the four rules of number, fractions and decimals, simplifying and solving in algebra, perimeters and areas, ratio and proportion, percentages, averages, graphs, angles and basic geometrical constructions. The book contains step-by-step worked examples with explanatory notes together with over two hundred practice exercises.

## Geometry in Ancient and Medieval India

*Dr. T.A. Sarasvati Amma*

This book is very useful as a textbook of Hindu Geometry. It deals with the mathematical parts of Jaina Canonical works and of the Hindu Siddhāntas.

In their search for a sufficient good approximation for the value of $\pi$ Indian mathematicians had discovered the tool of integration, which they used equally effectively for finding the surface area and volume of a sphere and in other fields. This discovery of integration was the sequel of the inextricable blending of geometry and series mathematics.

# वैदिक गणित

अथवा

## वेदों से प्राप्त सोलह सरल गणितीय सूत्र

### *जगद्गुरु स्वामी श्री भारतीकृष्णतीर्थ जी महाराज,*

### *अनुवाद : एअर वाइस मार्शल विश्वमोहन तिवारी*

*वैदिक गणित पर लिखित इस चमत्कारी एवं क्रांतिकारी ग्रंथ में एक नितान्त नवीन दृष्टिकोण प्रस्तुत किया गया है। इसमें संख्याओं एवं राशियों के विषय में जिस सत्य का प्रतिपादन हुआ है वह सभी विज्ञान तथा कला-विषयों में समान रूप से लागू होता है।*

*यह ग्रंथ आधुनिक पश्चिमी पद्धति से नितान्त भिन्न पद्धति का अनुसरण करता है, जो इस खोज पर आधारित है कि अन्तःप्रज्ञा से उच्चस्तरीय यथार्थ ज्ञान प्राप्त किया जा सकता है। इसमें यह प्रदर्शित किया गया है कि प्राचीन भारतीय पद्धति एवं उसकी गुप्त प्रक्रियाएँ गणित की विभिन्न समस्याओं को हल करने की क्षमता रखती हैं। जिस ब्रह्माण्ड में हम रहते हैं उसकी संरचना गणितमूलक है तथा गणितीय माप और संबंधों में व्यक्त नियमों का अनुसरण करती है। इस ग्रंथ के चालीस अध्यायों में गणित के सभी विषय— गुणन, भाग, खण्डीकरण, समीकरण, फलन इत्यादि— का समावेश हो गया है तथा उनसे संबंधित सभी प्रश्नों को स्पष्ट रूप से समझाकर अद्यावधि ज्ञात सरलतम प्रक्रिया से हल किया गया है। यह जगद्गुरु श्री भारतीकृष्णतीर्थ जी महाराज की आठ वर्षों की अविरत साधना का फल है।*

100 YEARS

1903

MOTILAL BA

PUBL